Narrative Innovation and Cultural Rewriting in the Cold War and After

Narrative Innovation and Cultural Rewriting in the Cold War and After

MARCEL CORNIS-POPE

palgrave

NARRATIVE INNOVATION AND CULTURAL REWRITING
IN THE COLD WAR AND AFTER
© MARCEL CORNIS-POPE, 2001
Softcover reprint of the hardcover 1st edition 2001 978-0-312-23837-7

First published 2001 by
PALGRAVE™
175 Fifth Avenue, New York, N.Y. 10010 and
Houndmills, Basingstoke, Hampshire RG21 6XS.
Companies and representatives throughout the world.

PALGRAVE is the new global publishing imprint of St. Martin's Press LLC Scholarly
and Reference Division and Palgrave Publishers Ltd (formerly Macmillan Press Ltd).

ISBN 978-1-349-63182-7   ISBN 978-1-4039-7003-9 (eBook)
DOI 10.1007/978-1-4039-7003-9

Library of Congress Cataloging-in-Publication Data
Cornis-Pope, Marcel
Narrative innovation and cultural rewriting in the Cold War era and after / Marcel
Cornis-Pope.
    p.   cm.
    Includes bibliographical references (p.) and index.

   1. American fiction—20th century—History and criticism.   2. Experimental
fiction, American—History and criticism.   3. Pynchon, Thomas—Criticism and
interpretation.   4. Sukenick, Ronald—Criticism and interpretation.   5. Federman,
Raymond—Criticism and interpretation.   6. Morrison, Toni—Criticism and
interpretation.   7. Postmodernism (Literature)—United States.   8. Cold War in
literature.   9. Narration (Rhetoric).   I. Title.

PS374.E95 P66   2001
813'.540911—dc21

                                                             2001031734

A catalogue record for this book is available from the British Library.

Design by Letre Libre, Inc.

First edition: December 2001
10   9   8   7   6   5   4   3   2   1

Permissions

CONTENTS

Preface

Begun before 1989 in an effort to correct the formalistic focus prevalent in discussions of postmodern fiction with a rhetorical and sociocultural reexamination, this book has undergone a major reorientation subsequently. The momentous events of 1989 have inspired a reevaluation of the impact of postmodern innovation not only during the Cold War era but also in the current post–Cold War restructuring. I have expanded my corpus of texts to include post-1989 works that demonstrate the new opportunities and challenges of innovative fiction after the fall of the Berlin Wall. I have also reconceptualized the discussion of specific directions in innovative fiction through new categories that better account for its contribution to the articulation of a polysystemic, multicultural environment. What started as a critical rereading of several types of innovative fiction has turned into a comprehensive rethinking of the goals of narrative innovation and cultural rewriting during and after the Cold War.

Narrative Innovation and Cultural Rewriting undertakes a systematic study of narrative responses to the polarized ethos of the latter half of the twentieth century that held cultures hostage to a confrontation between rival ideologies abroad and a clash between champions of uniformity and disruptive others at home. Considering a range of narrative projects and approaches (from polysystemic fiction to surfiction, postmodern feminism, and multicultural/postcolonial fiction), this book highlights their solutions to ontological divisions (real vs. imaginary, worldly and otherworldly), sociocultural oppositions (of race, class, or gender), and narratological dualities (imitation vs. invention, realism vs. formalism). Stepping back from traditional polarizations, innovative novelists have tried to envision an alternative history of irreducible particularities, excluded middles, and creative intercrossings.

Chapter 1 calls into question the prevailing tendency in criticism to dissociate two complementary sides of postmodern innovation, opposing deconstruction to rearticulation and self-reflection to representation. The "agonistic" consciousness imputed to postmodern experimentation is to a large extent a reflection of criticism's own divided response to norm-breaking art. Against this agonistic view of artistic change, chapter 1 proposes a transactive model of innovation that emphasizes the interventive-reformulative role of narrative imagination. Section 3 of the introduction considers the appropriateness of this "interventive" model in postmodern historical writing. Especially when confronted with traumatic historical events, innovative fiction produces radical disturbances in our representational frameworks, foregrounding those repressed human potentialities and alternative histories that never made it into the dominant History. The last section of chapter 1 addresses the role of narrative innovation and cultural rewriting in the post–Cold War transition. Against the tendency prevalent in both postcommunist Eastern Europe and in the United States to push postmodern innovation to the margin, I argue that innovative fiction can continue to play a significant role in the post–Cold War restructuring. Coover's *Pinocchio in Venice* (1991), Federman's *To Whom It May Concern* (1990), Morrison's *Jazz* (1992) and *Paradise* (1998), Pynchon's *Vineland* (1990) and *Mason & Dixon* (1997), and Sukenick's *Doggy Bag* (1994) and *Mosaic Man* (1999) give ample proof that contemporary writing has not surrendered its commitment to transformative thinking.

Chapter 2 examines three modes of narrative innovation that have challenged the Cold War "narratives of containment," freeing up and diversifying the channels of narrative exchange (polysystemic fiction), reinventing the rules by which "reality" is projected (surfiction), and "remember[ing] or mend[ing] what violence tears apart" (Joplin 51) with an imaginative patchwork of retrieved voices and stories (postmodern feminism). By pursuing narrative articulation in a self-critical way, displaying its underlying assumptions and constrains, these modes achieve a clearer understanding of the extent to which fiction can and does rewrite reality. The *polysystemic novels* of Robert Coover and Thomas Pynchon simultaneously build and question systematic narration, allowing dissonance to disrupt their totalistic claims. Their mistrust of totalization reflects both at the level of the story, which exploits the tensions between variegated systems of information and events, and at the level of discourse, which arbitrarily juxtaposes fact and fiction, undermining the conventions of historiographic reference. Intent on seeing "past the fiction of continuity, the fiction of cause and effect, the fiction of a humanized history endowed with 'reason'" (Pynchon, *V.* 286), Pynchon's and Coover's novels have created significant rifts in the Cold War ideologies, inserting in them the other's retrieved perspective.

While the fiction of Coover and Pynchon remains concerned with system analysis and transgression, working from inside the master narratives of Western culture, *surfiction* and the *postmodern feminist novel* pursue more radical modes of narrative and cultural rewriting. As practiced by Walter Abish, George Chambers, Raymond Federman, Kenneth Gangemi, Madeline Gins, Marianne Hauser, Steve Katz, Clarence Major, Paul Metcalf, Ursule Molinaro, Gilbert Sorrentino, and Ronald Sukenick, "surfiction" involves two related tasks: The first is disruptive, a polemical dismantling of the mimetic traditions of fiction; the second is reconstructive, shifting focus from the "experience of life" to the writer's effort to "reinvent" reality in a self-conscious process of narrative articulation. When surfictionists break away from a controlling plot, derailing narration with self-reflexive digressions or typographical designs, they assert their need to *unwrite/rewrite* their life stories, looking for imaginative ways to rearrange ordinary existence and open up the culture's available narrative plots.

The task of "unwriting" the dominant narrative tradition is central also to postmodern feminism. Like its experimental male counterpart, the self-reflexive work of feminist writers has called into question the conventional order of realism that has involved a systematic misrepresentation of women. But in addition to denouncing the traditional mechanisms of narrative representation, feminist theorists and fiction writers have been seeking new models of narrative articulation that rewrite cultural reality from alternative viewpoints. In the work of African American women writers, this revisionist impulse is amplified through a postcolonial strategy that Henry Louis Gates Jr. has called "signifying": a parodic appropriation of the language of the master in order to cajole, mock, trump the master's discourse, eventually changing the subject and its significance (*Figures in Black* 237–41). Resorting to what Morrison has called acts of "rememory" in *Beloved*, postmodern feminist fiction has tried to "*intervene* in history rather than *chronicle* it" (Marshall 150 [Unless otherwise noted, all emphases appear in the original.]), retrieving ignored events and shifting attention from "winners" to ordinary or historically displaced folk.

Polysystemic fiction, surfiction, and the innovative feminist novel challenge simultaneously the order of reality and the rules of narrative, "repatriat[ing]" them "from the realm of determinism to that of potential" (Sukenick, *Blown Away* 114). Their rewriting violates not only the "order of fate" but also the traditional order of fiction, forcing a reevaluation of the novel's epistemological and cultural domain. To paraphrase Jacques Derrida, in order to understand postmodern "rewriting" as a species of "writ[ing] differently," we "must reread differently" (*Of Grammatology* 87). The following chapters of my book attempt such reevaluative readings of Pynchon, Sukenick, Federman, and Morrison. Each chapter rereads the corpus of a writer's work from a new theoretical-analytic perspective: polysystem theory and postmodern cartographies in the Pynchon chapter, postmodern representational theories and ethnic criticism in the Sukenick chapter, performative-dialogic theories of discourse in the Federman chapter, trauma psychology and postcolonial theory in the Morrison chapter. In addition to exploring each writer's specific responses to recent cultural history, chapters 3 to 6 and the epilogue intercross a number of themes common to these four writers and others analyzed more briefly in chapter 2 (Acker, Coover, Didion, Hauser, Katz, Kingston, Major, and Walker). Among these themes are the critique of the Manichaean imagination of the Cold War, the search for alternatives to languages of power, and the dialogization of history to include ignored alternatives and excluded "others."

Narrative Innovation and Cultural Rewriting is focused primarily on North American examples, but these are reconsidered in the context of international postmodern theories and practices of fiction. For example, the introduction relates oppositional American postmodernism to an East European postmodernism of "resistance." As I have argued in a recent book on postmodern practices behind the Iron Curtain (*The Unfinished Battles: Romanian Postmodernism Before and After 1989*), the breakup of the Cold War structure was to a great extent a product of the increased cultural-philosophic dialogue across the ideological divide in which postmodern theories and practices played a catalytic role. Since the end of the Cold War, the "meaning" of the world has no longer been located in the imperial rivalry between superpowers, emanating instead from that vast middle ground between West and East, dominant and peripheral, released by the collapse of our polarized worldview. That middle ground can be found in many places: in the reemergence of Central Europe, in diasporic literatures, in multicultural or postcolonial identities, but also in the "third zones" located in a United States and Russia hybridized through the cross-fertilization between West and East, North and South. However fragile, these median areas are being continually replenished by the products of an innovative literature that emphasizes "interference" among the participating systems rather than assimilation or reified differentiation.

Acknowledgments

Of the many people who have encouraged this project over the years, special thanks are due to Ronald Bogue, Matei Calinescu, Janice Carlisle, Raymond Federman, Monika Fludernik, Jerry Klinkowitz, Paul Maltby, Larry McCaffery, Jerry McGuire, Brian McHale, Virgil Nemoianu, George Perkins, Brian Richardson, Mihai Spariosu, Henry Sussman, and Ronald Sukenick. Their comments on various independent articles and sections of my manuscript have helped sharpen my argument. The annual SSNL conferences have provided an ideal environment for testing out early drafts of several chapters and refining my conceptual framework. I am also most grateful to Gabriele Schwab, who has scrupulously reviewed my entire manuscript, and to Karen Wolny, senior editor of Palgrave, Amy McDermott, production editor, Meg Weaver, associate production manager, Annjeanette Kern, assistant production editor, and copyeditor Debby Manette for their expert guidance in preparing this manuscript for publication. This book would not have been possible without support from Harvard University in the form of an Andrew W. Mellon faculty fellowship (1987–88), which helped launch this project, and a semester research leave from the English Department at Virginia Commonwealth University that provided the final impetus for the completion of this manuscript. I also owe a world of gratitude to my daughters, Anca and Oana, who have had the temerity to read some of the fiction their father likes, and to my wife, Micaela, without whose continuous encouragement and gentle nudging this book would not have reached a conclusion.

Earlier drafts of certain sections have been previously published, but they are here substantially revised and expanded. The author thanks publishers for permission to use material from the following sources: *Critique: Studies in Contemporary Fiction* 29.2 and 42.1 for material included in chapters 2 and 5; *Pynchon Notes* 28–29 for material included in chapters 1 and 2; *The American Book Review* 13.1 and 14.3 for material incorporated in chapter 1; *Mimesis, Semiosis and Power* for material incorporated in chapters 1 and 2; *Style* 28.3 for material incorporated in chapters 1 and 5; *Narrative and Culture* for material used in chapters 1 and 2; *Violence and Mediation in Contemporary Culture* for material included in chapters 1, 4 and 6; *International Postmodernism* for material incorporated in chapter 1; *Symploke* 5.1–2 for material incorporated in chapter 3; *Engagement and Indifference: Beckett and the Political* for material incorporated in chapters 2 and 5; and *Federman A to X-X-X-X: A Recyclopedic Narrative* for material incorporated in chapter 5. (For a full citation of these previously published articles, see the preceding Permissions page and the Works Cited list at the end of this book.)

Richmond, Virginia

CHAPTER I

Postmodernism's Polytropic Imagination
Unwriting/Rewriting the Cold War Narratives of Polarization

There is a love of complexity, here in America, [. . .] pure Space waits the Surveyor, —no previous Lines, no fences, no streets to constrain polygony however extravagant, —especially in Maryland, where, encourag'd by the Re-survey Laws, warranted properties may possess hundreds of sides, —their angles pushing onward and inward, —all Sides zigging and zagging, going ahead and doubling back, making Loops inside Loops, —in America, 'twas ever, Poh! to simple Quadrilaterals.

—Thomas Pynchon, *Mason & Dixon* (586)

Beyond [the] literary and formalistic rupture, which critics have been discussing and analyzing for quite some time now, there is another form of rupture relevant to the understanding of the self-reflexive novel that has become visible in retrospect, and that is the rupture that occurred during the 1960s between the official discourse and the subject. [. . .] By the end of the Watergate crisis, all the official versions dealing with the Cold War, the McCarthy era, the Korean War [. . .], CIA activities in various parts of the world, the Vietnam war of course, and so on are being mistrusted, questioned and challenged not only in political writing [. . .] but also in the novel as it establishes a new relation with reality and with history, a relation based on doubt.

—Raymond Federman, *Critifiction* (23, 25)

I. SAYING "POH! TO SIMPLE QUADRILATERALS": INNOVATIVE FICTION AND THE QUEST FOR AN ALTERNATIVE NARRATIVE AND CULTURAL IMAGINATION

In his 1988 reappraisal of "Self-Reflexive Fiction," Raymond Federman described narrative innovation as a form of *resistance* against official constructions of reality during the Cold War era. The "New Fiction" took shape in the 1960s to fill "the linguistic gap created by the disarticulation of the official discourse in its relation with the individual" (*Critifiction* 25). Its main effort was to "place into the open, in order to challenge it, the question of representation in fiction, especially now that the line between the real and the imaginary" had been erased by the official discourse (25).

Coming of age in the "decade after Hiroshima," when mainstream fiction—as Thomas Pynchon also recalls—was "paralyzed by the political climate of the cold war and the McCarthy years" ("Is It O.K. to Be a Luddite?" 41), John Barth, Donald Barthelme, Richard Brautigan, Robert Coover, Ursula LeGuin, Thomas Pynchon, Ishmael Reed, and Kurt Vonnegut denounced the ideological myths on which official representations relied, retelling history in a satirical key. By replacing a conventional mimesis of "content" with a subversive mimesis of "form" that foregrounded the culture's symbolic systems, early postmodern fiction disrupted the traditional socialization of reality, disentangling the discourse of the subject from the discourse of the "Establishment" (Federman, *Critifiction* 23). Other writers emerging after 1968 (Walter Abish, Raymond Federman, Kenneth Gangemi, Madeline Gins, Steve Katz, Clarence Major, Gilbert Sorrentino, and Ronald Sukenick) rejected "mimetic realism and mimetic pretension" altogether (31), denouncing their "silent agreement with the official discourse of the State" (28–29). Their "critifictional discourse" (31) attacked—out of political necessity—"the vehicle that expressed and represented that reality: discursive language and the traditional form of the novel" (32).

Federman's retrospective ascribes to innovative fiction a strong reformulating function, committing it to a poetics of "divergence" (18). For Federman, this poetics begins by extricating fiction "from the postures and impostures of [traditional] realism" (21), expanding the available modes of narrative and cultural articulation. For Toni Morrison, it also begins by challenging

> statist language, censored and censoring. [. . .] Whether it is obscuring state language or the faux-language of mass media; whether it is the proud or calcified language of the academy or the commodity-driven language of science; whether it is the malign language of law-without-ethics, or language designed for the estrangement of minorities, hiding its racist plunder in its literary cheek—it must be rejected, altered, and exposed. ("Nobel Lecture 1993," 268–69)

But once the violence of these "policing languages of mastery" is exposed, the novelist looks for a narrative alternative whose aim is to liberate "human potential" (268) and "limn the actual, imagined, and possible lives of its speakers, readers, writers" (270).

The stakes of this search for a new narrative and cultural imagination were especially high during the protracted Cold War period that identified the "discourse of America" with a Manichaean rhetoric of confrontation. Deriving "its logic from the rigid major premise that the world was divided into two monolithic camps, one dedicated to promoting the inextricable combination of capitalism, democracy, and (Judeo-Christian) religion, and one seeking to destroy that ideological amalgamation by any means," the Cold War discourse of America gave rise to what Alan Nadel has described as a "containment culture" (3). America's policies of "deterrence" against the Soviet bloc were matched by narratives of containment at home, emphasizing conformity "to some idea of religion, to 'middle-class' values, to distinct gender roles and rigid courtship rituals" (4). Common to both external and the internal versions of containment was the effort to retain well-defined boundaries between Other and Same—whether the Other was the Soviet Union, a "series of recalcitrant Third World others" (Spanos 65), or heretics at home.

Across the ideological divide, the Soviet Union produced powerful counternarratives of containment, from Joseph Stalin's aggressive pursuit of a defensive cordon of satellite countries and stabilization of power through brutal purgings, to Leonid Brezhnev's program of "social imperialism" in the Third World. Developed over a period of several decades, the self-legitimizing narrative of Soviet power combined co-option, exporting to the Third World a primarily Eurocentric concept of socioeconomic revolution, with repression, quashing labor movements and national uprisings in the satellite countries. At home, this narrative pursued relentless industrialization, bureaucratic rationalization, and terror against the internal "enemy" who refused conversion to *homo Sovieticus.* On both sides of the Iron Curtain, Cold War ideology worked as a version of the "cooling systems of dissuasion and extermination" described by Jean Baudrillard (*The Evil Demon of Images* 24), annihilating historical memory and exorcising alternative views.

Not surprisingly, the proliferation of conflicting narratives of containment caused in time an ideological implosion, with these narratives "eventually split[ting] one another asunder" (Nadel xi). In the United States, the failure of containment as a foreign policy triggered the drama of the Vietnam War; the failure of containment as a discursive practice marked the beginnings of American postmodernism (Nadel 67), a theoretical and artistic movement that called into question the containment paradigm itself. The emergence of protest movements and oppositional literature contributed to the breakup of the "(neo)imperial structure of American/European modernity," foregrounding cultural possibilities utterly precluded by it (Spanos 67, 68). Challenging the dichotomous imagination of the Cold War and of modernism itself, postmodernism proposed polysystemic cartographies that mediated among differentiated subjects and cultures according to an extravagant geography of "zigging and zagging [sides], going ahead and doubling back, making Loops inside Loops" (Pynchon, *Mason & Dixon* 586). In communist Eastern Europe, an ideological system regarded as stable and all-pervasive created in much the same way conditions for its own dispersal. By turning "language and discourse [. . . into] the ultimate means of production" (Verdery 91), communism colonized the political imaginary of its people but also generated a cultural surplus, more symbolic work than the ideological system could absorb and control. This semiotic excess favored to some extent the system's opponents, allowing them to "reconquer a space for the production of genuine cultural values and to create counter-institutions to protect these" (Verdery 110). As I have argued elsewhere ("Critical Theory and the *Glasnost* Phenomenon" 131–56; *The Unfinished Battles* 7–58), the *Glasnost* phenomenon was a direct product of the increased cultural-philosophic dialogue across the ideological divide in which the postmodern theories and practices played a catalytic role, encouraging the development of a post–Cold War consciousness. Postmodernism emerged in Eastern Europe in the phase of late communism, which brought about the collapse of controlled economies, as a critical response to "aberrant political conditions, anachronistic social difficulties, and artificial cultural obstructions" (Cârneci 91, 96). Terms like "pluralism," "posttotalitarianism," "diversity," "recuperation," "reality," and "simulation" figured with increasing frequency in the East European (mostly unofficial) debates, suggesting a larger struggle against state-imposed fabrications of reality. As Mikhail Epstein reminds us, "Long before Western video technology began to produce an overabundance of 'authentic' images of an absent reality, this problem was already being solved in Russia by our ideology, by our press, and by statistics that calculated crops that would never be harvested to the hundredths of a percentage point" (qtd. in Berry 343). Against this "simulational logic operat[ing] at all levels of society" (Berry 343), Eastern European writers promoted a *postmodernism of resistance* that replaced monologic concepts with a dialogic understanding of reality based on creative disagreement rather than on blind consensus. The postmodern poetry of Jacek Bierezin, Volker Brown, Mircea Dinescu, Agnes Gergely, Viktor Krivulin, Ewa Lipska, Adam Puslojić, Tomaz Salamun, Tatyana Shcherbina, or Desző Tándori functioned as a vehicle of "moral resistance," opening "the doors and windows of poetry to life, reality, to history, to street language" (Iorgulescu 61). The innovative fiction of Gabriela Adameşteanu, Péter Esterházy, Danilo Kiš, Milan Kundera, Mircea Nedelciu, Liudmila Petrushevskaia, Tatyana Tolstaya, Christa Wolf, and Alexandr Zinoviev attacked the foundation of communist hyperreality, valorizing heteromorphous and marginalized messages.

The postmodern challenge to the Cold War narratives of containment included thus a reconceptualization of the culture from the margins, retrieving the experience of subaltern groups and excluded others. As Vincent B. Leitch explains, "[t]o conceptualize culture from the vantage point of social margins is to foreground violent omissions, repressions, and contingencies of traditional reason, representation, knowledge, and high culture. Put in other terms, postmodern cultural analysis characteristically entails an ethicopolitics of inclusiveness, multiculturalism, polylogue, social transformations, antiauthoritarianism, and decentralization" (133). Informed by the retrieved interests of women, African Americans, pretechnological societies, and so on, postmodernism has sought more inclusive models of cultural exchange that valorize the capacity of subjects to participate in a variety of "social logics" (Mouffe 142). In the area of race relations, postmodernism has questioned the master narrative of modernism predicated on "the rural/urban and the European/non-European" oppositions (Hogue 2), which has for the most part prevented people of color from sharing the benefits of modernization, welcoming a thin middle class at the expense of maginalizing the rest of the black population into urban ghettoes (Hogue 10, 19). In its place, postmodernism has proposed a "recategorization of social practices, identities, symbols, and experiences" that cuts across distinctions between "marginalized [and] hegemonic cultural communities," seeking forms of "transracial exchange" (Hogue 21). The polysystemic and multicultural fiction that has emerged since the mid-1970s mediates cultural and psychological otherness, turning its textual spaces into "borderlands" where "two or more cultures edge each other, [. . .] where under, lower, middle, and upper classes touch, where the space between two individuals shrinks with intimacy" (Anzaldúa, preface).

If the current geopolitical scene is more hospitable to "the cultivation of multilingualism, polyglossia, the arts of cultural mediation, deep intercultural understanding, and genuinely global consciousness" (Pratt 62), some credit is due to the revisionist imagination of postmodernism that has deconstructed the polarized cartographies of the Cold War era, replacing them with polysystemic mappings. And yet innovative literature has not received adequate recognition for the fact that "imagination has now acquired a singular new power in social life" and "[m]ore persons in more parts of the world consider a wider set of 'possible' lives than they ever did before" (Appadurai 197). Its multicultural agendas have been misconstrued as naively universalizing and narrative self-reflection has often been judged as a ruse concealing a lack of vision. The few recent attempts to foreground the connection between postmodern innovation and the deconstruction of Cold

War ideologies have left many questions unsettled. For example, Thomas Hill Schaub's *American Fiction in the Cold War* relegates the discussion of postmodernism to a short epilogue, leaving unresolved the question whether experimental writing is a viable response to, or a mere reflection of, "the politics of paralysis in the post-war period" (190). According to Alan Nadel's *Containment Culture* (1995), the failure of containment as a discursive strategy coincided with the beginnings of American postmodernism, but Nadel's book does not consider the specific ways in which postmodern innovation contributed to the breakup of the confrontational ideologies of the Cold War. The postmodern works he discusses in some detail (Didion's *Democracy*, Heller's *Catch 22*, and Walker's *Color Purple*) view pessimistically their chances for breaking out of the containment paradigm. Finally, Paul Maltby's pioneering book, *Dissident Postmodernists: Barthelme, Coover, Pynchon* (1991), identifies a "dissident" trend in American postmodernism but limits it to one group of writers and to mostly thematic concerns.

Critics who dismiss postmodernism as "self-reflexive gaming" ignore the complex cultural filiation of self-reflection that has emerged at the conjunction of a "Lacanian psychoanalytic moment, with its emphasis on specularity and the problematics of self and Other; the deconstructive moment, with its emphasis on textual rhetoric and self-reflexiveness; and the feminist moment, with its political and theoretical emphasis on the exclusion of women from traditional discourse" (Suleiman, *Subversive Intent* xv). They also fail to take into account the "extraliterary" causes of narrative self-reflection. In Jerome Klinkowitz's view, the "inhibiting conditions of living, loving, teaching, writing" in postindustrial America and the "wooden convention of genre-based and industry-controlled publishing" have had as much impact on the practice of innovative fiction as other specifically literary circumstances ("The Extra-Literary in Contemporary American Fiction" 19–20, 28). The response of innovative writers to these extraliterary conditions has been aesthetic as well as political: Concurrently with interrogating the prevailing codes, they have engaged in a process of narrative and cultural rewriting, pursuing versions of a multicultural "post-European novel" (Sukenick, "Unwriting" 27), such as Sukenick's generative fiction informed by a "law of mosaics" (27) or Anzaldúa's border narratives.

A reexamination of the best experimental fiction published since the mid-1960s bears out Paul Maltby's conclusion that postmodernism has undertaken a systematic critique of "hegemonic values—technocratic, bureaucratic, militaristic, business-oriented" (*Dissident Postmodernists* 25). Postmodern fiction has located the roots of our conflictual ethos in the polarizing imagination inherited from the Age of Reason and in the hypercompetitive animus of capitalism (E. L. Doctorow, Thomas Pynchon, Kurt Vonnegut). The main focus of the new fiction has fallen on the twentieth-century avatars of this divisive spirit, as embodied in the imperialistic confrontations of World War I and II (Walter Abish, Raymond Federman, Joseph Heller, Jerzy Kosinski, Thomas Pynchon, Kurt Vonnegut), the Cold War paranoia of the 1950s (William Burroughs, Robert Coover, Don DeLillo, Gilbert Sorrentino, Raymond Federman, Ursula Le Guin, Thomas Pynchon, Ronald Sukenick), the civil rights and gender confrontations of the 1960s (Clarence Major, Toni Morrison, Joyce Carol Oates, Johanna Russ, Ishmael Reed, Alice Walker), the Vietnam era unrest (Donald Barthelme, Steve Katz, Joyce Carol Oates, Thomas Pynchon, Ishmael Reed), cultural narcissism and fragmentation in the 1970s (Marianne Hauser, Thomas Pynchon,

Kurt Vonnegut), Reagan's "cowboy capitalism" in the 1980s and the new "culture wars" in the 1990s (Kathy Acker, Robert Coover, Don DeLillo, Steve Katz, Clarence Major, Ronald Sukenick). Each narrative critique is also a remapping of culture, enlarging it with alternative voices and values repressed by the narratives of capitalist modernity. The retrieval and reintegration of the "other" is central to postmodern narrative histories that trace the transatlantic "crossing," slavery, and the postemancipation life of African Americans (Toni Morrison, Ishmael Reed, Alice Walker), the patterns of dislocation and assimilation experienced by Native Americans (Leslie Marmon Silko, Louise Erdrich, Gerald Vizenor), or the difficulties of establishing a successful identity for ethnic minorities (Maxine Hong Kingston, Jamaica Kinkaid, Theresa Hak Cha) and sexual "others" (gay and lesbian fiction). Feminist writers have been busy fulfilling Virginia Woolf's "ambitious beyond [...] daring" suggestion to "re-write history" (47), "add[ing] a supplement to [it...] so that women figure there without impropriety" (39). Challenging the exclusionary logic of the "real" articulated by the masculine symbolic, feminist writers have tried to envision a "completely other history: a history of paradoxical laws and non-dialectical discontinuities: a history of absolutely heterogeneous pockets, irreducible particularities, of unheard of and incalculable sexual differences" (Cornell 2, 83).

We must ask, therefore, why innovative fiction has often been misconstrued as a self-indulgent formalism rather than as a meaningful response to a historical and literary crisis. Innovative writers share with their critics some responsibility for their misrepresentation, overstating the playful, self-canceling aspects of their work:

> The elements of the new fictitious discourse [...] must be digressive from one another—digressive from the element that preceeds and the element that follows. [...] Rather than being a stable image of daily life, fiction will be in a perpetual state of redoubling upon itself. It is from itself, from its own substance that the fictitious discourse will proliferate—imitating, repeating, parodying, retracting what it says. (Federman, "Surfiction—Four Propositions" 11)

Similarly, Gilbert Sorrentino's early *Kulchur* essays advertised a new type of "signalless American novel" that functioned as a primary reenactment rather than an interpretation of the processes of life (*Something Said* 13–48). In a more philosophic vein, William H. Gass has insisted that the story "is taking place the only place it could take place: on the patient page, in among the steadfast words, the metaphors of mind and imagination" (*Habitations of the Word* 78). Sidestepping what André Gide called the "tyranny of resemblance," literary discourse emphasizes the "existential productivity" (84) of its own metaphors: "We watch a thought in the process of composing itself. [...] Nothing is being represented. A thought, instead, is being *constructed*—a memory" (81, 83).

For Sukenick even these descriptions are "encumbered [...] by undigested leftovers from other theories." Gass's notion of fiction as a mental "model of the world [...] reintroduces the schizoid split between art and reality that one gets rid of in speaking about art as an addition to reality [...]. This is basically a subtler kind of imitation theory in which continuity between art and experience is broken because art is seen as a mode essentially different from experience" (*In Form* 23). Sukenick's own "digressions on the act of fiction" have called for a more radical version of nonmimetic fiction whose qualities "are abstraction, improvisation, and opacity" (211).

By making its language partly opaque, fiction resists the game of make-believe, calling attention to its phenomenological and technological reality, to "the truth of the page" (212).

This reorientation toward the material reality of the novel has been regarded by unsympathetic critics as an act of ideological "recoiling," creating "a new kind of flatness or depthlessness, a new kind of superficiality in the most literal sense" (Jameson, *Postmodernism* 9). Yet it is clear from other, more considered statements that the point innovative writers have been making is not that the world exists "wholly within the word" (Wilde, *Middle Grounds* 124), but that our versions of reality depend on perceptual and discursive systems that offer biased representations under the guise of a "natural" order of the world. Therefore, it is the writer's obligation to challenge naturalized conventions of representation, imagining better aesthetic and sociocultural syntheses. Disruption becomes in this context as important as rearticulation. The innovative writer's refusal to commit to a reified notion of History, "without rupture or break" and "oblivious to otherness, difference, quiddity" (Hassan, "The Aura of a New Man" 166), should not be taken as a proof of "ahistoricalness." On the contrary, through reclamation and revision the innovative novelist puts forth alternative histories that resist easy "recovery, use, assimilation" (166). For example, Sukenick's fiction avoids "the factuality of history," offering itself as *re-creation*, that is, an imaginative act of articulation "not different from that of composing one's reality" (*In Form* 206, 208). Postmodern feminists also oppose patriarchal histories that weave a "powerful story of male adventure and female suffering, of male freedom and female restrictions" (Gregory 141), turning fiction into a space of "fantasy and pleasure" within which "imaginary formations" more empowering for women can be conceived (Kristeva, "Women's Time" 197, 207).

Both projects move beyond the formalistic "autoreferentiality" imputed to them, reinventing the rules by which reality is projected. Though male and female postmodernists have not always agreed on the "gender" of traditional codifications (for Barthes the classical readerly text is "feminine"; for Cixous and Irigaray the realistic novel is "phallic"), both groups have singled out the realistic tradition as a worthy critical target:

> The fairy tale of the "realistic" novel whispers its assurance that the world is not mysterious, that it is predictable—if not to the characters, then to the author—that it is available to manipulation by the individual, that it is not only under control but that one can profit from this control. The key idea is verisimilitude: one can make an image of the real thing which, though not real, is such a persuasive likeness that it can represent our control over reality. This is the voodoo at the heart of mimetic theory that helps account for its tenacity. (Sukenick, *In Form* 3–4)

How entrenched this tradition still is can be gauged from a comment such as John Aldridge's that imputes postmodernism the lack of an unspoiled, Hemingway-type "relationship with the concrete objects." Whereas "Hemingway's works can be read as a series of instruction manuals on how to respond and comport one's self in the testing situations of life," those of postmodern writers are steeped in "a gratuitous [. . .] state of distortion and aberration" (*The American Novel and the Way We Live Now* 140, 150). Much criticism of innovative fiction has been articulated from similarly conservative positions that hold onto the very notions challenged by postmodernism, such as Aldridge's "dependable rules of feeling and conduct," De Villo Sloan's "traditional notion of the human" (37), or Charles Newman's trust in "a non-relative vantage point for observing the world" (5).

Critiques of postmodernism on the left have not been more helpful. The prevailing view among theorists like Jean Baudrillard, Fredric Jameson, Terry Eagleton, Hal Foster, Christopher Norris, and Alex Callinicos is that postmodernism remains an art of pastiche and simulation that renders history "reified, fragmented, fabricated—both imploded and depleted" (Hal Foster 123). The "neutral practice of [parodic] mimicry, [. . .] amputated of the satiric impulse, devoid of laughter and of any conviction that [. . .] some healthy normality still exists" (Jameson, Postmodernism 17), provides little room for transformation. Ironically, while denouncing postmodernism for blunting the critical edge of modern art, neo-Marxist critics remain oblivious to their own marginalization of postmodernism's adversarial potential. Jameson's benchmark analyses have focused primarily on forms of art that give support to his deterministic thesis that postmodernism is complicitous with the socioeconomics of late multinational capitalism (see Maltby, Dissident Postmodernists 8–14; McHale, "Postmodernism or the Anxiety of Master Narratives" 17–33; and Curtis White, "Jameson Out of Touch?" 21, 30). While theorizing the need for a more reflective "cognitive mapping" in art, the examples Jameson gives (the fiction of E. L. Doctorow, that "epic poet of the disappearance of the American radical past" [24], the abstract "hyperspaces" [38] of John Portman's Westin Bonaventure Hotel in Los Angeles, or the "new wave" cyberpunk science fiction described as a symptom of an "ultimate historicist breakdown" [231]) reinforce his pessimistic ahistorical belief in the capacity of capitalism to neutralize oppositional art. Jameson deems any form of artistic resistance a priori doomed, "an integral and functional part of the system's own internal strategies" (Postmodernism 203). The novel receives Jameson's most skeptical appraisal, as "the weakest of the newer cultural areas" by comparison to "its narrative counterparts in film and video" (298) and even to postmodern consumer items in "boutiques and fashionable little restaurants" (377). What Jameson ignores is the fact that the methods he envisions for the future—such as "global cognitive mapping" (54) that will enable us to understand the great "multinational and decentered communicational network in which we find ourselves caught as individual subjects" (44), or "transcoding" that will interplay different systems and establish subtle linkages in contemporary culture—have been successfully tested in polysystemic, feminist, and hypertextual fiction.

It is tempting to say with Jean-François Lyotard that behind the diverse critiques of postmodern experimentation "there is an identical call for order, a desire for unity, for identity, for security, or popularity. [. . .] Artists and writers must be brought back into the bosom of the community, or at least, if the latter is considered to be ill, they must be assigned the task of healing it" ("Answering the Question" 73). Not surprisingly, criticism has had as many problems with postmodernism's attack on verified models and grounds as with its self-questioning spirit. For every complaint of self-reflexive indulgence levied at postmodernism we can cite countercriticisms of excessive cultural disruption. The "agonistic" consciousness imputed to postmodern innovation is to a large extent a reflection of criticism's own divided response to norm-breaking art.

2. LEARNING TO LIVE WITH POSTMODERNISM'S
SUBVERSIVE DEMON: FROM AN AGONISTIC
TO A TRANSACTIVE MODEL OF NARRATIVE INNOVATION

Pynchon's metaphor for optimizing (narrative) communication, we recall from *The Crying of Lot 49*, is "Maxwell's Demon," the clever sorting device hypothesized by James Clerk Maxwell in 1871 to test the strength of his second law of thermodynamics:

> As the Demon sat and sorted the molecules into hot and cold, the system was said to lose entropy. But somehow the loss was offset by the information the Demon gained.
> [. . .]
> "Communication is the key," cried Nefastis. "The Demon passes his data on to the sensitive, and the sensitive reply in kind. There are untold billions of molecules in that box. The Demon collects data on each and every one. At some deep psychic level he must get through. The sensitive must receive the staggering set of energies, and feed back something like the same quantity of information. To keep it cycling." (105)

Maxwell's original demon was supposed to work like an intelligent distributor in an idealized heat engine, feeding "hot" and "cold" molecules through a trapdoor between two chambers in such away that, in violation of the law of increased entropy, he would produce an information-based order that contradicted the probable distribution of molecules in random mixtures. But as Leo Szilard and Leon Brillouin (see his chapter 13) have shown since, the violation of the second law of thermodynamics is only apparent: The decrease in thermodynamic entropy through the information fed by the demon into the system is offset by the "process of acquiring the necessary knowledge" about molecules, which creates entropy at other levels (J. Kerry Grant, *A Companion to "The Crying of Lot 49"* 85–86). Though aware of the fact, Pynchon's version of the Demon capitalizes on the *metaphoric coincidence* between the loops of energy and information, proposing to control the entropy resulting from the informational flow in a way similar to that imagined by Maxwell for the thermodynamic process. The success of communication depends on the existence of a "sensitive" respondent at the other end, capable of sorting "hot" and "cold" informational signs and inputting the system with new energy in order to reduce its spontaneous drift toward "equivocation." Maxwell's Demon works best as a literary analogy (the photograph on the Demon's box shows Clerk Maxwell with "hands [. . .] cropped out of the photograph. He might have been holding a book" [*The Crying of Lot 49* 107]), representing accurately the process of fiction. Like Maxwell's original device, fiction exploits the metaphoric coincidence between the world of experience and the narrative world constructed through textual information. The "two equations" only happen to "look alike" (106), yet their intersection produces a phenomenological world that is richer than both. With a receptive interpreter reading creatively the Demon's messages and feeding back her own patterned responses, the narrative machine can submit the random flow of signs to new, imaginative configurations.

In *The Crying of Lot 49*, Oedipa is at first overwhelmed by the intricate "message from the grave" she receives from Pierce Inverarity in the form of a testamentary assignment. As the novel's main puzzle-solver, Oedipa lacks modernism's faith in successful narrative synthesis. No single center of meaning, no dominant perspective is allowed to emerge. But while neither Inverarity's will nor Oedipa's reinterpretation of it can contain the entropy that threatens post–World War II technological society, both set in motion a process that converts "waste" into extravagant patterns of meaning. "Trystero," the secret organization Oedipa discovers by accident as she tries to make sense of Pierce's enigmatic legacy, is a good metaphor for the workings of postmodernism itself: Trystero lives off the entropy it creates, turning waste into new, insidious orders. Postmodernism similarly allows two different emphases to interact: a deconstructive, improvisational impulse with a re-creative, negentropic one.

Criticism has often dissociated these two sides of innovation, describing the postmodern project as overly fanciful and hopelessly disruptive. Underlying these descriptions is a nostalgic notion of narration as an effective integrative machine. Within this model, epistemological and compositional concerns are either overlooked or willingly subordinated to the "positive socializing function of literature" (Charles Newman 173). What the contemporary novel needs, according to Newman, is "neither amusement, nor edification, but the demonstration of real authority which is not to be confused with sincerity, and of an understanding which is not gratuitous" (5–6). By endorsing such dubious notions of narrative authority, mainstream critics reinstate the "phony duality" realism/formalism they elsewhere denounce (see Aldridge, "The New American Assembly-Line Fiction" 26). Radical sociocultural critics have also opposed realist/modernist epistemologies to the fractured, antiauthoritarian poetics of postmodern fiction. According to Terry Eagleton, postmodernism is a "joke" at the expense of the "revolutionary avant-gardism" of Tzara or Mayakovsky, dissolving "art into the prevailing forms of commodity production" (131). Even when postmodernism is recognized as a phase of the avant-garde, as in Charles Russell's *Poets, Prophets, and Revolutionaries* (1985) or in Andreas Huyssen's *After the Great Divide* (1986), this filiation is carefully qualified. For Russell, postmodern art can be oppositional in the avant-gardist sense but "fails to offer any principle upon which a significantly different art or social vision might be created" (238). Huyssen also regards postmodernism as an "endgame of the avant-garde and not as the radical breakthrough it often claimed to be," lacking the original "consciousness that social change and the transformation of everyday life were at stake in every artistic experiment" (168, 170).

In their troubled response to postmodernism's disruptive poetics, both Marxist and neoconservative critiques illustrate our guilty conscience about artistic innovation. As René Girard explains, the "negative view of innovation is inseparable from a conception of the spiritual and intellectual life dominated by stable imitation. Being the source of eternal truth, of eternal beauty, of eternal goodness, the models should never change. Only when these transcendental models are toppled, can innovation acquire a positive meaning" (11). Imitation and innovation appear incompatible, caught in a psychodrama that pits a quasi-theological fear of heretical change against a "terroristic" promotion of innovation for its own sake. And yet, as Girard reminds us, the "tendency to define 'innovation' in more and more 'radical' and anti-mimetic terms" (16) is spurred by the competitive ethos inherent in mimesis itself. For Girard the "mimetic rivalry unleashed by the abandonment of transcendental models," and the anti-innovative resentment that blocks "external" and "internal mediation" are equally harmful. Both are expressions of a conflictive mentality that fails to see imitation and innovation as complementary (14). The precondition for "real innovation is a minimal respect for the past, and a mastery of its achievements" (19). Successful innovation challenges tradition from inside, transforming it.

Though Girard's concept of innovation may appear overly cautious, excluding radical shifts as "meaningless agitation" (18–19), it does offer us a way out of traditional dualistic relationships (mimesis/innovation, formalism/realism). We need, however, to expand his definition to account for stronger forms of experimentation ("L=A=N=G=U=A=G=E" poetry, surfiction, postmodern feminism, postcolonial writing) that deconstruct the opposition realism/self-reflection. By Girard's own admission, radical artistic innovation "has a fecundity which has no parallel in science, technology or economic life," even though it may have lost of late some of its impetus fed by individualistic ideologies of creativity (17–18). We must also take heed of the recent critiques of the emphasis on "individualism" and "originality" inherent in our concept of innovation, which postmodernism itself has partly deconstructed. A historicist approach to experimental fiction will have to replace the agonistic-individualistic model of innovation with a transactive one that emphasizes the reformulative role of narrative imagination. This will allow us to confront two further biases in contemporary criticism: on one hand, the hypostatization of innovation as the irreconcilable "other" of tradition, with little cultural purchase beyond that of disrupting the established discursive structures; on the other, the collapsing of innovation and tradition into an easily digestible form of mainstream postmodernism. In what follows, I will discuss both aspects in some detail.

a. Relocating Postmodern Innovation: Contemporary Culture's Critical Interlocutor or Incommensurate "Other"?

Innovative fiction has often been regarded as a collection of antireferential procedures that estrange narration from reality. The prevailing critical terminology has prevented a more balanced understanding of narrative innovation as both disruptive and reconstructive. Used all too often to describe recent forms of narrative, terms such as "metafiction," "antinarrative," "pure fiction," "invention," "parody," and "pastiche," have reduced postmodernism to an either/or logic that pits deconstruction to articulation and "invention" to "imitation." But as we have seen, these terms cannot exist as absolute opposites. Viewed as totally "new, different, nonexistent [. . .] invention is in fact impossible, since to truly invent one would have to reinvent invention from a different episteme, an episteme that does not as yet exist. And this reinvention, because it is a *reinvention*, would ultimately have to be predicated on the original episteme, thus rendering it no more status than a repetition or a refiguration" (Desai 120).

Does this mean that invention/innovation are unavailable as cultural strategies? Certainly not, especially if we take into account the varied etymological and historical embodiments they have taken: invention as "coming upon" or "discovery" (from the Latin *venire*), invention as "composing" (from the rhetorical term *inventio*), or invention as the devising of "other worlds through corollary processes" of "rearticulation" (Desai 121, 122). In this latter sense, innovation has been critical not only to groups of oppressed people, who need to "reinvent" themselves, but also to novelistic imagination in general. As Derek Attridge puts it, literature has always been about the encounter with an "other": an alternative world, the "singular otherness of the other person," a "new existent that cannot be apprehended by old modes of understanding" (22, 24). Innovation in this sense is an ethical act: Only through refashioning the norms whereby we understand/represent the world and other human beings can fiction achieve a "creative, responsible responsiveness to the other" (29).

Contemporary innovative writers have understood the need to move beyond a binary logic better than some theorists, refusing both a naive experiential stance and nihilistic self-deconstruction. The narrative strategies employed in surfiction, for example, seek to radicalize fiction's rapport with reality, opening the self-contained "system of language up to experience beyond language" (Sukenick, *In Form* 11). The four sections in Steve Katz's *Moving Parts* (1977) may look like experiments in "pure writing," with "the quality of music, riffs and jams" (Klinkowitz, *The Life of Fiction* 106). A closer examination will find them concerned with the narrative and ideological frames that construct "reality," throwing a "cataract of dogma" over "your perceptions of things as they are" (Katz interview in LeClair and McCaffery 226). What they finally suggest is that fiction can rewrite reality, establishing "uncontrollable and mysterious resonance[s]" with experience: "Art prepares the bed of contingencies from which reality sprouts, ripens, and is harvested" (75). In similar ways, postmodern feminist art challenges our expectations of a "transparent" reality while simultaneously reshaping the space "between writing and experience," fact and fiction (Hunt 199). Recognizing that people are trapped not only in history but also in language, having internalized a system of conventions that no longer allows them to experience reality as a whole, the innovative writer is interested in "reinventing" both: "I think storytelling [. . .] serves a function in human intercourse. And I want to stretch the bounds of the potential of storytelling, which always has to be refitted to the times, reexamined, reinvented. In other words, the structure of the exchange has to be reimagined" (Katz interview in LeClair and McCaffery 231). The pressing task of redefining the novel as a vehicle of cultural exchange ascribes innovation an important cultural function. Viewed in this light, innovation means more than "playing with the mechanics of and approaches to storytelling": It means "exploring, discovering, maybe inventing the forms appropriate to my understanding of the world" (221).

According to its own practitioners, innovative fiction should thus be viewed as a critical interlocutor of culture. The alleged "unreadability" of experimental fiction—by contrast to the "legible, pleasing [. . .] reassur[ing]" mainstream novel—can play a corrective role in contemporary culture, disorienting us "in relation to ourselves" and the world's "reality" (Federman, *Critifiction* 70–71). The comfort of readability comes from the ease with which it allows us to play our "mental cinema of realism beyond language" (72), recognizing "*righteously* [. . .] how coherent, continuous, whole, rational, logical, how secure we are in our culture" (71). By contrast, innovative fiction is disruptive, forcing us to reexamine our relation with culture (73); it is also re-creative, capturing existence in its "pluridimensionality" that for Federman includes "power, economy, war, peace, sex, violence, all the great tricks of reality" (74) but also the language that expresses them, "for language too is *a* reality" (73).

The analyses of innovative fiction published over the last ten to fifteen years have made somewhat clearer the connection between a writer's struggle to rearticulate her "life story" and her critique of the prevailing modes of narrative and cultural articulation. Recent criticism has reassessed the postmodern agenda with more sophisticated theoretical tools (feminist, new historicist, poststructural, postcolonial), departing from an earlier view of experimentation as politically uninvolved. Linda

Hutcheon's work is a case in point: After several books dedicated to the formalist poetics of metafiction (*Narcissistic Narrative: The Metafictional Paradox*, 1980; *A Theory of Parody*, 1985; *A Poetics of Postmodernism*, 1988), the critic has refocused attention on the political drive of postmodernism, arguing in *The Politics of Postmodernism* (1989) that self-reflection is inextricably bound up with a critique of power and domination. In her discussion of *Saints and Scholars* (1987), Hutcheon credits Terry Eagleton with a "return to history and politics through, not despite, metafictional self-consciousness and parodic intertextuality" (61). But even when revised, many of the concepts associated with self-referentiality (metafiction, parody, pastiche, intertextual appropriation) continue to create evaluative problems for criticism. A term like "metafiction" is essentially a misnomer, positing a problematic locus outside and above the fictional process, whence an effective critique of narrative models can be attempted. As Hutcheon concedes, a critical framework that emphasizes the metafictional reappropriation of forms of the past may miss postmodernism's potential for both "de-toxifying" (153) and changing the dominant systems of meaning and value.

While appropriately rehistoricizing innovative fiction, locating literary aesthetics within the framework of a cultural politics, recent reassessments continue to waver between two theoretical descriptions of innovation: One explains the novelist's task in deconstructive terms as a "purification" of language by "rendering [it] seemingly incoherent, irrational, illogical, and even meaningless" (Federman, *Critifiction* 33); the other emphasizes the socially relevant task of reformulation, arguing that "the techniques of parody, irony, introspection, self-reflexiveness directly challenge the oppressive forces of social and literary authorities," offering "a new idea of history" (32). The first evaluation tends to limit postmodern fiction to a "disruptive complicity" (33) with the culture; the other credits innovative fiction with a rearticulative capacity. At the root of this theoretical hesitation is a dissociative model of narration in which deconstruction and rearticulation, invention and reflection are often at odds. Derived from a simplified application of Jacques Derrida's deconstruction, or of Michel Foucault's and Jean-François Lyotard's critiques of the falsely universalizing discourses of modernity, this model emphasizes divisiveness and incompleteness in narration. The role of postmodern writing, as Lyotard saw it, is to lay bare the modes of ordering that societies resort to in order to minimize risk and unpredictability. These modes situate and legitimize "first-order" practices of inquiry within a totalizing metadiscourse that makes "an explicit appeal to some grand narrative such as the dialectics of Spirit, the hermeneutics of meaning, the emancipation of the rational or working subject, or the creation of wealth" (*The Postmodern Condition* xxiii). Postmodern writing is called upon to undercut these powerful metanarratives, resisting

> the simple and naive exchangeability of things in our world. [. . . T]o write is necessarily to allude to something else which is not easily communicated [. . .] to advance or want something that is not clear, and to discover a means of giving testimony of that which is precisely not yet included in the circulation of commodities [. . .]. ("A Conversation with Jean-François Lyotard," Olson and Hirsh 173, 176)

When it does not simply suggest an economy of the "unpresentable" (*The Postmodern Condition* 82), Lyotard's theory of resisting writing translates into "discontinuous, catastrophic, nonrectifiable, and paradoxical" *petits récits* that recognize "the heteromorphous nature of the language games" and of reality (60, 66). These discontinuous local stories promote experimentation and differentiation, but they also prevent a comprehensive view of the cultural system they are part of. Therein lie both the strength and the limitations of the postmodern order envisioned by Lyotard.

The shortcomings become readily apparent when Lyotard's concept of nontotalizable language games is applied uncritically to postmodern practices. One consequence is the denial of any self-reflexive perspective from which a critique of narrative and cultural articulation may begin. To quote the neopragmatist wisdom that best represents this position, traditional interpreters believed in a "master narrative" that provided a model of "general rationality" predicated on an equally general ignorance of contingencies (Fish 120). Today's interpreters are "never quite able to take themselves seriously because they are always aware that the terms in which they describe themselves are subject to change, always aware of the contingency and fragility of their vocabularies and thus of their selves" (Rorty 73–74). But while appropriately questioning the totalizing claims of a traditional hermeneutics, the new pragmatists deny also a more flexible form of a cultural critique that encourages leaps outside one's own perspective. In the absence of a serious examination of, and exchanges between, local narratives, we are left with a "*socius* of separate communities possessing no control or power over each other" (Leitch 112).

Lyotard's own position in this respect is more complex, allowing for intersubjective and interdiscursive negotiation. He describes society as an "interweaving of heteromorphous classes of utterances" in which individuals participate as nodes that mediate several discourses (*The Postmodern Condition* 65). Since there is no "language game or genre or discourse which is able to encompass all the different discourses or genres," postmodern writing remains engaged in a negotiation with other discursive systems, expressing its own *différend* ("A Conversation with Jean-François Lyotard," Olson and Hirsh 180). But this negotiation takes agonistic-competitive forms in Lyotard, even if the "agonistic" is not between people but rather "between language games or genres of discourse" (179). Lyotard's opposition between "grand" and "smallish" narratives further accentuates the divisiveness of the postmodern project, organizing it around "different ways of narrating [. . .] stories" (181).

The task of defining a cooperative rather than competitive model of signification that would reconcile the disruptive side of postmodern innovation with the goal of rearticulation has been pursued more vigorously by postdeconstructionist, postcolonial, and postmodern feminist theorists. Feminism's challenge, according to Sandra Harding, is to articulate a gender-specific epistemology as a defense against male claims of "objectivism/universalism," on the one hand, and self-denying relativism, on the other (87). For this particular task Lyotard's "agonistic theory of language and paralogistic theory of legitimation cannot serve as basis" (Benhabib 122). The tendency of Lyotard's version of postmodernism to put everything "under erasure" undermines important cultural concepts, such as those of knowing subject, gendered agent, female experience. As Nancy Fraser and Linda J. Nicholson insist, feminism still needs conceptual "narratives about changes in social organization and ideology, empirical and social-theoretical analyses of macrostructures and institutions, [. . .] critical-hermeneutical and institutional analyses of cultural production" (26). In order to remain effective, feminism must distinguish "master narratives," which subsume the interests of different groups, from "grand narratives" (Aronowitz and Giroux 70), which articulate big stories such as the subordination of women in patriarchy or the exploitation of native cultures in colonialism. The latter are still useful provided they remain "pragmatic and fallibilistic" (Fraser and Nicholson 35).

But the task of articulating larger narratives about women is not without pitfalls. As long as they stay within the framework of essentialist realism, feminist theory and fiction risk perpetuating some of its totalizing gestures: the logocentric search for "ultimate causes," the subordination of contingency to abstract generalization, and teleological plotting. In writing a coherent narrative of women's struggle against male domination, feminism may overlook the "historical specificity of different societies and groups," forgetting that "there is no unitary idea of identity, and gender is only one kind of social division among others—race, class, ethnicity, age, and sexual preference" (Przybylowicz 298). Therefore, feminism will profit from broaching its conceptual categories with a postmodern awareness of their manifold and provisional nature. Postmodernism and feminism can collaborate toward "an epistemology and politics which recognizes the lack of metanarratives and foundational guarantees but which nonetheless insists on formulating minimal criteria of validity for our discursive and cultural practices" (Benhabib 125). The resulting postmodern feminism can be *experimental,* testing new modes of writing, but also *collaborative-transactive,* replacing the agonistic environment posited by Lyotard and by the entrenched tradition of "social Darwinism" that presents "life as battle, everything in terms of defeat and victory: Man versus Nature, Man versus Woman, Black versus White, Good versus Evil, God versus Devil" (LeGuin 36) with a cooperative environment in which the relationship between object and subject, self and other, identity and difference is continually renegotiated.

The concept of difference itself has been reevaluated in recent postmodern feminism and certain postcolonial projects, being removed from the earlier emphasis on a "single concept of 'otherness' [that] has associations of binarity, hierarchy, and supplementarity [. . .] in favor of a more plural and disprivileging concept of difference" (Hutcheon, *A Poetics of Postmodernism* 65). For Rey Chow, the indiscriminate embrace of a pristine, dehistoricized difference is no less problematic that the traditional reification of identity. Both rely on a process of idealization (of self or other) that "is always associated with violence" (xxxiii). Therefore, the play of cultural differences should be invoked not in order to enhance division but rather to create a more responsive environment for intercultural translation and negotiation. The interstitial "third space" between global and national cultures or between Self and Other imagined by Edward Soja, Homi Bhabha, or Rey Chow liberates difference from the Western binary logic, allowing it to create "new forms of meaning, and strategies of identification, through processes of negotiation where no discursive authority can be established without revealing the difference of itself" (Bhabha, "DissemiNation" 313). Translating this space in feminist terms, Donna Przybylowicz proposes to replace the "ideological division of the world" with a "notion of intersecting and relating realms" (268) that encourage interaction without absorbing "differences into a universalizing discourse" (261). The deconstruction of the dominant cultural-linguistic system is a necessary stage in this project, but it is complemented with a "utopian moment," a search for a "language [that] can change society" (290) and mediate between "antagonistic voices" (261).

A similar attempt to integrate these conflicting tasks into an alternative model of cultural production can be found in Mihai Spariosu's work, from *Literature, Mimesis and Play* (1982), *Dionysus Reborn* (1989), and *God of Many Names* (1991) to his more recent

The Wreath of Wild Vine (1997). His books can be read as stages in a larger project that attempts to redefine the "relationship of literature to other modes of discourse in Western thought by focusing on the notions of play, mimesis, and power" (*God of Many Names* x). The history of these interrelated terms is recapitulated in the form of a bifurcated narrative that has as one plot the gradual suppression/regulation of an-archic play through mimesis. As redefined by Plato and Aristotle, mimesis subordi-nates the "spontaneous, free, [and] arbitrary" power of poetry to a principle of "reason, Ideal Form, and Eternal Order" (*Literature, Mimesis and Play* 21–22). A coun-terplot emerges subsequently with Kant and German idealism that begins the "un-even, and by no means irreversible process of restoring play to its pre-Platonic high cultural status," as "an indispensable cognitive tool" (9, 22). Spariosu urges us to see these plots as polytropic and complementary. Thus play can lead to the "creation and establishment of certain [. . .] power configurations" (*Literature, Mimesis and Play* 10), but it also problematizes our power mentality, diffusing conflict into heterogeneous forms such as "agon, chance, necessity, play as freedom, mimicry, and simulation" (*Dionysus Reborn* xiii). Mimesis, in turn, seeks to convert the "immediate power con-figurations [of anarchic play] into mediated, representational ones"; but the repres-sion/mediation of an "unshamed and violent" prerational power by a "rational mentality" creates a "loss of presence and a yearning for (absolute) authority" (11). Literary mimesis is thus doubly constituted, reflecting the "conflictive [. . .] nature of Western humans and their play: gentle, reasonable, and peace loving on the one hand, and competitive, intractable, and warmongering on the other" (xiv).

Even from this brief synopsis, it is obvious where Spariosu's intellectual sympa-thies lie. His comparative archeology of Western thought has included from the be-ginning a critical dimension, interrogating the continuous resurgence of the "power principle" in our cultural practices and proposing in its place a model of semiosis based on mediating play. *The Wreath of Wild Olive* sharpens Spariosu's critique of our conflictive "dialectic of identity and difference" (155). Calling into question both the totalistic notions of identity inherited from the rationalist tradition as a whole, "which erase, reconcile, or subordinate all difference" (82), and the agonistic con-ceptualization of difference as "unique, singular, and irreconcilable" (82) in Niet-zsche and his twentieth-century heirs (Heidegger, Gadamer, Derrida, Foucault, and Deleuze), Spariosu urges us to embrace an "irenic mentality," that is, a "mode of thought, behavior, and pathos grounded in the principle of peace" (*The Wreath of Wild Olive* 274, n. 1). Through "ludic liminality and irenic transformation" (xii), Spariosu hopes to move us beyond both a stubborn mimetic tradition, which subordinates the play of difference to a totalizing order of Reason, and the agonistic reassertion of difference, which endorses violence as inescapable.

Spariosu's search for alternatives to mimetic thinking that would engage other-ness in the frame of "irenic, responsive understanding" (xv) has relevance for our discussion of innovative fiction. In Spariosu's view, modernism and postmodernism rehearse and partly transcend "the relentless, but inconclusive, agon between archaic and median values that has been central to Western civilization" (211). Modernism appears to support median values of reason and morality, but mostly in a "parodic, and problematic fashion" (211). Postmodernism recovers the "archaic," prerational values in Western culture, but its work is both "archeological" and critical, denounc-ing the subordination of all cultural production to a concept of power dissimulated as "reason, knowledge, morality, and truth" (245). Its work is pluralistic, translating into a "ludic-liminal discourse of the Other and otherness" (214).

The three authors chosen by Spariosu in support of his argument—Nabokov, Lowry, and Orwell—make the demonstration somewhat difficult by falling between the modernist and postmodernist paradigms, even though it could be argued that it was their very "in-betweeness" that allowed them to subvert boundaries, problematize identities, and open themselves to the experience of "an irenic alternative world" (225). Spariosu's demonstration could have benefited from the inclusion of more recent postmodern writers who have exposed the violence of cultural polarization, seeking nonantagonistic definitions of sexual, racial, and socioethnic difference. The solutions advanced by the polysystemic novel, surfiction, and postmodern feminism parallel the current theoretical search for a model of nonconflictive signification. Keenly aware of the "economy of violence" inherent in patriarchal discourse (Derrida, *Writing and Difference* 117), postmodern feminism has tried to recover those alternative semiotic relations (antiabstract, multiple, collaborative) repressed by the "war within discourse." Similarly, surfiction has used improvisational techniques to destabilize hierarchical codes, replacing a power-oriented representational language (figured in Sukenick's work as an "art of rape") with a process-oriented language of experiencing.

The products of this experimental imagination mediate between identity and difference, self and other, signification and play. The representational paradigm is not rejected but revised, converted into a form of "liberated" mimesis that includes the "free constructive activity of the inventive faculty" (Spariosu, *Dionysus Reborn* 246).

b. "Interventive" Narration or "Midfiction"?

We must distinguish the imaginative mediation pursued by innovative fiction, which reconfigures a culture's narratives, from the amalgam of metafiction and traditional realism advocated beginning in the early 1980s in response to postmodern deconstruction. "Midfiction"—as Alan Wilde called this mainstream version of experimentation—"negotiate[s] the oppositional extremes of realism and reflexivity (both their presuppositions, and their technical procedures)," subordinating metafictional techniques to a mimetic poetics (Wilde, *Middle Grounds* 44–45). The limitations of this new narrative category are made clear through Wilde's notion of "suspensive" irony. In contrast to both a constructive modernist irony, which projects a stabilizing aesthetic order on the modern anarchy, and the "chaos-drunk" irony of surfiction (141), the suspensive irony of "midfiction" works obliquely, "by way of some strategic *écart* or swerve in its fabric," to allow us to glimpse "the moral perplexities of inhabiting a world that is itself, as 'text,' ontologically ironic, contingent and problematic" ("Strange Displacements of the Ordinary" 192). Other cases of postmodern irony discussed by Wilde (e.g., the "generative" wit of Coover and Barthelme) show more "willingness to live with uncertainty, to tolerate, and, in some cases, to welcome a world seen as random and multiple, even, at times, absurd." But even this irony is carefully qualified, praised for creating "anironic enclaves of value in the face of—but not in the place of—a meaningless universe" (*Middle Grounds* 148).

The defense of "midfiction" is usually accompanied by the claim that distinctions between innovative and traditional narratives have collapsed in the 1980s. According to John Barth's much-publicized "Literature of Replenishment" essay (1979), postmodernism can survive only if it transcends the opposition between "modernist" (experimental) and "premodernist" (traditional) literature, keeping "one foot always in the narrative past [...] and one foot in, one might say, the Parisian structuralist present" (*The Friday Book* 204). Even a steadfast innovator like Clarence Major concedes "that the spirit of radical experimentalism and innovation gradually mellowed out during the seventies and eighties and are now finding their way into the mainstream of American writing. [...] Subliminally their influences are there throughout just about the whole spectrum of American fiction today—so much so that we don't notice that they are present in a more diffuse way in the culture and in American fiction writing" (interview in McCaffery, *Some Other Frequency* 259). While Major's comment still admits a useful cross-fertilization between mainstream fiction and radical innovation, other critics have expressed doubts concerning the impact of experimental fiction in the neoconservative 1980s and 1990s. As the editor of a *Chicago Review* double issue on "{In/Re}novative Fiction" noted, experimental techniques such as the use of the present tense, self-reflexive metaplots, second-person point of view, or the merging of reality and fiction "have been assimilated into standard practice and are used by mainstream writers [...] not so much because they seriously serve for 'making strange' any longer, but as signs of allegiance to the tradition which spawned them" (Lauzen, "This New Text" 7). And yet this special 1981 issue and others published since prove that postmodernism continues to offer innovations that—by Sarah E. Lauzen's own admission—represent a welcome alternative to the "Victorian appearance" (5) of the contemporary house of fiction. The revisionist poetics of innovative fiction still provides a better chance for negotiating cultural polarities than bland amalgams like "midfiction."

Admittedly, some forms of "midfiction" are daring enough in their use of revamped avant-garde techniques to make a difference. The examples usually cited are the "New Wave" science fiction of William Gibson, Samuel Delany, Johanna Russ, and Tom Disch (McCaffery, "The Fictions of the Present" 1167–70; McHale, *Constructing Postmodernism* 225–68); also the "new young thugs of innovative fiction" Mark Leyner, Mark Amerika, William Vollman, Eurudice, and Criss Mazza (Federman, *Critifiction* 131). Still, with the exception of Delaney, Russ, and a few well-established innovators added by Larry McCaffery to this group (McElroy, Atwood, Piercy, DeLillo), the "new wave" science fiction does not take its "nontraditional notions of 'realism'" ("The Fictions of the Present" 1170) too far. Even Gibson's prototypic cyberpunk novel *Neuromancer* (1984) could not find liberating narrative alternatives to the cyberspace of "consensual hallucination" accessed by its characters via computer jocks and "simstim" implants. Except for the periodic disruptions and enlargements of the narrative perspective through the appropriation of somebody else's viewpoint, the narrative proceeds conventionally, illustrating for Brian McHale the "centered, centripetal" model upon which postindustrial global capitalism continues to rely (*Constructing Postmodernism* 260).

Computer-created hypertext fiction has a better claim to revolutionizing our reading and writing habits, but even this mode runs—according to a supporter— "the risk of being so distended and slackly driven as to lose its centripetal form, to give way to a static low-charged lyricism—that dreamy, gravityless, lost-in-space feeling of the early sci-fi films" (Coover, "The End of Books" 25). Though more genuinely dialogic than printed fiction, hypertext fiction inevitably limits its interactive creativity. We could, in fact, argue that a printed novel provides the reader with the experience of a "nearly endless narrative" more easily than an electronic hypertext whose complicated navigation and pressures of "randomness and expansiveness might come to feel as oppressive to fiction writers [and readers] as linearity and closure did for modern and postmodern writers" (Travis 108).

The emergence of "midfiction" in the 1980s has narrowed the boundary between experimentation and traditional fiction but has not erased it entirely. Some blurring of distinctions is, of course, not only inevitable but also desirable in this age of hybridization and boundary crossing. As Larry McCaffery notes, generic differences are today

> even more difficult to maintain than they were only a quarter of a century ago. Should rock videos by Madonna, Peter Gabriel, or Laurie Anderson be considered mainstream simply because they are enormously popular—even though they employ visual and poetic techniques that twenty-five years ago would certainly have been considered highly experimental? Is William Gibson's "cyberpunk" novel, *Neuromancer,* "avant-garde" since it employs unusual formal techniques (the use of collage, cut-ups, appropriation of other texts, the introduction of bizarre new vocabularies and metaphors)? Or does the publication by the genre science-fiction industry establish it as pop? ("The Avant-Pop Phenomenon" 216)

Experimental postmodernism has contributed directly to this scrambling of codes. But while the polysystemic fiction of Coover and Pynchon, the surfiction of Federman and Sukenick, or the postmodern feminism of Acker and Russ have pursued the deconstruction of traditional genres as part of a larger revisionistic project, recent hybrids such as cyberpunk fiction, "avant-pop," or "fusion fiction" have largely given up literature's ambition to function as an oppositional discourse. Despite its affinities with earlier experiments, avant-pop departs from oppositional postmodernism, trying to reconcile innovation with techno-pop culture. On the surface, the project of "combining Pop Art's focus on mass culture with the avant-garde's spirit of transgression" (McCaffery, "Reconfiguring the Logic . . ." 1) sounds attractive. Yet the desired synthesis remains lopsided as long as

> Avant-Pop shares with Pop art the crucial recognition that *popular* culture, rather than traditional sources of high culture [. . .] is now what supplies the citizens of postindustrial nations with the key images, character and narrative archetypes, metaphors, and points of reference and allusion that help us establish our sense of who we are, what we want and fear, and how we see ourselves in the world. Thus, the content of Pop Art and Avant-Pop overlap to the extent that they both focus on consumer products. (McCaffery, "Avant-Pop" xviii)

Other critics have found this overlap questionable, indicating a surrender of fiction's iconoclastic power. As Geoffrey Green argued in the 1988 Brown University symposium on innovative American fiction, the "critical functionality of the term 'postmodernism'" (*Unspeakable Practices* 262) and of the strategies associated with it are preempted when they become staples of the entertainment industry. The flaunting of self-referentiality in rock videos, TV shows, and advertising does not make these works culturally "experimental." By contrast to Kathy Acker's politically charged punk novels or to the self-questioning films of Woody Allen, Robert Altman, Lizzie Borden, Terry Gillian, David Lynch, or Yvonne Rainer, the sophomoric postmodernism of *Moonlighting, Rock, Wayne's World,* and *Beavis and Butt Head* accomplishes little beyond a "self-reflexive savagery" that moves us "from Semio-text to semiliterate butt jokes, all in one lifetime" (Leland 61).

McCaffery himself dissociates this spurious form of "avant-pop," resulting from the repackaging of avant-garde techniques by popular culture, from a critical version that parodies and subverts mass market genres. In the first instance, the term "avant-pop" simply registers the unprecedented expansion of pop culture in the age of "Hyperconsumer Capitalism," when "pop culture has not only displaced nature and 'colonized' the physical space of nearly every country on earth, but [. . .] has also begun to colonize even those inner, subjective realms that nearly everyone once believed were inviolable" ("Avant-Pop" xii, xiii). In the second instance, "avant-pop" represents a critical response to a culture of "disposable consumer goods, narratives, images, adds, signs, and electronically generated stimuli" (xiv). This response ranges from an ironic recycling of the archetypes of pop culture, to more disruptive strategies that call into question the "multidimensional *hyperreality*" (xiv) of consumer capitalism. With few exceptions, the writers included by McCaffery in *After Yesterday's Crash: The Avant-Pop Anthology* (1995)—from Coover, DeLillo, Federman, Sukenick, and Vizenor to Mark America, Eurudice, Rikki Ducornet, and Curtis White— reconfigure pop archetypes as part of a broader engagement with the cultural ideologies that construct people's lives and identities.

What these writers have in common is the "interventive, i. e., aggressive, interactive" impulse of their fiction, "leading to [cultural] action" (Sukenick, "The Rival Tradition" 4). In contrast to the "accidental avant-garde" discounted by John O'Kane because it "inadvertently registers the crisis of postmodern society, always trying to sell what it is in fashion" (179–80), "interventive" fiction reconfigures the given narrative and cultural codes. In that sense, innovative fiction exemplifies not an "aesthetic deadend endgame," as Sacvan Bercovitch has proposed, but rather the type of risk-taking "middle game" he associates with ideological criticism: "The middle game [. . .] concerns positionality, inventiveness, and strategic combinations. It specializes, we might say, in potentiality. It works through indirections, circumventions, and possibilities." Bercovitch invites us to imagine "a perennial middle game, one that requires you, at each decisive juncture, to reconfigure your strategy—in effect to start anew—within a complex situation-in-process [. . .]. What would that game look like?" (38). It would look much like the "game" of innovative fiction.

c. A "Cunning Balance": Innovation as Rewriting

In Clarence Major's inspired description, innovative fiction "takes chances; defies categories; barks at the moon; is generally difficult and cranky but often sweet and sour, bitter and profound and funny. It's never comfortable in the company of well-mannered fictions; it's impolite and even criminal at times; it's an outlaw sometimes watched by the guardians of proper culture" ("A Meditation" 163). A path-breaker and an "outlaw," innovative fiction works both inside and outside mainstream culture to redefine relationships between margins and centers, representation and invention. What innovative fiction seeks is a "cunning" balance (Federman, *Double or Nothing* 9) between its interrogation of narrative and cultural orders and its need to define new semiotic possibilities. This transformative dialogue starts at the level of language but engages the logic of social sense as well.

A difficulty we continue to confront in discussions of innovative fiction is the absence of concepts flexible enough to account for its "double-coded politics" (Hutcheon, *The Politics of Postmodernism* 101), interested in denaturalizing and revising the narrative configurations of culture. The twofold focus on narrative and cultural *(re)articulation* I have been proposing beginning with my articles "Postmodernism Beyond Self-Reflection: Radical Mimesis in Recent Fiction" (1991) and "Narrative Innovation and Cultural Rewriting: The Pynchon-Morrison-Sukenick Connection" (1993) may be a step in the right direction. This conceptual focus valorizes the sociocultural reformulation that goes on in self-reflexive art, replacing a more limited understanding of postmodern innovation as mere subversion; it also acknowledges the self-problematized nature of the postmodern revisioning. Innovative fiction interrogates traditional modes of narrative and cultural articulation, seeking to reconnect us with ignored aspects of individual and collective experience. As Sukenick put it, before "we write fact into language again," we need to "unwrite the book of life till the difference between language and the rest of experience is as clear as possible" ("Unwriting" 4). The novelist's effort begins with an unwriting of "what has been formulated as experience," but continues with an imaginative rewriting that allows "a new sense of experience" to develop (26).

In the words of Maurice Blanchot, quoted by Federman in *Critifiction* (124), "to write is always first to rewrite, and to rewrite does not mean to revert to a previous form of writing, no more than to an anteriority of speech, or of presence, or of meaning. To rewrite is a form of undoubling which always precedes unity, or suspends it while plagiarizing it." This type of "pla(y)giaristic" rewriting—to use a choice Federman term—invents "new sets of rules by which the familiar pieces could be rearranged." It also "liberates" difference within cultural reality (*Critifiction* 125–26), allowing us to "re-vision" it. As defined further by postmodern feminism, "re-vision" is both an "act of looking back, of seeing with fresh eyes"—hence an "act of survival" for those who have been misrepresented (Rich, *On Lies, Secrets, and Silence* 35)—and an act of looking forward, shifting attention from the given representations "to the axis of vision itself—to the modes of organizing vision and hearing which result in the production of [those] image[s]" (Doane, Mellencamp, and Williams 6).

Innovative fiction has approached the problem of rewriting with a certain degree of trepidation. Such postmodern titles as *The Unnamable, Texts for Nothing, Unspeakable Practices—Unnatural Acts, Amalgamemnon, Lost in the Funhouse, The Death of the Novel, The Ticket that Exploded, The Voice in the Closet, The Exagggerations of Peter Prince, My Amputations, The Crying of Lot 49, Mulligan Stew, Long Talking Bad Conditions Blues,* and *The Burning Book* testify to the difficulties of articulating a new vision in a fragmented, self-reproducing culture. Still, innovative fiction has found ways to overcome its own impossibility and self-advertised death. At the end of the 1960s, Ronald Sukenick drew a sizable collection of stories from the idea of "the death of the novel," noting that the contemporary novelist "who is acutely in touch with life" had no other alternative but "to start from scratch: Reality doesn't exist, time doesn't exist, personality doesn't exist. God was the omniscient author, but he died; now no one knows the plot, and since our reality lacks the sanction of a creator, there's no guarantee as to the authenticity of the received version" (*The Death of the Novel* 41). Raymond Federman also wrote his first novel in the "funerary climate" of the self-questioning 1960s, confronting simultaneously "the impossibility and necessity of writing. [. . .] It was by doubting history, society, politics, culture, as well as his own art, [. . .] that the writer somehow managed to do his work" (*Critifiction* 114–15). Faced with the "Cold War dogma" and the "threat of the nuclear apocalypse," the postwar novelist felt, in Coover's words, "that so much of the trouble we found ourselves in was the consequence of not being imaginative enough about the ways out" (interview in LeClair and McCaffery 66). In order to become possible, postmodern fiction had to create an imaginative breach in the Cold War narratives of containment, turning a "crisis of representation" into a successful narrative production. In the end "the story did get told" (Federman, *Critifiction* 116), but the "holes," "gaps," "closets," and "precipices" (86) turned out to be as meaningful as what got written—a fitting testimony to the "impossibility and the necessity of the act of writing in the Postmodern/Post-Holocaust era" (87).

3. Rewriting History's "Ghoststories": The Bifurcated Focus of Innovative Fiction

As a fabric of events "lodged in the room of the infinite possibilities they have ousted," history is for Stephen Dedalus not too different from a "ghoststory" (Joyce, *Ulysses* 9.148). Derrida also speculates that any narrative negotiation of the unsettled space between life and death, past and memory, "can only *maintain itself* with some ghost, can only *talk with or about* some ghost" (*Specters of Marx* xviii). A "logic of haunting" (10–11) weighs upon historical narrative, forcing it repeatedly to come to terms with the "ghost" of the past that "never dies" (75). The spectral presence of "numerous phenomena that still remain unexplained" makes "[p]eriodic revisions—literally new visions and perceptions—[. . .] necessary" (Southgate 65–66, 67). Rewriting is thus very much part of the process of historical construction, but its role remains ambivalent, both liberating and limiting. Imaginative re-visioning can disrupt the linearity of a given system of "truths," regarded as the "soul" of traditional History, opening it up to the promise of new "event-ness." But as Pynchon's *Vineland* suggests, rewriting also functions as a tool of cultural appropriation, subordinating contingencies to retrospective plots. No act of historical interpretation can escape this contradictory dialectic that generates both exciting new contours and distressing flatness. Therefore, to borrow Robert E. Spiller's recommendation in his introduction to the 1948 *Literary History of the United States,* "each generation should produce at least one [. . .] history of the United States, for each generation must define the past in its own terms" (vii). Postmodern fiction has welcomed this challenge, participating in a major critical revisioning of modern history. Alert to the ambiguities of the term "history," which covers both the totality of past human actions and the accounts we construct of them, postmodern fiction has concerned itself concurrently with historical events and with the "processes of historical thinking, the means by which history in the second sense is arrived at" (Walsh 16). More often than not, the reinterpretation of history is carried out from a contemporary perspective, highlighting—in Georg Lukács's terms—"the *prehistory of the present,*" those "historical, social, and human forces that have in the course of a long evolution molded our life, giving it the shape we now know" (337–38). It is also performed from the counterbalancing subject positions of women, people of color, gays, nomads, and sundry "others" who enrich history with their "ghoststories" and new present relevance. A novel like Robert Coover's *The Public Burning* (1977), which exposes the self-promotional superpower narratives of the 1950s, has even more relevance today, at a time marked by the emergence of new nationalistic and ethno-centric discourses. The faith that Uncle Sam—Coover's mouthpiece for the ideology of confrontation—has in unrestrained masculinist power resonates with the ideologi-cal platform of the new right: "To him, a closed frontier was like a hardened artery and too much government, too much system, too much political theory, was a kind of se-nility" (*The Public Burning* 205). Eisenhower's championing of U.S. economic interests in the world, summed up by Coover as a Western plot pitting "tall, handsome, blue-eyed" Ike (240) against foreign "swarthiness" (238), has found new applications in the Gulf War. Familiar is also Sister Fowler's complaint that "[w]e have refused to live under God's control, and now live under guvvamint control! [. . .] The tithes we re-fused God we must now pay in taxes!" (418).

Such forward-looking analogies should not be viewed as breaches of perspective, reducing all "history to contemporary history" (McEwan 4, 183). While Neil McEwan justifiably questions the "Whig view" of history "which makes the past a prologue to the present and distorts it by hindsight" (4), he too hastily ascribes this "lack of men-tal perspective" (6) to all postmodern historical "rewriting." Upon closer examina-tion, "rewriters" of history like Coover, Morrison, Pynchon, or Reed, prove to be just as interested in examining the past in its own terms, acknowledging "attitudes which are unlike ours" and "granting the difference in similarities" (McEwan 6). Polysys-temic fiction, surfiction, postmodern feminism, and postcolonial literature are con-cerned with recovering the differential potential of the past, reclaiming those stories that have never made it into the dominant History. In the words of a Morrison re-viewer, they all tell "ghost stor[ies] about history" (qtd. in Samuels and Hudson-Weems 135). The "ghosts" are revenant figures, "the very content of historical erasure" that return to haunt us (Felman and Laub 267); but also devenant figures, products of a changing understanding of past and present. The postmodern "ghost stories about history" seek not only to locate the roots of our *Zeitgeist* but also to rearticulate our cultural imaginary in response to the challenges of the past.

Retrieving the "ghoststories" of our past creates "effects of diversifying, exploring, experimenting, undoing, disorienting, and dehabituating" in our conventional un-derstanding of history (Connor, *The English Novel in History* 6). Readers have often found this type of historical rewriting discomforting. A characteristic complaint is that postmodern fiction nullifies our sense of truth, replacing "reality [. . .] not by the veracity but by the voracity of language" (Charles Newman 91). For Newman, such novels as Barth's *Letters*, Coover's *The Public Burning*, Gaddis's *J. R.*, McElroy's *Lookout Cartridge*, and Pynchon's *Gravity's Rainbow* "disclaim external evidence, eschew the reciprocal, the reportorial, the historical, and at times even the felt—but all with a prodigious verbal activity which insists that they are *creating* everything they tell us" (90–91). Similarly, Richard Andersen concluded his section on *The Public Burning* by charging Coover with "pillag[ing] history for material that [he] can process into show-off exercises that call attention mostly to the narrative designs of their author" (133). What Coover's critics objected to was not only the controversial use of a his-torical figure (Nixon) as the novel's raisonneur, but also the broader emphasis on history as construction. Coover's Nixon understands all too well the "power of the Word" to make history:

What was fact, what intent, what was framework, what was essence? Strange the impact of History, the grip it had on us, yet it was nothing but words. Accidental accretions for the most part, leaving most of the story out. We have not yet begun to explore the true power of the Word. (*The Public Burning* 136)

History rewards those skilled at "play[ing] games with the evidence, manipulat[ing] language itself, [making] History a partisan ally" (136). But Coover does not condone this type of historical fabrication performed in his novel by Nixon, Uncle Sam, or the Cold War press. His own rewriting works against such crude revisionism, retrieving the "ghosts" of a past ousted or misrepresented by it.

What innovative fiction calls into question are not documented historical realities but their narrative and conceptual representation. To borrow Edward M. Bruner's terminological distinctions, postmodernism applies its critical revisionism to "life as told (expression)," and to some extent to "life as experienced," in order to reclaim "life as lived (reality)" (7). While reminding us that experience is refashioned through storytelling, innovative fiction proposes alternative narrations meant to correct the distortions perpetrated by traditional history. This kind of re-visioning is essential to a healthy polycentric culture interested in pursuing "many different accounts of the past—none claiming any special privilege" (Southgate 8–9). Therefore, the right to raise "critical, methodological, epistemological, philosophic questions" about the way history "is thought, written, or established" must be defended incessantly against those who confuse critical thinking with vulgar revisionism:

> Whoever calls for vigilance in the reading of history, [. . .] or demands a reconsidera-
> tion of the concepts, procedures, and productions of historical truth or the presuppo-
> sitions of historiography, and so forth, risks being accused today, through
> amalgamation, contagion, or confusion, of "revisionism" [. . .] A very disturbing his-
> torical situation which risks imposing an *a priori* censorship on historical research or on
> historical reflection whenever they touch on sensitive areas of our present existence. It
> is urgent to point out that entire wings of history, that of this century in particular, in
> Europe and outside of Europe, will *still* have to be interrogated and brought to light,
> radical questions will have to be asked and reformulated without there being anything
> at all "revisionist" about that. Let us even say: on the contrary. (Derrida, *Specters of Marx*
> 185–86, n. 5)

Traditional historiography sought scientific assurance based on a few operational articles of faith, such as the claim that "Truth is absolute; it is as absolute as the world is real. It does not exist because individuals wish it to any more than the world exists for their convenience. Although observers have more or less partial views of the truth, its actuality is unrelated to the desires or the particular angles of visions of the viewers" (Handlin 405). Postmodernism calls into question such absolutes, suggesting that the past comes to us not as a unified story but rather as an "archival discourse consisting of multiple voices and heteroglot languages" (Leitch 141). Taking its cue from the philosophies of history developed immediately before and after World War II (the "negative dialectic" of the Frankfurt School that challenged totalizing visions of history, Heidegger's emphasis on the gap between "historicity" and "real history," Sartre's questioning of history's intelligibility in terms of the aspiration to individual freedom, Foucault's critique of the unified notions of historical agency and memory), postmodernism has foregrounded the problematic nature of all historical representation that relies on the power of narration for "truth." But while subverting history's grand plots, postmodernism also calls attention to other, "lesser narratives" that have emerged in their place, with equally self-serving agendas tied to nationalistic, ethnocentric, or corporate interests. A critical focus on the narrative and ideological articulations of history is, therefore, even more necessary today as cultures become divided by contending *petits récits*. Postmodernism can provide a new critical understanding, uncovering the subtle links between our narrative explanations of "history" and the "centers of power [. . .] that collect and order information about what is going on" (Vattimo 134). It can also offer imaginative rewritings of history, reclaiming details of everyday life that do not fit into easy patterns. Like the "new historicist," the innovative fiction writer wants to retrieve a "minor" social history from behind "major" political events, compensating for "the impossibility of fully reconstructing and reentering the culture" of the past by "constantly returning to particular lives and particular situations, to the material necessities and social pressures that men and women daily confront" (Greenblatt 5).

Critics who decry the ahistoricity of postmodernism disregard the outpouring of narrative texts that explore issues of memory, temporality, and historical representation. In pursuing a "desacralized" vision of events (Vattimo 140), innovative fiction has alternated between a retrospective-revisionist and a proleptic-transformative task. For example, Pynchon's *Gravity's Rainbow* and *Mason & Dixon*, Coover's *Public Burning*, DeLillo's *Libra*, and Reed's *Mumbo Jumbo* call into question the ideological frameworks that obfuscate individual experience. As examples of Hutcheon's revisionist "historiographic metafiction" (*The Politics of Postmodernism* 62–92), these novels disrupt official representations, retrieving the human potential suppressed by history's grand plots. On the other hand, Federman's *The Twofold Vibration*, Morrison's *Sula* and *Paradise*, and Sukenick's *98.6* and *Mosaic Man* adopt an "exploratory" form of historical rewriting, approaching self and humanity "from a potential point of view, preremembering the future rather than remembering the past" (Federman, *The Twofold Vibration* 1–2). Common to both approaches is a focus on the *process of articulation*, both in the narrative-experiential and in the cultural sense. Innovative fiction interrogates traditional modes of articulation, seeking new strategies to reconnect us with ignored aspects of individual and collective experience.

a. Focus on Narrative Articulation: Experiential Recovery and Improvisation

Innovative fiction employs self-reflexive procedures to call attention to the conflicting processes of narrative realization involved in the construction of historical or present-day "reality." Realism is its favorite critical target because of the role it continues to play, according to Jean-François Lyotard, in reconfirming a reality that modern socioeconomic orders—whether informed by the "power [. . .] of the party" or the "power [. . .] of capital"—have already emptied of meaning. In the destabilizing context of modernity, in which familiar objects and institutions no longer evoke "reality" except as nostalgia or mockery (*The Postmodern Condition* 75–76), realism takes on the task of "preserv[ing] various consciousnesses from doubt," reconfiguring the referent around a sharable point of view that "endows it with a recognizable meaning" (74). Postmodern art needs to challenge the rules by which these "reassuring" visions are constructed (74), concurrently with inventing "other realities" (77).

Both tasks have been successfully pursued by innovative fiction. Its self-critical, multivoiced narrative performances challenge the "great system of constraints by which the West compelled the everyday to bring itself into discourse" (Foucault,

Power, Truth, Strategy 91), foregrounding the myths upon which realism relies: the privileging of an authorial perspective, the logocentric faith in expressive plenitude, the presupposition of a canonical pre-text (Society, Truth, Reality) that the novel has to imitate. Federman's "surfiction," for example, generates a surplus of voices and retellings that cannot be easily resolved. While discrediting conventional representation, Federman's texts propose alternative modes of narrative articulation that project a paradoxical story "as it was, as it were, as it happened" (*The Twofold Vibration* 50). Likewise the all-inclusive and disorderly novels of Coover, DeLillo, Pynchon, and McElroy upset the interpretive habits enforced by realism (selectivity, searching for buried secrets, the expectation of meaningful closure). Polysystemic fiction encourages us to overread through a technique that attaches "greater or more elaborate significance to some object, event, or character than seems reasonably warranted" (Lauzen, "Men Wearing Macintoshes" 62). By inviting and at the same time frustrating a reading for minute significance, polysystemic fiction forces us to ask not "What does [the text] mean?" (60) but rather "where to look for meaning, and what to regard as significant in the evolution of innovative forms?" (62). Charles Newman rightly argues that such "profusion of discourse" threatens the "continuity of the legislative power in the political" and aesthetic orders (33). Innovative fiction interrogates not only Newman's expectation of a unified perspective, but also the novelist's effort to capture the "wholeness" of human experiences: "I lose them when I try to bring them back, even for description. The image tries to represent something it is not [. . .]. Twisted repressed moments fill their place" (Major, *Reflex and Bone Structure* 107–8). But the innovative writer's response to these questions is never simply negative: While denouncing the "extremely sophisticated system of illusions" created by traditional realism (Federman interview in LeClair and McCaffery 140–41), the novelist keeps trying to extricate a "real story" from the "edifice of words" (Federman, *The Voice in the Closet* 1, 18; my pagination).

All postmodern narrative participates, according to Lyotard, in a paradoxical dynamic that combines the task of "perpetually flushing out" the artifices of representation with that of reinventing the familiar rules and seeking a new "realization" (*The Postmodern Condition* 79, 81). This process of narrative (de)realization is especially complicated in historical fiction, requiring repeated establishment procedures that produce an incomplete, dilemmatic encoding of reality. For Lyotard, as the self-questioning historian of a post-Auschwitz/post-Gulag world, reality "is not what is 'given' to this or that 'subject,'" but "a state of "the *differend*" (*The Differend* 4, 9)—that is, a radical point of difference among the languages that try to negotiate events of the past. The "differend" emerges whenever history's victims or excluded have to define themselves in the language of the victors/victimizers, a language in which their suffering has no meaning. Since historical representation always involves totalizing narratives that suppress the story of the "other," the task of the postmodern historiographer is to listen "to that which does not speak through a tradition, or that which speaks through a silenced tradition" (Levi 52). The postmodern historian must perform simultaneously the role of witness, storyteller, and critical interpreter, fully aware of "the irreducible alterity of the past" yet trying to "recover that alterity in as close an approximation of 'how it actually was' as possible" (Spiegel 196).

The traditional historiographer pursued what he believed to be a depersonalized body of truth that existed independently of his acts of construction, and which nevertheless authenticated itself through the tautness of form. Thus George Saintsbury understood his task to be that of "intelligently comprehend[ing] all the accessible and important documents on the subject," followed by an effort to "digest" and "order" this information so that "not merely congeries of details, but a regular structure of history, informed and governed throughout by a philosophic idea, should be the result" (626). The historian was expected to produce "from the literary as well as historical side, [. . .] an organic whole composed in orderly fashion and manifesting a distinct and meritorious style" (626). By linking their notion of historical truth to a principle of rhetorical unity, traditional historiographers unwittingly called attention to the poetic side of their production. Postmodern historical writing foregrounds this relationship with a vengeance, exploiting "the metaphorical similarities between sets of real events and the conventional structures of our fictions," or between historical narratives—"the contents of which are as much *invented as found*"— and literature (Hayden White, *Tropics of Discourse* 82, 91). Like the novelist, the historian relies on "explanation by emplotment," through which "a sequence of events fashioned into a story is gradually revealed to be a story of particular kind" (White, *Metahistory* 7). André Bleikasten (9) expands on this comparison:

> [O]ne "does history" the same way one does literature, and that act of doing (*poïesis*) belongs to somebody, implicates a productive agent operating under particular circumstances according to specific modalities which leave their imprint on what is being produced. That is to say, history is necessarily historio-*graphy*: it is written and, as such, unfolds as a discourse, charged with intentions and aimed at producing certain effects.

Historical "poïesis" involves three broad constitutive acts: "a *mise en mots*, according to communicable linguistic and rhetorical codes; a *mise en scène*, in accordance with representational procedures that are likewise codified; and a *mise en intrigue* (*mythos*, in Aristotle's sense, or 'emplotment,' to borrow Hayden White's term)" (9). These acts simultaneously constitute and problematize history: "All recounting carries forth a fiction, is transformed into fiction by the word that engenders it. But all narrative, even a fairy tale, is also a denied fiction: upon hearing it or reading it, *one must believe it*" (9).

With its contradictory dynamic of narrative realization, postmodernism raises further questions concerning the status of events and of agents. In postmodern vision, history is composed not only of what achieves prominence or comes to pass but also of what does not: potential events, arbitrary coincidences, unplotted contingencies. The regime of events is left unsettled also by the ambiguity of historical and narrative agency (of doers and tellers). As a rule, the narrating persona in historiographic fiction adopts the perspectives of "historian" (diegetic-retrospective), "painter" (pictorial-reconstructive), "memorialist" (autobiographic), or "performer" (mimetic-testimonial; see Fleischman 61–63). In traditional historiography these perspectives are often mutually exclusive or in conflict. As a participant/witness in the sequence of events, the writer does not yet have, strictly speaking, a historical consciousness; he is at best a "chronicler, [. . .] always on the verge of participation, or at least of a presence in the action that is in effect the presence of a witness" (Genette, *Narrative Discourse Revisited* 104). As a historian, the writer becomes, in Gérard Genette's apt phrase, "a *subsequent witness*" (80), speaking from a dissociated extradiegetic perspective: the composite perspective of reconstructed history. Postmodern historical fiction, on the other hand, often intermingles these positions: The narrative voice can be simultaneously retrospective-authorial, dealing with verifiable "facts," and self-implicated/revisionistic, acknowledging that these narrative reconstructions are neither privileged nor complete. For example, Pynchon's narrators typically waver between the present tense, which suggests the position of a witness involved in history as process, and a foreboding future tense that suggests a fatalistic view of history. In Doctorow's fiction, new historical truths emerge from a "nonfactual witness which doesn't destroy the facts or lie about them or change them, but in some peculiar way illuminates them" (Doctorow, "Ragtime Revisited" 8).

Especially when confronted with the "unspeakable" historical events (the Nazi Holocaust, ethnic genocides, social catastrophes), innovative fiction problematizes every component in its narrative situation: "mode," "person" or "voice," and "perspective" in F. K. Stanzel's terminology (46–62). "Voice" is especially problematic in this type of fiction. A "'dialogization' *of the narrator's own voice*" takes place that involves not only Bakhtin's notion that the narrator speaks for a "polyphony" of voices but also a more radical "plurality of centers of consciousness irreducible to a common denominator" (Fleischman 220). The communicative circuit is also split between the various narrative relays and their intended narratees, trapping the reader in a continuous "narrational crossfire" (Coste 174). Even when the addressee remains problematic (which is often the case in recent fiction), postmodernism maintains a strong focus on the function of "addressivity" (Connor, *The English Novel in History* 11–13, 42–43), struggling to provoke a response, to build a new narrative contract. In lieu of the objectified, authoritative voices of traditional historiography, postmodern fiction prefers inclusive, dialogic voices such as the second person that accommodate "the implied reader in their referential field," placing "permanently on the actual reader the onus of defining himself in relation to the text and its enunciator" (Coste 176). As Federman's *To Whom It May Concern,* Morrison's *Beloved* and *Paradise,* and Pynchon's *Gravity's Rainbow* and *Mason & Dixon* suggest, history can only be (re)articulated through dialogic modes of narrative, its burdens shared.

The scandalous subjects of history can be better mediated if the "Cyclopean space" of traditional fiction is converted into a dialogic tangle of "writerly/readerly levels, vertical, horizontal and sagittal manners of perceptions and conceptions" (Major, "A Meditation" 165). A number of innovative techniques have been employed toward this goal. Inherited from avant-garde art and earlier mavericks like Rabelais and Sterne, the technique of *improvisation* has contributed significantly to the breakup of traditional plot structures, converting text-time into a "fluid montage of many possibilities" and bringing "the past and the present and the future together" in the "now-as-I-read experience" (Major, "A Meditation"168). Through improvisation, fiction "break[s] free of controlled storytelling," spinning new configurations that "exist *independently* of the strings which [. . .] connect them to the real world" (170). But this freedom is not unqualified. As Major is quick to point out, improvisation grants us "the power of doing the unexpected, of making huge reversals, turning societies, cultures around or up-side-down" (176), but even the most inventive work can be read "from left to right, so to speak, from experimentation to convention" (176). Therefore, innovative fiction must remain critical of its "truths," combining cultural revision with "an honest and unrelenting investigation of its own processes" (171).

Formalization and *invention* work in similar ways to open up the space of fiction to new, experimental configurations. The goal of formalization, as explained by Jean Ricardou (119–23), is to "denaturalize" fiction, foregrounding its articulatory principles. This self-ironic use of formalization is well represented in Barth, Coover, DeLillo, and Pynchon, whose fictions foreground a "geometric" mode of thinking obsessed with charting and patterning; but also in Kathy Acker's parodistic use of cut-up techniques and montages of citations that taunt the boundary between anarchy and order. Every Coover novel has at least one character who thinks that stories "somehow always had to do with numbers, numbers and sequence, and maybe this was God's game, too, having started maybe with one and two and set them humping" (*John's Wife* 52). Coover's own interest in games, numerological combinations, and repetitive patterns has a philosophic rationale. As the writer argued in an interview, under the present "conditions of arbitrariness" created by the "breakdown of religious structures and of many of the principles of the Enlightenment which have supported our institutions," it is incumbent upon art to put "the random parts together in an order which provides a pattern for living" (Gado 153). Art is called upon to negotiate a subtle balance between randomness and arbitrary orders. Pynchon's novels work in similar ways, pitting a dogmatic commitment to design (such as Herbert Stencil's in *V.*) to a phenomenological view that recognizes no pattern. His latest epic *Mason & Dixon* exposes the rationalistic roots of our geometric imagination, opposing to it the fluid (anti)cartography of liminal cities and "borderland" worlds.

Other writers have used formalization more systematically, replacing the plot-driven structure of narrative with a combinatorial matrix of formal elements. The phonic and syntactic permutations pursued by the *Ouvroir de Littérature Potentielle* (OuLiPo) group (Italo Calvino, Georges Perec, Raymond Queneau, and Harry Mathews) function as narrative "generators," engaging the reader in subtle rhetorical and cultural explorations. Thus, the absence of the most frequent French letter "e" in George Perec's novel *La Disparition* (1969) triggers a discourse about cultural and linguistic "disappearances." *La Vie mode d'emploi* (1978; *Life: A User's Manual*, 1987) by the same author resorts to two different sets of formal constraints—an expanded chess board and a complicated topological figure, the "bi-carré latin orthogonal d'ordre 10"—to dramatize the obsession of Percival Bartlebooth (whose name recalls both the mythic quester and Melville's "scrivener" Bartleby) with order-making. By contrast to OuLiPo's "art in a closed field" (McHale, *Constructing Postmodernism* 28), Maurice Roche's *Codex* (1974) undermines narrative intelligibility with its unstable visual-linguistic frames that alternate fiction, commentary, and self-citation. Roche's model of (dis)assemblage has attracted Federman's attention, suggesting to him strategies for "de-realizing" a conventional historical plot.

The American surfictionists have taken this experiment in a different direction, moving beyond the "cul-de-sac" formalism of earlier metafiction or the OuLiPo writers. "Through puns or through arbitrary devices, through the imposition of odd schemes, non sequiturs, even through improvisations" (*In Form* 111), Sukenick's fiction attempts to crack the protective shell of language and release its transformative energies. Its ambition is to reconnect us with "the mute world beyond language from which language above all separates us, and which, therefore, it has the power to restore" (*In Form* 39). Therefore, Sukenick and his colleagues have preferred a freer model of improvisation derived from jazz to the systematic combinatorial art of the *nouveaux romanciers*. As Sukenick explains, his fiction

has never been much concerned with form, unless to break down other forms, thereby creating a vacuum inviting new material. OuLiPo methods do the same thing, it's true, but then, a convention established, repeat programmatically, where an antiformalist like myself would go on to improvise as in a jazz solo, picking up on the last phrase and going from there. ("Unwriting" 4)

In surfiction, "[f]orm, like language, [comes] after the fact" ("Unwriting" 4), but it is not entirely spontaneous or open-ended. Like jazz, surfiction draws on the "thrilling tension between the 'freedom' of blowing and the imperatives of order" (Katz interview in LeClair and McCaffery 223).

Innovative fiction emerges thus from a paradoxical interplay of imaginative freedom and formal constraints. In most models of innovative fiction, some combination of approaches and tasks has been proposed. Sukenick's "generative fiction" is predicated on an odd blend of improvisation and abstraction, following simultaneously an experiential and a conceptual-formalistic route (see chapter 4). Postmodern feminism has been similarly divided between a deconstructive impulse (a desire to dismantle the traditional cultural-linguistic system) and a "utopian" search for a "language [that] can change society" (Przybylowicz 290). The various propositions of innovative fiction appear thus contradictory, difficult to integrate. Perhaps they need not be integrated but rather perceived as competing claims within the postmodern project. By emphasizing the tension between experiential improvisation and critical reflection, postmodernism can keep better track of the "large and small facts of our daily lives" (Sukenick, *In Form* 242), expanding our ontological and cultural purview.

b. Focus on Cultural Articulation: From "Resistance" to Reformulation

When innovative fiction problematizes narrative articulation, it also engages us in broader reexamination of the cultural conditions that allow one story to prevail over others. Its own narrative dynamic suggests that the "construction and choice of one story over others" is governed less by a relation to truth than by "a will to power," "a desire *not* to hear certain other voices or stories" (Flax 195). Therefore, postmodern fiction provides an ideal locus where claims of narrative/cultural authority are examined and reformulated (Ormiston and Sassower 15, 16). The "classical-canonic" view of authority governed by a transcendental "correspondence between narrative and reality" (99) is replaced in innovative fiction by an emphasis on the self-legitimizing power of narration, which can make and unmake historical worlds. As William H. Gass put it succinctly, "What we remember of our own past depends very largely on what of it we've put our tongue to telling and retelling. [. . .] Historians make more history than the men they write about [. . .]" (*Fiction and the Figures of Life* 126).

As long as power determines truth, "[w]ho claims Truth, Truth abandons. History is hir'd, or coerc'd, only in Interests that must ever prove base. She is too innocent to be left within the reach of anyone in Power"—so comments a young savant in Pynchon's eighteenth-century epic *Mason & Dixon* (350). The solution he suggests is proto-postmodern: History needs "to be tended lovingly and honorably by fabulists and counterfeiters, Ballad-Mongers [. . .] Masters of Disguise [in order] to provide her the Costume [. . .] and Speech nimble enough to keep her beyond the Desires [. . .] of Government" (350). The duty of the historian is to "seek the Truth, yet [. . .] do everything he can, not to tell it" in a way that would reify it, turning his account into a "Chain of single Links" (349). In Foucauldian terms, the postmodern historian needs to write an "effective" rather than an "objective" history, focused on "the reversal of a relationship or forces, the usurpation of power, the appropriation of a vocabulary turned against those who had once used it" (*Language, Counter-Memory, Practice* 154). This type of work would recognize the problematic nature of all representation, without giving up its effort to articulate alternative perspectives in which history would feature as a web of contending links.

A thorough rereading—such as I am attempting in this book—of some of the best narratives published in the United States since the mid-1960s reveals the fact that innovative fiction has been from the beginning concerned with redefining the relationship between history and fiction, narrative and cultural articulation. As an imaginative response to the "conditions we live in," innovative fiction has taken upon itself "to rescue experience from any system [. . .] that threatens to devitalize or manipulate experience" (Sukenick, *In Form* 11). The self-reflexive focus on narrative articulation has provided the necessary basis for a stronger project of sociocultural reformulation. The difference between parodistic metafiction and revisionist postmodernism boils down to the role they ascribe to historical re-visioning. If for John Barth, art cannot "redeem the barbarities of history," being content to "sustain, refresh, expand, ennoble and enrich our spirits along the painful way" (*Chimera* 17), for Coover, Federman, Morrison, Pynchon, and Sukenick art must engage the causes that produce historical traumas, imagining better sociocultural and narrative futures. Applying "the Utopian radicality of post-structuralist difference and alterity" to the "here-and-now," postmodern feminism is interested in "creating and transforming history. [. . . F]eminist critical theory [and fiction] aligns historiography and critical theorizing with the making and doing of history" (Radhakrishnan 193, 202–3).

The crucial connection between technical innovation and cultural reformulation has been acknowledged by theorists of the avant-garde, but its efficacy in contemporary art is still a matter of debate. The "greatest aesthetic and social challenge" of today's experimental art resides for Charles Russell at the level of formal constraints. Having outlived the individualistic ideals that animated the historical avant-garde, contemporary experiments "arise self-consciously—self-reflexively—from within their culture and discourse" ("Subversion and Legitimation" 55). Self-reflection becomes the instrument of a critique from within, focused on the sociocultural parameters that art "sets itself against and which it would change," but which also "limit or deny its activist desire" (56). Hence the ambivalence of this critique: While postmodernism fails to break up the dominant system, it still manages to denaturalize its ideological structures, illuminating "the arbitrary nature of all discourse, and by extension of all social systems of meaning and power" (57).

In an earlier essay devoted to fiction, Russell granted a stronger sociocultural function to postmodern innovation. Acting on the knowledge "that all perception, cognition, action, and articulation are shaped, if not determined by the social domain," innovative fiction employs its "new aesthetic and social configuration" oppositionally, to challenge the dominant meaning system ("Individual Voice in the Collective Discourse" 34). Innovative fiction adopts characteristic avant-garde techniques ("the aggressive disruption of language and of the literary text through fragmentation, aleatory structures, strained metaphors, collage, 'cut-ups,' self-reflexively arbitrary formal structuration") to alter the "reader's literary expectations and his habitual acceptance of social discourse" (34). The purpose of these techniques is often explicitly political:

For example, Pynchon interrogates the "confusing multiplicity of meaning systems," seeking some space for personal expression in the collective language that speaks us all; and Sukenick responds to environments that remain "chaotic, ambiguous, and antagonistic" by creating a "personal reality in the absence of acceptable social totalities" (35). To add one more example conspicuously absent from Russell's essay, innovative feminist writers wrestle with the dominant discourses, exposing their limitations and seeking out those "constitutive elements which can be appropriated for an alternative nonhegemonic critical practice" (Przybylowicz, Hartsock, and McCallum 7–8). In all cases, innovation acquires sociocultural significance precisely because of its capacity to disrupt and rewrite the prevailing meaning systems.

In order to properly assess this sociocultural role, we need to consider the fact that innovative fiction has redefined the notion of commitment for us. The programmatic essay added by Raymond Federman to the 1981 edition of his *Surfiction* collection distinguished between two forms of artistic commitment: one referential, the other self-referential. After quoting Sartre's plea for politically committed literature, Federman argued that contemporary literature must confront its own crisis of knowledge and language before it can address the historical crisis that concerned Sartre. The writer should begin by freeing fiction from "the kind of knowledge that is received, approved, determined by conventions. [. . .] The new novel affirms its own autonomy by exposing its own lies: it tells false stories, inauthentic stories; it abolishes absolute knowledge and what passes for reality" ("Fiction Today or the Pursuit of Non-Knowledge," *Surfiction* 300). But the novel is not content to function as a version of the "LIAR'S PARADOX" (308), giving the "lie" to the rest of culture. It also applies its reflexive incisiveness to the culture's narratives, "searching [. . .] within the fiction itself for the meaning of what it means" to construct fictions (300). This latter focus translates Sartre's concept of commitment into a concern with the writer's "freedom of speech [. . .]. This freedom to explore the possibilities of saying and writing everything is [. . .] as crucial and as subversive as what Sartre proposed some thirty years ago" (305). Innovative fiction seeks to "reinstate things, the world, and man in their proper place—in a [new truth]. That [. . .] is also a form of literary commitment" (306).

The fact that Federman's model of literary commitment draws more on Beckett than on Sartre, should not be regarded as a diminishment of fiction's social relevance. As Marc Chénetier has argued in a discussion of postmodernist "écriture engagé," the "commitment to [innovative] writing" is as important as political commitment (220). By contrast to a "text whose 'commitment' is reduced to ideas that are asserted in a monosemic production that makes impossible the reader's freedom of intervention," "polyvocal" experimental writing can liberate us from the "regime of monologic ideas and better resists didactic usage" (218). For Federman, Beckett represents the ideal commitment to writing—one that, without being openly political, has mapped better than any other project our postwar "journey to chaos" (to borrow the title of Federman's 1965 book on Beckett). Federman's understanding revises the customary view of Beckett as a writer of negation and impossibility, the very antithesis of Joyce (see Kenner 33–34; Hassan, "Joyce, Beckett, and the Postmodern Imagination" 183). In Federman's rereading, Beckett's confrontation with impossible hurdles (a world of absurd postwar divisions, a complex but impotent language, a process of signification reduced to silence or noise) did not result in a mere preempting of the narrative process, but rather in a putting forward of the "unpresentable in presentation itself," which is the task assigned by Lyotard to all postmodernism (*The Postmodern Condition* 81). This task requires the cooperation of contradictory narrative modes:

> The emphasis can be placed on the powerlessness of the faculty of presentation, on the nostalgia for presence felt by the human subject, on the obscure and futile will which inhabits him in spite of everything. The emphasis can be placed, rather, on the power of the faculty to conceive, on its "inhumanity" so to speak [. . .] since it is not the business of our understanding whether or not human sensibility or imagination can match what it conceives. (79–80)

The purpose of the first mode is to deconstruct "artifices of presentation which make it possible to subordinate thought to the gaze and to turn it away from the unrepresentable" (79). By contrast, the second mode draws on "the power of the faculty to conceive," putting forth a new realization both in the sense of a *mise en ouvre* and a new *mise en scène* of reality (81). Beckett's fiction bears out Lyotard's description, exemplifying a complex process of narrative (de)realization in the sequence *Molloy* (1951), *Malone meurt* (1951), and *The Unnamable* (1953), where the narrators of each subsequent novel claim to have authored the previous ones, opening up these narratives (and, indeed, Beckett's entire fiction) to competing enunciations. What emerges is not an aesthetics of "failure" but rather an aesthetics of (self-)revision. Beckett's "decreative" procedures are political at least by implication, blocking the culture's master narratives and putting "in question all aspects of narrative validity" (Hayman 113). As Beckett explained in a letter to Axel Kaun, the goal of today's writer is to "bore one hole after another in [language], until what lurks within it— be it something or nothing—begins to seep through" (*Disjecta* 132). In Beckett's case, what seeps through is the psychodrama of articulation, the feverish encounter with an elusive reality that confers a provisional status to everything said.

For the innovative novelist, confronted with the difficulty of articulation in a post-Holocaust, (post -)Cold War era, Beckett has proven a better guide than Sartre. Much innovative fiction revolves around the impulse to "reinstate" things in a "new truth." This impulse has created an unstilled oscillation between disarticulation/rearticulation, saying and unsaying. Like Beckett, the experimental novelist proclaims the "impossibility of saying the world," but mainly in order to reveal within it "the incredible possibility that everything can be said now, everything is on the verge of being said ANEW" (Federman, *Surfiction* 306). Even when more radical models than those offered by Beckett are used, innovative fiction still struggles with the issue of articulating an effective mediating story in an age dominated by forms of sociocultural and political division. For example, women's fiction has had to deal with the question of how to develop strategies of resistance within today's "multiple and by no means coherent field of allegiances" and discriminations (Przybylowicz, Hartsock, and McCallum 6). Responding to the heterogeneity of "subjugated social locations which women occupy" and to the awareness that woman's own identity is "plural, heterogeneous, and fractured" (6, 7), postmodern feminism has embraced a new kind of "negative hermeneutics," "vigilantly negotiat[ing woman's] contradictory positioning within a network of race, class, and gender" (6, 9). All innovative fiction works as a "negative hermeneutics," deconstructing the ruling narrative practices and seeking to develop a new cultural consciousness. As a result, innovative fiction is caught between conflicting *narrative logics*.

Propositions about the "real" world fall, according to the classical logic, under the regime of necessity. Propositions in fiction, "on the contrary, are governed by the modality of possibility; they require, in short, 'suspension of belief as well as disbelief'" (McHale, *Postmodern Fiction* 33). But it is this act of contradictory suspension that gives fiction its special "amphibious" status: neither true nor false, neither completely possible nor impossible.

Innovative fiction does not emerge from mere "impossibility," as D. W. Fokkema (81–98) and the early postmodern manifestoes proposed, but also from "hypothesis, provisional supposition," and revised projections (Sukenick, *In Form* 99). The two regimes, however, are kept in tension, as when the narrative material is introduced through what Matei Calinescu describes as the technique of the *palinode* (309–10): a common enough rhetorical strategy with the postmodernists who claim not to deal in plot details, while insinuating them nonetheless in the very act of their negation. Oftentimes the technique of the palinode dramatizes a difficult process of narrative enunciation, as in Toni Morrison's *Beloved* where the traumatic event of the slave woman's infanticide can be revisited only by Sethe, Stamp Paid, Baby Suggs, and Beloved herself, in silent narrations that may or may not find a narratee; or in Federman's fiction where a sequences of x's mark the unspeakable event of his family's disappearance in the gas chambers. Federman's narrators circle continually this radical erasure, knowing like Sethe that they "could not close the circle, pin it down for anybody who had to ask" (*Beloved* 163). Postmodern articulation, as Philippe Sollers suggests, must remain provisional, an amalgam of telling and retraction:

> [T]o *tell all* signifies something quite different today than a sort of indefinite, manifestly scandalous verbalization. To *tell all*, and this involves no paradox whatsoever, would be to refuse vigorously to state what have you, but *while stating* it, or again to reject such verbalization insofar as it continually conceals and justifies novelistic mystification. (192)

The palinodic, unresolvable text of postmodernism dramatizes the essential circularity of all fiction in its "encounter with something resembling the Real, while not being either the truth or its meaning. Fiction never ceases to repeat itself, to repeat the expectation of an encounter. An encounter perhaps always missed, whose failure it keeps describing for us. [. . .] That encounter is encircled on all sides by narratives, by descriptions, by symbols [. . .] that are set in motion the very instant a photograph [of reality] is taken" (Durand, "Image, récit" 94–95).

Far from being a mere formalistic fancy, the concern with "the circumstances of [fiction's] own possibilities" (Federman, *Surfiction* 292) engages the potentialities and limitations of social imagination. As a rewriting of the cultural order, postmodern fiction "re-invents the nothing new" (Federman, *Critifiction* 128), relocating it within more responsive structures. At the same time, literature "re-invents what [has] been banished, hidden, or expelled from individual or collective memory" (128). These tasks come into conflict with the given cultural field: While "everything can be said, and must be said, in any possible way" (Federman, "Surfiction—Four Propositions" 12), everything has already been said. Therefore, the innovative novelist must use a double strategy that interplays "unwriting" with "rewriting." The novelist's effort begins with a deconstruction of the "sanctioned descriptions of life" (Sukenick, "Unwriting" 26); but it is followed by an imaginative rewriting that creates "a new and more inclusive 'reality.'" Innovative fiction combines thus a poetics of *resistance* with one of *reformulation,* highlighting the process of continuous rearticulation that the novel performs on life. Though the goal of producing a "new reality" that did not pre-exist the work of fiction recalls the Romantic concept of autonomous creation, the innovative writer pursues it with an awareness of the ideological investments inherent in every narrative act. The process of rearticulation must go on indefinitely because it engages not only the "sanctioned descriptions of life" but also fiction's own "fossilized formulas of discourse" (Sukenick, "Autogyro" 294).

4. INNOVATIVE FICTION IN THE POST–COLD WAR TRANSITION: CHARTING A COURSE BEYOND THE MASTER PLOTS OF GLOBALIZATION AND "END OF HISTORY"

Even though the momentous events of 1989 have put an end to the confrontational world of superpower politics, they have not brought about an instant democratization of sociocultural exchanges. The post–Cold War period has freed the social imaginary of the vestiges of competing grand narratives, but it has often replaced them—and not only in the former Eastern European bloc—with narratives of a nationalistic or ethnocentric kind that promote invidious cultural distinctions. Much of this new ethnic fundamentalism has emerged in direct reaction to the pressure of the First World's "globalizing" ideologies that, far from being "deimperialized," reinforce the "international division of labor and appropriation [. . .] benefiting First World countries at the expense of Third World" (Ebert 286). While globalization promotes "the detachment of cultural material from particular territories and its circulation in repackaged heterogeneous, boundary-violating forms" (Buell 42), its arrangements favor on the whole the consumer cultures of the West. Hegemonic rather than truly multipolar, the "new world order" is "disturbingly reminiscent of the old" (Lazarus 95).

The new tensions between global interdependence and ethnocentric separatism, First World centers and Third World peripheries, speak to a state of ideological crisis typical of periods of transition. Some have blamed this crisis on the "weak thought" of postmodernism, even though its hybrid methodologies have been more helpful on the whole in negotiating such tensions than "stronger" and more polarized political projects. We still must ask whether postmodernism can continue to make a difference in a post-1989 world characterized by the erosion of oppositional discourses and—if we take Baudrillard seriously—by a more general breakdown of our sense of reality in the "hysteria of production and reproduction of the real" (*Simulations* 44). Are aesthetic "resistance" and narrative innovation viable strategies in the post–Cold War restructuring; can literary discourse mediate between the ethnocentric concepts of culture that have reemerged in many places? The dissident, antifoundational approaches developed by postmodernism in response to Cold War ideologies may look outmoded after the collapse of those normative metanarratives. As Michael Epstein has argued in an interview, the late twentieth century swallowed "the sublime forms and great ideas of previous epochs," including those of a utopian avant-garde that tried "to bring the future into the present, or to move the present into the future." In its place, a postutopian "rear-garde" has emerged that cultivates "loaded boredom": "Art is now tired both of realism, which tries to coincide with reality, and of the avant-garde, which rushes forward and leaves reality behind. [. . .] The rear-garde [. . .] falls behind deliberately, [. . .] collecting the remainders of the accelerated historical process, its trash, dust, rubbish" (Berry, Johnson, and Miller-Pogacar 116–17). And yet, as Epstein's own work has proven, it is possible to envision alternatives to both disruptive, antiutopian postmodernism and a postutopian rearguard of "neutral resistance." Epstein's involvement between 1982 and 1990 in several interdisciplinary improvisational groups in Moscow (the Club of Essayists, the Laboratory of Contemporary Culture) and his more recent American publications (*After the Future*, 1995; *Transcultural Experiments* [see Berry and Epstein], 1999) have encouraged a form of improvisational "dialogic thinking" that breaks through the dualities of self and other, public and private, intellectual and mundane, creating new dimensions for discourse.

While Michael Epstein and his colleagues are eager to explore new possibilities for cultural "reinvention" at the intersection of postmodernism and performance theory, other critics have been content to support a "rear-garde art" of ironic retrenchment. The prevailing trend after 1989, both in Eastern Europe and in the United States, has been to push postmodern innovation to the margin as part of a concerted reaction against those alternative paradigms responsible in the view of their critics for creating "a crisis of authority, power, identity, and values. [. . .] At stake [. . .] is an attempt by conservatives to dismantle those sectors of the arts that combine artistic freedom with social criticism" (Giroux 6, 8). This conservative temper has contaminated not only the popular press but also academic reappraisals of contemporary literature. In volume 5 of *The Encyclopedia of World Literature in the 20th Century*, Robert F. Kiernan notes that "postmodern metafictionists remain a significant part of the literary scene, but they ceased in the 1980s to enjoy the critical deference they received in the late 1960s and 1970s [. . . Their works] found only limited audiences, not just because they are inherently difficult books, but because their self-reflexive conceits and multidimensional, nonlinear combinations of fantasies, joke, and horror seem to many readers a throwback to the intellectual temper of the 1970s" (26). Critical attention has gone instead to "several novelists of the middle rank with distinctive but not distinctively postmodern signatures" (Louis Auchincloss, Joyce Carol Oates, E. L. Doctorow) and to "a galaxy of relatively new writers [. . .] associated with what is voguishly termed 'hick chic,'" which falls back on more conventional realism (25).

As this overview of contemporary fiction suggests, the swerving away from experimentalism has ideological motivation. Fiction is "rescued" from its "intense concentration on language" because formal experimentation is associated by the conservative *Kulturkampf* with the subversive ideologies of the 1960s and 1970s. The list of the 100 best twentieth-century novels assembled in 1998 by Random House (see "Modern Library/100 Best Novels") is a case in point. Few postmodern writers made the list (Kurt Vonnegut, Joseph Heller, E. L. Doctorow, John Fowles, Salmon Rushdie, and Muriel Spark). In an amnesic revision that took literature back one or two generations, the list reinstated the primarily white modernist male canon from Joyce and Fitzgerald to Bellow, Mailer and Dickey. By contrast, the writers featured prominently in the reception of American literature abroad are most often those left out by the Modern Library list: Abish, Barth, Barthelme, Coover, Federman, Gaddis, Hawkes, Pynchon, Sorrentino, and Sukenick (D'haen and Bertens 9); also postmodern feminist and ethnic authors such as Rita Dove, Marilyn French, Toni Morrison, Grace Paley, and Alice Walker. The Italian bibliography of Coover, Pynchon, Barthelme, Burroughs, and Purdy "is almost as rich as the American one"

(149). In Spain, the translations of Barthelme, Cisneros, DeLillo, Gaddis, Kosinski, Morrison, Pynchon, and Reed have run through several reprints (177, 178–81), and in Germany works like Abish's *How German Is It,* Coover's *The Public Burning,* Federman's *Double or Nothing* and *The Twofold Vibration,* or Pynchon's *Gravity's Rainbow* and *Vineland* have been "recommended to every serious reader of contemporary fiction" as "models" of innovation and social criticism (22, 27, 44, 56). Academic criticism has devoted monographs and essays not only to Pynchon and Vonnegut, but also to Gass, Federman, feminist metropolitan fiction, and postmodern native American literature, treating their works as "acts of defiance and critique," attempts to break open official history and develop "ironic and parodistic counter-histories of their own" (D'haen and Bertens 82, 88). The convergence of translation, popular criticism, and academic research has helped the "reception of innovative writers and experimental works" in all these countries (146).

The United States have lacked all but a negative convergence between popular criticism and academic studies. While working at cross-purposes, the former to discredit narrative and cultural experimentation, the latter to turn it into a cultist object of study, popular criticism and academic research have both contributed to the marginalization of postmodern fiction. The extent of this marginalization can be gauged in a comparison between two successive historiographic efforts, both under the editorship of Emory Elliott, the *Columbia Literary History of the United States* (1988) and *The Columbia History of the American Novel* (1991). The treatment of contemporary fiction in the *Columbia Literary History of the United States* is imaginative, highlighting both "dialogic oppositions" and "certain continuities." Complexity and divergence are maintained through a focus on directions that resist easy assimilation: "the powerful history of realism, along with a sequence of persistent and various disputes with its veracity, its philosophical possibility, its relevance" (Malcolm Bradbury, "Neorealist Fiction" 1127); or the revisionist effort of new fiction, liberating narration from self-perpetuating conventions of language and recovering suppressed aspects of experience (Catharine R. Stimpson, "Literature as Radical Statement"; Raymond Federman, "Self-Reflexive Fiction"; Larry McCaffery, "The Fictions of the Present"). Under his inclusive, political as well as interartistic definition of experimentation, Henry M. Sayre ("The Avant-Garde and Experimental Writing") discusses innovative poetry and fiction, poststructuralist critical theory, minimalist sculpture, conceptual art, post-Cageian music, and feminist photography. In Seyre's view— shared with the other critical reassessments written by Bradbury, Federman, Stimpson, and McCaffery—rumors about the "death" of experimentation are greatly exaggerated. The political metafiction of Coover, Heller, Pynchon, and Reed, feminist fiction, minimalism, the cyberpunk novel, and the literature of "radical statement" (Stimpson 1073) written by writers associated with the antiwar, ecofeminist, and gay movements are given as examples of the positive transformations that occur "when one form of power—the political, say—confronts another form of power— that of imagination" (Molesworth, "Culture, Power and Society," 1044).

By contrast, *The Columbia History of the American Novel: New Views* flattens the discussion of contemporary fiction, emphasizing forms of late modernism or "midfiction" over postmodern experimentation. Both Elliott's general introduction and Patrick O'Donnell's preface to the last section propose definitions of the novel that are baggy enough to include "most texts called 'novels' at the moment" (xi), but also uncommitted enough to collapse conventional and avant-garde approaches. The general editor's list of "renowned artists of the novel" emerging since mid-twentieth century adds to the dominant figures of the 1950s only two "metafictionists" (Barth and Pynchon), three more-or-less mainstreamed feminists (Morrison, Oates, and Kingston), and a qualified experimenter (DeLillo). The analytic essays on post-modern realism, gendered (gay) fiction, and the avant-garde expand this canon, but they are highly selective themselves, downplaying the cultural significance of narra-tive innovation. Though claiming not to evaluate the "collage of proliferating move-ments and subjects" (514), *The Columbia History of the American Novel* appears often monological in its treatment of contemporary fiction. The master narrative that runs through much of the last section describes postmodernism as an embattled cul-ture, trapped ideologically between "the atavistic and jingoistic mutterings of the cultural right" and the "uncritical tribalism [. . .] of many of the proponents of mul-ticulturalism" (Cornel West, "Postmodern Culture" 517). Poetically, postmodernism is said to waver between formal nihilism and an exaggerated faith in the subversive power of fiction. The literary reassessments that follow use concepts that contribute to a leveling of narrative projects, despite an otherwise laudable effort to emphasize their political value. Under the term "postmodern realism" (521–41), José David Saldívar discusses manifestations of "a larger politics of the possible and of resis-tance," but lumps together innovative "magic realists" like Carpentier, García Márquez, Morrison, Kingston, and Islas, whose "resistance" includes a critical rewriting of the conventions of historical realism, with "minimalists" like Carver, Vi-ramontes, Oates, Stone, and Doctorow, whose response to the culturally "reproduc-tive" rhetoric of realism is more duplicitous, often coupled with the "digestible, best-seller style." Ed Cohen's article "Constructing Gender" confronts the issue of cultural and narrative ideology more directly, arguing that gay fiction follows non-prescriptive paths of desire, upsetting the "normative versions of gender and sexual-ity" (545). But with one significant exception—Samuel Delany's deconstruction of the "natural" articulations of gender and sexuality—the fiction that Cohen reviews does not significantly transform the aesthetic ideology of the novel.

The two final entries, "Postmodern Fiction" (Molly Hite) and "The Avant-Garde" (Robert Boyers), promise to broach the issue of narrative innovation more directly, but the presence of two separate chapters—one treating twelve "major" writ-ers, the other discussing aspects of the "avant-garde"—already suggests a double process of canonization. Though critical of the confining definitions imposed on postmodernism by earlier scholarship, Molly Hite perpetuates some of them, calling postmodern fiction a "white male genre" (698) and leaving out innovative feminists such as Maxine Hong Kingston, Grace Paley, Marge Piercy, Toni Morrison, and Alice Walker—to name but a few. The only two women mentioned among the "major post-modernists" are Kathy Acker and Joanna Russ; by contrast, Hite's earlier book *The Other Side of the Story* (1989) focused entirely on feminist narrative innovation, making radical claims for its contributions to postmodern experimentalism. Robert Boyers's discussion of "The Avant-Garde" begins with some of the same "major" postmod-ernists (Hawkes, Barth, Coover) and with a vaguely defined opposition between an academic avant-garde (728), characterized by "visionary pedantry" and a "refusal to provide [. . .] continuity and closure," and a socially aware approach that assumes "that there is something for which to be grateful in a genuine avant-garde" (727). What that something is, we glean finally from Boyers's analyses, which valorize those "substantial and troubling" works that display less "wayward brilliance" and more "moral urgency" (Abish's *How German Is It* rather than *Alphabetical Africa*). This approach reinstates the division between "the signifiers of language" and "the things they describe," which otherwise Boyers decries (731), subordinating narrative innovation to paraphrasable thematic concerns.

Such rereadings of postmodernism are problematic for two reasons: On the one hand, they operate with one-sided definitions of innovation that create discriminatory hierarchies; on the other, they level the postmodern scene without much concern for the critic's own acts of appropriation. As Linda Hutcheon warns, it is impossible to write "anything enlightening about postmodernism, without acknowledging the perspective from which it is said, a perspective that will be inevitably limited" (*The Politics of Postmodernism* 15). The critical reconstruction of postmodernism should proceed in full awareness of the ideological frames imposed by the critic; but it should also interplay them with other perspectives, replacing a "foundational discourse" with a telling of alternative "stories of postmodernism" (McHale, *Constructing Postmodernism* 4), as McHale, Hutcheon, Maltby, Suleiman, and Christine Brooke-Rose (*Stories, Theories and Things*) have endeavored in recent years.

Any attempt at theoretical rewriting (mine no less) must start by acknowledging the paradoxical nature of the critical process that constructs and is in turn constructed by its object. In the case of postmodernism, the object itself is *in transition*, busily reconstructing itself under our own eyes. Criticism should catch postmodern innovation in its shift from a critique of the grand narratives of the Cold War era to the mediation of heterogeneous new interests. The collapse of the bipolar order of ideological metanarratives has made "consensus on world-views and values unlikely, [with] all extant *Weltanschauungen* firmly grounded in their respective cultural traditions" (Bauman, *Legislators and Interpreters* 5). Therefore, what the post–Cold War transition needs most is "specialists in translation between traditions" (5), capable of restoring conversation in places where it has failed—and there have been many in recent years. As a literature in radical transition, postmodernism can better respond to these needs, rearticulating differences in nonconflictive narratives that cut across borders and divisions.

However, as Raymond Federman cautions, a "literature in transition" may become a literature in danger of extinction, reduced to a "mere *supplement of [popular] culture*" ("The Last Stand of Literature" 191). Postmodern writers have become complete outsiders, but not necessarily by choice: "[M]any important and innovative books are prevented from being disseminated by publishers, editors, literary agents, critics, librarians, and even professors who refuse [. . .] to read, [. . .] discuss, teach the current literature, the literature being written today" (191). Still, this grim assessment of today's market-driven publishing needs qualification. Surely, not all innovation is dead, not all nonconventional writers (e.g., Coover, Ginsberg, Morrison, Pynchon) have been corrupted by the market. Federman concedes that, by virtue of its positioning "outside the literary establishment" (192), the post–Cold War literature can give "renewed legitimacy for questions of ethics and politics, for questions of action, intention, and meaning" (191). A "literature of transition" can respond more honestly to the "rhythms of change, of transition, of metamorphosis" (192). Freed from the "formalistic frivolities" of earlier postmodernism, innovative fiction can regain the "urgency of the act of writing," making literature matter again "in the arena of social changes and cultural production" (190).

Federman's guarded optimism about literature's potential for renewal in the post–Cold War transition is shared by other experimentalists. A 1990 collection of essays called *The Politics of Poetic Form* (edited by Charles Bernstein) made bold claims for literature as a form of "resistance," "provocation," "remaking" (Jerome Rothenberg), "writing as a *counter*-reading" (Bruce Andrews), active historical "recovery" (Susan Howe), subversive "ludic experience" (Nicole Brossard), "vital force" and "public plasma" (Charles Bernstein). In response to the contemporary "crisis of expression," this collection rallied innovative writers around an oppositional poetics designed to take art "beyond the speculation of skepticism to a critically active stance against forms of domination" (Hunt 198). This liberationist project may appear utopian in the rapidly shrinking space that literary discourse occupies in the post–Cold War high-tech mediascape. As Erica Hunt cautions, projects of poetic "opposition and resistance" fall prey to their own "nostalgia for a lost unity or richness of culture" (201). By taking "the ground of textuality of language as the stake and the prize, ignoring other, extratextual grounds and strategies" (Andrews 34), these projects reinstate a conflict between social demands and the demands of writing. The converse approach that overemphasizes literature's political role is also problematic, relying all too often on a crude concept of language that facilitates one-way translations between producers and receivers of messages, political agendas, and verbal articulations. While postmodernism has the opportunity to "abandon the dead-end dichotomy of politics and aesthetics which for too long has dominated accounts of modernism," the point as Andreas Huyssen states it is not to eliminate but to heighten "the productive tension between the political and the aesthetic, between history and the text, between engagement and the mission of art" ("Mapping the Postmodern" 52). Instead of relinquishing its marginality to join the mainstream at all costs, innovative literature can make its position both inside and outside the given discourse system critically productive. Its work can begin with an "internal" transgression of discursive limits in order to carve "out a place for the social to exist in some freer way inside the human individual being" (Piombino 233). Concurrently, innovative writing can reconfigure the larger sociocultural situation, offering its innovations "not only as alternative aesthetic conventions but also as alternative social formations" (Bernstein, "Comedy and the Poetics of Political Form" 243). The examples of "radical rethinking & reinvention of expression" that Hunt discusses in her own essay—Toni Morrison's *Beloved* and Primo Levi's *Survival in Auschwitz*—are culturally effective precisely because they respond to "extremely violent conditions" (211) by applying their capacity for reformulation both to intradiscursive relations and to the extralinguistic contexts that govern them.

The best literary products of the last decade have continued along this line, rethinking simultaneously the discursive and ideological frameworks within which cultural exchanges unfold. These frameworks have been significantly diversified by postmodernism's "heterological" disposition, interested in developing a "discourse on the other" in Michel de Certeau's sense of the phrase. Postmodernism's idea of otherness was initially indebted to "identity politics" understood both as the recognition of "various social labels used in inter-subjective relationships, [. . .] which individuals attach to themselves or are attached to them by others, and which identify them as members of a specific group"; and as "the militant assertion of their specific identities by various oppressed groups" (Callinicos, *Theories and Narratives* 181). More recently, postmodernism has submitted identity politics to an antifoundationalist critique, arguing that identities are "constructed rather than self-evidently deduced from experience" and that "access to our remotest personal feelings is dependent on social narratives, paradigms, and even ideologies" (Mohanty 42, 47–48). Identity is described as fluid and "dialogical" (Taylor 33) not just because it is built on how others perceive us, but also because each of us occupies multiple positions in the world. The figure of the self-conscious "nomad" has been proposed as a solution to the issue of "how to find an adequate symbolic language to describe fractured plural identities in migration and a fractured social world, a messy and disordered geography of plates, continents, or fractal zones, slipping, sliding, and skidding into, under, and over one another" (Probyn 13). Cultures themselves are being reconsidered as kinds of "hybrids" that make possible a "range of subject-positions from which people can struggle against racist ideologies and practices" (Giroux 21).

The efficacy of postmodernism's "discourse on the other" has been acknowledged even by its opponents who have engaged in a frantic counteroffensive to return American culture to its Eurocentric and patriarchal roots. "Political correctness" and "multiculturalism" have become neoconservative code words for "'left opposition' and 'people of color,' both ideal scapegoats now that the cold war has ended" (Shohat and Stam 46). It is also true that certain aspects of this discourse of otherness have lost their "innocence" or credibility in the post–Cold War transition:

> Things are not so simple: the idea of a postmodern paradise in which one can try on identities like costumes in a shopping mall ("I'm a happy cosmopolitan, you can be a happy essentialist, they can be ironists or defenders of the one and only Faith") appears to me now as not only naive, but intolerably thoughtless in a world where—once again—whole populations are murdered in the name of (ethnic) identity. (Suleiman, "The Politics of Postmodernism after the Wall" 54)

Saying that identities are socially constructed should not prevent us from taking them seriously and examining their ideological investments. Post–Cold War literature is called upon to develop a postessentialist/postrelativist approach that will regard identities neither as "self-evidently based on the authentic experiences of members of a cultural or social group," nor as "all equally unreal," but rather as products of "objective social location" and of "legitimate and illegitimate experience" (Mohanty 54).

Even with these qualifications, the heterologic narratives of postmodernism continue to raise questions among postcolonial and ethnic critics. Will otherness be simply reappropriated by the "benevolent" discourse of Western postmodernism, which "masquerades as the absent nonrepresenter who lets the oppressed speak for themselves" (Spivak 292)? Will cultural difference disappear along with notions of "sovereign powers, universal truths, [. . .] local authenticity" (Rabinow 258), swallowed by the globalizing pressures? bell hooks thinks so when she charges that white critics write "about black culture 'cause it's the 'in' subject without interrogating their work to see whether or not it helps perpetuate [. . .] racist domination, [and participate] in the commodification of 'blackness' that is so peculiar to postmodern strategies of colonization" (qtd. in Olson and Hirsh 116). She also reminds us that what borders inspire is not postmodern play but real encounters with "the terrorizing force of white supremacy" ("Representing Whiteness" 363). The emphasis on "multicultural" and "borderline" subjects represents an important challenge to the traditional ideologies of division, but its results remain ambiguous, translating into a paradoxical recentering of white culture in the very act of its marginalization. The recent tendency of white male culture to adopt the subversive strategies of liminality is a case in point. As the white runaway communities in Pynchon's novels from *Crying of Lot 49* to *Vineland* suggest, this can be seen both as a positive pluralization of identity and as an effort on the part of the white culture to reappropriate power from the margin, erasing cultural differences by embracing/incorporating them. Wendy Somerson exemplifies this with Russell Banks's *Rule of the Bone* (1995), a novel in which the white male turns into a "border being" in order to safeguard his relevance in the age of multiculturalism and the relevance of the United States in the cultural zone of the Caribbeans. For Robyn Wiegman the "proliferation of interracial male bonding narratives" works in similar ways, inserting the white man in the "alien (read: black) discourse" in order to recuperate the black man's "potential protest and challenge to white masculine power" (91, 94).

As Rey Chow shrewdly points out, between the effort of traditional colonialism to defend white consciousness against the subjectivity of the "barbaric" captives and the "*indiscriminate* embrace" of a "pure otherness-in-pristine-luminosity" (32) that has become fashionable in some academic circles there is a subtle similarity: in both cases "*what remains constant is the belief that 'we' are not 'them,' and that 'white' is not 'other'*" (31). Neither a globalist notion of multiculturalism that recognizes the "unqualified multiplicity of cultures without positing any ways for them to interact meaningfully" (Berry and Epstein 97) nor a defensive localism that promotes one's own culture or ethnicity is an adequate approach to the issue of otherness. The alternative proposed by Edward Soja, Homi Bhabha, Rey Chow, Michael Epstein, and others is the articulation of a "third space" of negotiation between self and other, native and foreign, global and local. Postmodern fiction can function as that "interstitial place where links occur between differentiated [cultural] subjects" (Travis 132), but also as an antiphonal chamber that interrogates the nativist features of the cultures that come into contact, encouraging them to read one another critically.

Narrative imagination must continue to bear also on our concept of history, opening it up to ignored possibilities. Postmodern fiction has successfully emphasized alternate histories not only in the "parallel worlds" science fiction of William Gibson and Bruce Sterling, but also in the political fiction of Coover, Doctorow, or Pynchon focused on the crossroads of history. For example, Gibson and Sterling's *The Difference Engine* (1991) reimagines Victorian England from the perspective of our cybernetic present, mobilizing a guerrilla force of computer hackers against the global power of British imperialism backed by Charles Babbage's 1829 invention of the first "steam" computer. E. L. Doctorow's *The Waterworks* (1994) considers, from the crossroad year 1871 in the history of New York and with the hindsight offered by the Reaganite 1980s, what American culture could have become had it not succumbed to the pressures of economic self-interest and various scientific-religious master plans to control civil society. Equally important in restoring depth to a history flattened by hegemonic narratives has been the innovative work of Toni Morrison, Alice Walker, Gayl Jones, and Ntozake Shange, which allows the ghosts of memory to disrupt official history, recovering the silenced stories of slaves and women. All these works bear out Baudrillard's warning that "otherness denied becomes a specter" and that "everything we thought left behind for ever by the ineluctable march of universal progress, is not dead at all, but on the contrary likely to return" (*The Transparency of Evil* 122, 138).

Our *fin-de-millénnaire* raises new and complicated questions for literary and cultural discourse. A host of contemporary problems—from continued economic and social inequality, to multiplication of ethnic wars and the emergence of profit-oriented virtual states organized by drug consortia—calls into question the triumphalist "end-of-history" vision that Francis Fukuyama and others have associated with the end of the Cold War. The 1989 revolutions marked for Fukuyama "the definitive triumph of Western liberal capitalism, but also, since this system no longer faced any serious competitors, the end of history" (Callinicos, *Theories and Narratives* 4–5). From a theoretical perspective, Fukuyama's thesis is problematic because it continues, instead of suspending, the grand narrative of Western liberalism and techno-informational progress that fed the Cold War. Inspired by Hegel, Fukuyama's view of history is strongly teleological, positing the final transfer of man from the realm of historical struggle to that of freedom and the satisfaction of consumer demands as a goal and condition for "mak[ing] all particular events potentially intelligible" (Fukuyama 56). Fukuyama's view is also polarized, explaining recent history in terms of the rivalry between two teleological visions with opposed notions about the "end of history" and the "end of man." In Fukuyama's reading of recent history, the post-1989 period brings about the defeat of one universalist vision (the Marxist-Leninist) by another, that of liberal capitalism. Though Fukuyama's interpretation of this denouement is ambiguous, proclaiming simultaneously the achievement of one "end-of-history" vision, and the end of "History [. . .] understood as a single, coherent, evolutionary process" (xii), it still suggests a closed concept of history.

As Sacvan Bercovitch explains, a "closed system ends in a coherent Reality. [. . .] An open system works in the opposite way. It begins and ends in a mixed state [. . .]: a provisionally (or apparently) unified configuration of concrete realities that's neither coherent nor incoherent because the absolutes that work to unify them also remain subject to question" (20). By positing the worldwide triumph of liberal capitalism as the denouement of the Cold War intersystemic contest, Fukuyama's "end-of-history" vision fits a closed system epistemology. To many of us, however, the post-1989 world suggests rather Bercovitch's open system, defined by "non-coherence" and a "mixed state." For those generations that "have had to learn to live with their repressed fears of a future that will be used up before it arrives—either by nuclear holocaust or by the damage perpetrated every day on the environment" (Francese 3), the post-1989 world is entropic rather than triumphalist. Likewise, for those of us concerned with fratricidal ethnic conflicts, the resurgence of an extreme right, and technological warfare that has turned mass killing into a postmodern art, the end of millennium "conjures up the image, not of the End of History, but of history as the endless repetition of disaster" (Callinicos, *Theories and Narratives* 13).

The "end-of-history" is just one of the many forms of "post-alization" that Theresa Ebert imputes to current thinking, whose effect is to arrest history and subvert the "project of emancipation" (181). The list of terms generated by this vision is endless: "postclass, postfoundationalist, postemancipatory, posthistory, postgender, postdialectical, postteleological, postpatriarchy, post-essentialist" (181) and—of course—post–Cold War. Postmodernism itself has played an ambiguous role in this regard, contributing to but also exposing this "post-festum" thinking. In critiquing the "master narratives" of traditional history, "which purport to disclose the 'Weltplan' or 'overarching meaning of history'[:] the classical concept of Fate, the Christian doctrine of Providence, the bourgeois notion of Progress, and the Marxist vision of the world-historical destiny of the Proletariat" (Hayden White, "Storytelling" 69), postmodernism has given the impression that no narrative integration of the past is possible any longer. It has been argued that the great "metarécits" of rationality, revolution, and democracy have been seriously challenged by Auschwitz and the Stalinist gulags (Lyotard, "Discussion" 581–84). But does this mean that contemporary discourse must move beyond "history," abdicating from any attempt to articulate and explain the past? After Auschwitz refuted Hegel, Budapest '56 refuted Marxism, and May 1968 refuted liberalism (Lyotard, The Differend 179), postmodernism no longer believes in "a possible rational course of history" or in a "unified point of view on history" (Vattimo, "The End of [Hi]story" 133). But is this tantamount with saying that we are witnessing the "end of history"? Both Lyotard and Gianni Vattimo tend to answer affirmatively, arguing that since historiography has become virtually impossible in the "society of communication," "we no longer live" or think historically (Vattimo134). The prolonged death of history is also one of Jean Baudrillard's favorite themes:

> The worst indeed is that there is no end to anything and that everything will continue to take place in a slow, fastidious, recurring and all-encompassing hysterical manner— like nails and hair continue to grow after death. Fundamentally, of course, all this is already dead and instead of a joyous or tragic resolution [. . .] we are left with a vexatious [. . .] outcome that is secreted into metastatic resistance to death. In the wake of all that resurfaces, history backtracks on its own footsteps in a compulsive attempt at rehabilitation [. . .]. History and the end of history are up for sale. Communism and the end of communism at bargain discount prices. [. . .] All the ideologies of the West are also up for sale; they can be purchased at a low price on all latitudes of the globe. ("Hystericizing the Millennium" 3, 5)

Thus postmodern and conservative thinkers find themselves in unexpected agreement over the end-of-history theme, though they approach it with different expectations and conclusions.

As Conrad understood at the end of the nineteenth century, we assume that history is ended and that anarchy has been ushered in whenever there is a sharp break with the values of the past (Zabel xii). The horrors of genocides, nuclear threats, and ethnic purgings have intensified the eschatological sensibility of literature through the twentieth century. Its most recent return to the end-of-history theme is paradoxical not only because it comes at a time that can be regarded as one of the most eventful in recent history, but also because it reinvents "the classics of the end" already experienced by the generation of the late 1950s and 1960s (Derrida, Specters of Marx 14–15, 56–63, 70–72). Postmodern fiction finds itself in the paradoxical position of repeatedly announcing the end of history only to question/surmount it in the act of its narration. "[L]iving in a ghostly aftermath, as though history were somehow already finished," writers from Beckett and Blanchot to Federman, Morrison, and Vonnegut make the ending "a principle of narrative generation," resuming "the possibility of history" (Connor, The English Novel in History 203–4; Connor's own examples are Anthony Burgess, Angus Wilson, Russell Hoban, and Doris Lessing). As a result of its paradoxical "inhabitation of history," postmodern fiction has impelled us to reexamine the ways in which we represent history. Thus what has ended is not our capacity to think historically, but rather our view of history as a unifying discourse, "whose function is to compose the finally reduced diversity of time into a to-tality fully closed upon itself" (Foucault, *Language, Counter-Memory, Practice* 152).

Against the constrictive horizon of a posthistory described by capitalist apologists like Fukuyama or Marxist envisioners of a uniform proletarian future, innovative fiction continues to imagine emancipatory human strivings that make history "pro-visional and revisable" (Butler, "Poststructuralism and Postmarxism" 8). The sug-gestion in Coover's *Pinocchio in Venice* (1991), Federman's *To Whom It May Concern* (1990), Morrison's *Jazz* (1992) and *Paradise* (1998), Pynchon's *Vineland* (1990) and *Mason & Dixon* (1997), or Sukenick's *Doggy Bag* (1994) and *Mosaic Man* (1999) is not that we have exhausted our capacity to think historically, but rather that our histor-ical imaginary is being choked by forces of stagnation and repolarization. The philosophers of the "Reign of Reason" have "cheerily dispose[d] of any allegations of Paradise," and people all over the world have "[i]n thoughtless greed, within pitiable brief generations, [. . .] devastated a Garden in which, once, anything might grow" (*Mason & Dixon* 135). And yet the innovative novelist has still been able to imagine an alternative journey not outside history but rather back to its basic metamorphic en-ergies and forward to a world wherein "all boundaries [and divisions] shall be eras'd" (406). Against the widespread view that postmodernism is anti- or postutopian (Jameson, *Postmodernism* 334–40, 401–6), these novels and others written by a younger and as-yet-unacknowledged "party of [narrative] Utopia" (Curtis White, "Jameson Out of Touch?" 30) give ample proof that contemporary innovative writ-ing has not surrendered its commitment to transformative thinking.

CHAPTER 2

Innovative Responses to the Metanarratives of Modern History

POLYSYSTEMIC FICTION, SURFICTION, AND THE POSTMODERN FEMINIST NOVEL

The church's pale façade [. . .] peers out [. . .] upon this shabby but bejeweled old tart of a city, the mystery of reason confronting the mystery of desire, and what it seems to be saying is: history, true, is at best a disappointment, [. . .] but it is also, in spite of itself, beautiful [. . .]. Not an easy idea for the old professor to accept, any more than that traditional Venetian notion of art as [. . .] discourse with time [. . .], a kind of ongoing dialogue between form and history.

—Robert Coover, Pinocchio in Venice (175–76)

[T]he book of life cannot be paraphrased, it cannot be prescribed, it cannot be predicted, it cannot be dictated, it cannot be imitated, it cannot resemble some other book, it cannot begin, it cannot end, it cannot be made up, it cannot be about major characters minor characters or any other characters other than those of the alphabet, it cannot be about the right ideas, it cannot be controlled.[. . .] There is only one thing you can do with the book of life: add to it.

—Ronald Sukenick, Mosaic Man (145, 197)

[The aim of white and black women] is the (trans)formation of this hierarchical oppositional grammar into an intense, sometimes conflictual, sometimes pleasurable and pleasureful commingling of opposites in a "free" zone or "borderland," where anxieties about identities (which we have when we must maintain difference or sameness) and the noncoincidence of self and other [. . .] are affirmed for the purpose of negotiation.

—Elizabeth A. Meese, (Ex)tensions (131)

I. WHEN THE "MYSTERY OF REASON" CONFRONTS THE "MYSTERY OF DESIRE": THE REARTICULATION OF HISTORY IN POLYSYSTEMIC FICTION

Confronted with the disappointing but beautiful "mystery" of history, innovative literature continues to articulate a totality of sorts, even if this is an "*agitated* totality, not a rested one" (Andrews 29–30). By pursuing new connections in the "uncontrollable and mysterious resonance" of experience, the innovative novelist converts mystery into mastery, making the "intricate surface of reality" depend on complex narrative "designs" (Katz, *Moving Parts* 3). But this mastery is not left unquestioned. Responding to the fundamental paradox of all acts of order-making that remain "tentative" and incomplete and yet demand an ever "sharper eye or a surer, better-articulated language" (Foucault, *The Order of Things* xix), innovative fiction interrogates not only the prevailing socializations of reality but also its own "rewritings," submitting them to an ongoing process of revision.

The question of narrative/cultural *mastery* and its critique is central to what Tom LeClair has called the "systems novel" (see *In the Loop*, 1987; *The Art of Excess*, 1989). The richly layered works of John Barth, Robert Coover, Joseph Heller, Ursula Le Guin, Joseph McElroy, and Thomas Pynchon are directly concerned with socio-discursive macrostructures, submitting the personal and the local to a systemic evaluation. By taking an integrative approach, the "systems novels" achieve, according to LeClair, a triple "mastery of the world in which they were written, mastery of narrative methods and mastery of the reader" (*The Art of Excess* 5). At the same time, they foreground the sources of their mastery, calling attention to the structuring operations they perform. Instead of concealing the integrative mechanism of fiction, systems novels expose it with a vengeance, providing "information about information," "discourse [that] represents other kinds and behaviors of discourse" (19). The "mastery" they claim is finally revealed as a synecdochal illusion that allows the "parts [...] selected, structured, proportioned, and scaled" to "suggest, not exhaust, the whole of discourse" (18). The result of this dialectic of "excessive mastery" is a species of problematized process-text that both illustrates and violates our deep-seated need for stable orders.

An immediate problem with the systems novel described by LeClair is that its persistent focus on global systems translates on a narrative level in extensive patterning. In its effort to "master" various information systems (including its own narrative conventions), the systems novel ends by reinforcing their dominance. LeClair argues, in fact, that entrenched cultural systems can be challenged only through confident acts of novelistic mastery, that in the presence of self-reproducing power institutions "only extraordinarily knowledgeable and skilled works of literature—masterworks—have the kind of power that asserts the efficacy of literature and leads readers to contest and possibly reformulate the mastering systems they live with" (1). I find this argument problematic not only because it compromises the postmodern critique of totalization by returning us to an integrative notion of literary "masterpiece" (according to this line of thought, *Gravity's Rainbow* is a more "significant work than *The Color Purple*" since it "masters a set of global conditions that *The Color Purple* does not address, conditions and systems in which all readers—black or white, female or male, old or young—are imbricated" [3]), but also because LeClair does not allow literary dissension to go far enough. He reduces it to such procedures as overloading, excess, and structural *complication*. In his view, systems novels increase the entropy of the narrative process, creating an "informational density" that cannot be "naturalized" by the reader (18–19), but they do not significantly challenge "the power systems they exist within and are about" (6).

LeClair's approach remains faithful to the positivistic bias of systems theory, lacking a sharper ideological critique of the contemporary technologies of information as provided by Derrida, Foucault, and Jameson, or an alternative model of nontotalizing connectivism as found in stochastics and polysystems theory. A significant shortcoming of systemic approaches is their tendency to treat open-ended biological or sociocultural mechanisms in terms of closed physical ones. As commonly employed in the 1960s and early 1970s, the term "System" denoted "a static, functionalist model of society rather than an unstable, dialectical one" (Maltby, *Dissident Postmodernists* 166). Yet clearly, the processes at work within even partially open systems complicate the universal laws posited by general systems theory, allowing for nonassimilable behaviors at both a local and a global level. While biological and cultural systems tend, according to Anthony Wilden (131–38) toward over-organization, they prevent the threat of rigidification by allowing meaningful variations or randomatic turbulence into the system. The work of the German sociologist Niklas Luhmann has further refined our understanding of a "social system," emphasizing the autonomy of the subsystems that constitute a functionally differentiated society. Social systems are self-organized but heterarchical: While they ensure the self-regulation of the system as a whole, they do not allow any single subsystem to dominate. Applying these propositions to literary culture, Siegfried J. Schmidt describes literature as a dynamic self-organizing social system, encouraging multiple "memberships" in various subsystems. Each subsystem develops its own model of reality, but these partial models do not add up to a compulsory "*ortho*-model." Still, in Schmidt's description the literary system conserves its autonomy and stability, resisting external perturbations and absorbing "noisy" components (413–24).

Postmodernism challenges the alleged autonomy and stability of cultural systems, calling into question their processes of autopoïesis. All literature involves, according to Joseph Natoli, a "struggle of order and disorder" (2), producing transformative "entanglements" in society's "regimes of signs" (4, 16), but postmodernism openly cultivates the "difference of disorder" (203) or simulates orders primarily to critique them. Even when it adopts a systemic perspective, postmodern fiction thwarts our interpretation with "abstruse information," "extended metaphors," "parodic parallels," and intricate knots of myths (Kuehl 108, 111, 113). These "maximalist" techniques (209) enact with a vengeance the condition attributed by William Paulson to all literary discourse, that of a "a noisy transmission channel that assumes its noise so as to become something other than a transmission channel, [. . .] a perturbation or source of variety in the circulation and production of discourses and ideas" (viii-ix). Postmodernism uses its informational and structural disorder to ideological advantage, turning noise into a radical form of communication that expands the structures of our understanding. As Edward Mendelson notes, encyclopedic novels such as *Gravity's Rainbow* do not simply incorporate "the full range of knowledge and beliefs of a national culture," but also expose "the ideological perspectives from which that culture shapes and interprets its knowledge" ("Gravity's Encyclopedia" 162).

Postmodern encyclopedic fiction is concerned both with system analysis and system transgression, with narrative mastery and its undoing. This type of novel challenges inherited modes of representation in at least two ways: (1) by revising the referential system of traditional fiction, breaking down its norms of selectivity, unity, and verisimilitude; and (2) by exacerbating and at the same time questioning our need for explanatory orders. "Men live by fictions," Coover reminds his interviewer Larry McCaffery, and the effort to "isolate little bits" in the flow of life and "make reasonable stories out of them" is not unreasonable. The problem emerges when some stories "start throwing their weight around," reducing life to abstract general-izations. The novelist's task is to denounce these transcendental explanations, "un-dermin[ing] their authority a bit, work[ing] variations, call[ing] attention to their fictional natures" (LeClair and McCaffery 68). Conversely, since the "world itself [is] a construct of fictions, [. . .] the fiction maker's function is to furnish better fic-tions with which we can re-form our notions of things" (interview with Gado, *First Person* 149–50). The iconoclastic writer refuses to just tell stories, asking "[h]ow are stories told," "[w]ho's telling the story," and "how does the making of that story ac-tually embed certain values" (Coover's comment in *Unspeakable Practices* 242).

Coover's early work suggests both the advantages and the limitations of this type of systemic critique. Several stories collected in *Pricksongs and Descants* (1969) focused on the effects of semiotic and political manipulation on a social body all too eager to play the assigned cultural game. In Coover's stories, quiz games, TV shows, Holly-wood movies, the popular mythology of spectator sports all contribute to the social-ization of individual imagination, preventing transformative cultural responses. Equally deadening are narrative conventions and habitual modes of perception. Coover's following two novels explored this closing of social and narrative imagina-tion in all its implications. *The Origin of the Brunists* (1966) and *The Universal Baseball As-sociation, Inc., J. Henry Waugh, Prop.* (1968) are concerned with the construction of a subjective system (a millennial religion in the former, a game world in the latter) that is subsequently enforced upon the world as a strategy of division and control. Henry Waugh is interested not only in designing a parlor baseball game that, like the war games he admires, seeks a "perfect balance between offense and defense" (*The Universal Baseball Association* 19), but also in recording the history of his invented Base-ball League in order to enforce "some pattern or other" on it (211). His "lust for pat-tern" (230) is shared by the townsfolk of West Condon (*The Origin of the Brunists*), who create a new religion around the survivor of a local mining disaster, turning a senseless catastrophe into an end-of-the-world epiphany (211). Both the Brunist re-ligion, which has the ability to absorb all contestation, and Henry Waugh's Baseball Association evolve toward increasing complexity and closure. What starts as a game of dice and charts designed by a bored middle-age accountant acquires a mysterious life of its own that reproduces the patterns of conflict from the larger world. Several critics have read Coover's novel allegorically, either as a rewriting of Christian his-tory in which J. Henry Waugh plays Jehovah, his favorite rookie, Damon Rutherford, the daemon, Jock Casey, a version of Jesus Christ, and the eight chapters of the book "loosely correspond to the seven days of creation plus an implied apocalyptic mo-ment" (McCaffery, "Robert Coover" 111–12; see also Andersen 60–61, and Cope 37–41, 44); or as an elaborate political parable, with the UBA initials of Henry's as-sociation standing for the USA, the Rutherfords for the New Frontier, the chancel-lors of the Baseball Association for various recent presidents, and the sequence of fatal events that claim the lives of several chancellors and key players as a represen-tation of the traumatic 1960s (Cope 42–43).

Henry's own desire to wrest some sense of historical order from "the terror of eternity" (238) aggravates the tensions inside his imaginary baseball league. Con-fronted with the threat of a "history [which] shows all signs of coming to an end" (Coover interview with Hertzel 25–26), Henry intervenes in the events of his as-sociation, taking sides and upsetting the balance of "strategy and luck, accident and pattern, power and intelligence" (45). His meticulous archiving of the association's first hundred years further rigidifies its history. But Henry's approach is innocuous by comparison to the methods employed by the association's fictional chancellor, Fenn McCaffree. He plants agents within the rival parties and uses his elaborate information machinery to police the life of the association. The difference in method between McCaffree and the association's previous chancellors comments on America's transition during the Cold War from a communitarian emphasis to corporate control. McCaffree's totalitarian approach contrast with Henry's understanding of history as provisional, always open to "a new ordering, perspective, personal vision, the disclosure of a pattern because [. . .] perfection wasn't a thing, a closed moment, a static fact, but *process,* [. . .] and the process was transformation" (211–12).

Even if not particularly hopeful about man's capacity to control—rather than be controlled by—his fictional devices, Coover's early works reasserted periodically Henry's re-creative approach to history against a dogmatic perspective such as McCaffree's. Dedicated to Cervantes, Coover's first collection of stories urged the "use of the fabulous to probe beyond the phenomenological, beyond appearances, beyond randomly perceived events, beyond mere history" (*Pricksongs and Descants* 78). Beginning with "The Door: A Prologue of Sorts" and moving through "Seven Exemplary Fictions" and the three stories in "The Sentient Lens," Coover's texts upset conventional definitions of reality with their discordant voices and perspectives. Even stories like "Morris in Chains," that remain trapped in the technocultural system, manage to insert paradoxical discourse and thinking in the "game plan," expanding "that vibrant space between poles" (interview in LeClair and McCaffery 72). What emerges in all instances is a plurivocal vision and structure of narrative.

While Bakhtin's concept of dialogics describes adequately some aspects of this poetics (see Cope 10–11, 16, 59–113), polysystem theory better highlights Coover's effort to accommodate alternative histories and voices in his fiction. As a radical development of the "dynamic functionalism" theorized by Russian formalism (Jurij Tynjanov and Boris Eikhenbaum, in particular) and of the subsequent multi-system theories of Bakhtin, Bertalanffy, Prigogine, and J. Dubois, *polysystem theory* describes— in the words of its main proponent—"system[s] of various systems, which intersect with each other and partly overlap, using concurrently different options, yet functioning as one structured whole, whose members are interdependent" (Even-Zohar, "Polysystem Theory" 11). Itamar Even-Zohar's basic assumption is that systematicity does not preclude multiplicity or heterogeneity, the concept of "system" being useful to literary studies as long as it is conceived as an "open structure consisting of several [. . .] concurrent nets-of-relation" (12). Nor does systematicity exclude conflict and mutual readjustment, as Iurij M. Lotman's cultural semiotics makes abundantly clear. The cultural framework within which texts are continuously (re)written is pictured by the Tartu scholar as an intercrossing of variously positioned discursive universes rather than as a stable hierarchy of thought systems. Divisions can emerge between the different sets of codes that a text carries, or between these and the repertoire of cultural associations embedded in the "language of the reader" (see *Analysis of the Poetic Text;* also "On the Semiotic Mechanism of Culture"). Likewise, in Even-Zohar's concept of "stratified heterogeneity" tensions intervene at all levels of the "polysystem": between functions and subsystems, canonized and peripheral positions. A polysystem maintains itself not by staying "untouched," but through intersystemic struggle and change. The "insufficiency" felt within one system encourages "interference" with other systems, resulting in "innovative transfers" of items, features, and functions (Even-Zohar, "Interference in Dependent Literary Polysystems" 81, 82).

In traditional unisystem theories peripheries were often excluded as extrasystemic. By contrast, polysystem theory refocuses attention on peripheries, allowing them to create tension within one system and become centers of other adjacent systems. We are urged to think of polysystems not "in terms of *one* center and *one* periphery, since several such positions are hypothesized" (Even-Zohar, "Polysystem Theory" 14), but as dynamic stratifications that continually redefine center and periphery. This theoretical position is quite useful in considering those postmodern narrative practices that redefine the rapport between margin and center, dominant and secondary structures. Unlike other systemic perspectives that have appeared resistant to issues of marginality and minority positions, remaining essentially "system building" approaches and thus "exposing the problem of theory transfer and Eurocentrism," polysystem theory allows a focus on alternative experiences (Tötösy de Zepetnek 31, n. 11). Polysystem theory can help us rethink the postmodern project, moving beyond an earlier critical emphasis on the disruptive function of postmodern practices to a more balanced view that takes into account postmodernism's effort to reintegrate excluded voices and cultural peripheries. We could thus argue that postmodernism employs strategies of systemic disruption as part of a transformative agenda that converts closed, hierarchized systems into dynamic polysystems.

Anticipations of a polysystemic mode of thought can be found in nineteenth-century literary and theoretical discourse. Consider Nathaniel Hawthorne's conclusion to "Wakefield" (1835):

> Amidst the seeming confusion of our mysterious world, individuals are so nicely adjusted to a system, and systems to one another, and to a whole, that, by stepping aside for a moment, a man exposes himself to a fearful risk of losing his place forever. Like Wakefield, he may become, as it were, the Outcast of the Universe. (*Twice-Told Tales* 140)

While the rough outline of a (poly)systemic view of literature and culture can be found in the theoretical texts of Friedrich Schlegel and Goethe (see Dimić 50), structural examples of a polysystemic approach are harder to come by, unless one considers Melville's multilayered, centrifugal narratives that have influenced postmodernists like Coover, or Dostoevsky's polyphonic works that inspired Bakhtin's own theory of narrative heteroglossia. Most other nineteenth-century narratives stop short of a polysystemic approach, looking for ways to adjust individuals to a social "system and systems to one another." Hawthorne, Dickens, Eliot, Flaubert, Balzac, Turgenev, and James focused their attention on this process of socialization, endorsing its community- and nation-building function while also critiquing some of its excesses, such as the punitive exclusion of those who "step aside" (deviants, minorities, women, children). But, as Homi Bhabha has argued, the "demand for a holistic, representative vision of society could only be represented in a discourse that was *at the same time* obsessively fixed upon, and uncertain of, the boundaries of society, and the margins of the text" (*The Location of Culture* 144). What the authors of the "baggy" Victorian novels discovered was that the "scraps, patches and rags of daily life" could not be easily turned "into the signs of a coherent national culture" (145). Their socio-narrative integration remained superficial, enhancing the "liminality" of both the people inscribed in the discourse and of the discursive structures themselves. The nineteenth-century novel was thus *liminalized*, "*internally* marked by the discourses of minorities, the heterogeneous histories of contending people, antagonistic authorities and tense locations of cultural difference" (148).

While the novelists of the previous era tried for the most part to mask this discursive liminality in the totalizing poetics of realist narrative, postmodern innovators have turned liminality to advantage, exploiting the subversive potential of sidestepping and the free circulation across systems. The play of the postmodern text makes genuine otherness conceivable by transmitting "both the *mots d'ordre* of established order and identity and the *mots d'ordre* of difference and otherness, of disorder" (Natoli 21, 10). The "'motley' society macroimage" that Joseph Natoli recognizes in Pynchon, Barth, Sorrentino, and others allows us to see "what a social order cannot see but has already been made 'see-able' within the motley model" (124). Likewise, Kathy Acker's "transgressive carnivalesque" challenges "the empire of language and meaning" (141, 144), intruding unruly messages in it. But making otherness "conceivable" is not enough: Postmodernism needs to rearticulate our lifeworlds, turning "the monologues of [. . .] historical determination into a dialogic whereby we hear previously unheard voices and resee difference" (150).

This is also Coover's understanding. As he explains in a comprehensive review of contemporary fiction, a more significant form of systemic disruption can be obtained when fiction "rejects mere modifications in the evolving group mythos [. . .] and attacks instead the supporting structures themselves, the homologous forms," making room for a "disruptive, eccentric, even inaccessible" narrative voice ("On Reading 300 American Novels" 37–38). The relationship between dominant and eccentric narrative voice, self and other is fundamental to Coover's own fiction. Robert Morace has suggested that "the basic plot of Coover's plays, stories, and novels" pits "the pattern-keeper, who accepts the determinacy of a teleological universe, versus the pattern-breaker, who embraces indeterminacy and imaginative freedom" (193). Although Morace has focused on Coover's concern with "the tyrant Other" as pattern-keeper and oppressor of individual imagination, I want to argue that equally important in Coover's fiction is the identification of the other with the excluded and repressed. Coover's protagonists seek out these excluded "others" because they function as expanding mirrors for the self, promising the experience of lost alterity. Coover prods his characters to engage these figures of replenishing otherness, but he also discloses the contradictory dynamic that underlies this effort, with the self desiring both to master and be mastered by an other. The first impulse leads to the incorporation of the other; the second, to submission to an idealized other that functions as "transcendent or absolute pole of address, summoned each time that a subject speaks to another subject" (Boons-Grafé 298). As Coover's fiction shrewdly suggests, this idealized figure of otherness may end up looking very much like the "tyrant Other." In Jacques Lacan's terms, they are both versions of the *grand-autre* (68), the great Other in whose gaze the subject seeks its identity.

Coover's early stories approached traditional plots from "the other side," recovering the silenced or victimized point of view: of Noah's sacrificed sibling in "The Brother," of Joseph, the bewildered and disconsolate husband of Virgin Mary in "J's Marriage," of Snow-White's wicked mother/queen in "The Dead Queen." Still, the artificiality of these voices points to the problematic nature of the recuperation. The space of narrative potentiality created through alternative retellings dissolves in self-contradiction or is absorbed back into a violent popular imaginary, as in "The

Babysitter" or "A Pedestrian Accident." *The Public Burning* (1977) makes a more convincing attempt to free individual imagination from dogmatic concepts of reality and reinscribe the other in history. The novel is too complex and polysystemic for a mere political reading, encouraging alternative anthropological-narratological concerns with the circus imagery and rituals of scapegoating (LeClair, *The Art of Excess* 106ff), metafictional critiques of the "language of power" (Mazurek 30), or the mixture of high and low styles in the Cold War "narrative[s] of containment" (Nadel 159–60). Still, these issues cannot be properly understood without some reflection on Coover's revisionist politics of history that reconsiders relationships between centers of power and margins, selves and others. Coover's novel foregrounds the confrontational narratives of the Cold War, tracing their origin back to a Manichaean collective ideology embodied in Uncle Sam—the novel's official reflector and "tyrant Other." At the same time, Coover's novel attempts to destabilize official history, denouncing it as a performance benefiting those in power and opening it up to excluded voices.

Jackson Cope has rightly argued that *The Public Burning* makes a notable effort to move from monologic to dialogic discourse: "The novel incorporates at times a dialogue about history (Ethel's operetta with Ike), but more usually supplies an immense cacophony of views, overlapping of voices that wedge Coover out as a bit singer in his own chorale. Nixon has no more privilege than the author, though he is so desperately trying to listen into history's conversation" (71–72). *The Public Burning* stops short of creating a truly "heteroglossic" structure, capable of distinguishing "a plurality of independent and unmerged voices" and a polyphony of "consciousnesses with equal rights and each with its own world" (Bakhtin, *Problems of Dostoevsky's Poetics* 6). In spite of its structural heterogeneity, the novel remains trapped in the ideological "chorale" that shapes America's narrative. As Coover explains, "I was striving for a text that would seem to have been written by the whole nation through all its history, as though the sentences had been forming themselves all this time, accumulating toward this experience. I wanted thousands of echoes, all the sounds of the nation" (interview in LeClair and McCaffery 75–76). Few of those echoes are radically dissonant, capable of putting forth effective alternatives to the official narrative. The public performance in Times Square that ends with the execution of the Rosenbergs is dominated by the composite voice of Uncle Sam. Coover's creation is the unchallenged master of "a lot of styles" (*The Public Burning* 89), a "pieced-together semiotic Frankenstein" (LeClair, *The Art of Excess* 128). But his voice is pseudodialogic as long as Uncle Sam admits no alternative viewpoints and has "nothing to believe in except himself. An audience of one" (*The Public Burning* 233). In spite of his protean, folksy vocabulary, Uncle Sam is too class and race conscious to represent genuine American "heteroglossia." His pronouncements make clear that he is "not partial to Jews, Muslims, Hindus, Buddhists, Voodooists, or Romanists. If he had any favorites at all, they were among people like Ezra Benson's Mormons, the eccentric, evangelical, and fundamentalist sects nurtured here on this soil" (345). Unlike his political incarnations (such as President Eisenhower) who pursue a more hypocritical version of American fundamentalism, Uncle Sam proclaims not only "America's election" but also her right to annihilate its cultural others ("those who expect to reap the blessings of freedom must, like men, undergo the fatigue of [. . .] massacreein'" [7]).

Alternative voices such as those of the Rosenbergs or their defenders are not absent from Coover's novel, but they have difficulty in breaking through Uncle Sam's polarizing discourse. Although *The Public Burning* pursues a historical vindication of the other, the counternarrative of Julius and Ethel Rosenberg remains uncertain and mediated. On the rare occasions they are allowed to take center stage, the Rosenbergs alternate between dignified silence and rhetorical statements that are immediately discredited by Uncle Sam and Nixon as political "grandstanding" (101). The traditional interlocutors of power (the press, creative writers, political philosophers) are also subdued. With the exception of a few references to Arthur Miller's politically charged play "The Crucible" (1953), which had the McCarthy era as much in mind as the Salem witchcraft trials on which it was based, to Einstein's statements in defense of the Rosenbergs, W. E. B. DuBois's writings on race that brought him the charge of being a Soviet agent, and the novels of Steinbeck, Farrell, Caldwell, and Moravia denounced in Congress for their "filth, perversion, and degeneracy" (215), Coover presents high culture as a reinforcer of the power structures. Likewise, popular entertainment provides not only the metaphors but also the texture of political conformism in the 1950s (Eisenhower's confrontation with the "swarthy" A-bomb "rustlers" follows the script of Fred Zinnemann's classic western, *High Noon*, and the Times Square pageantry is directed by none other than Cecil B. De Mille, informer of the House Un-American Activities Committee and winner of an Oscar for staging the "greatest [circus] show on earth" [281]). In Coover's sweeping perspective, all political and cultural institutions are responsible for maintaining the Manichaean narrative that pits America against its cultural and ideological others. *Time* magazine, personified as "The National Poet Laureate," displays a "great poetic affinity for War" (323), reducing history to linear confrontations that reinforce "deep tribal" prejudice (328). The *New York Times* embodies the "Spirit of History" in a more raw, more manipulative form. Its newspeak reports promote "arbitrariness as a principle," replacing logical relations with "randomness as design" and objectivity with "a willful program for the stacking of perceptions" (191).

And yet the very discourses that repress the humanity of the other end up by being disrupted or contaminated by it. The vindication of the other takes place within the dominant discourse that is rendered uncertain and pluralized. Under Uncle Sam's very nose, a "rash of evil doings" spreads around the world and the Times Square stage undergoes subversive changes: A manikin dressed like Uncle Sam with a Hitler mustache is strapped into the electric chair, and the luminous slogan "America the hope of the world" is distorted to read successively "dope," "rope," "rape," "rake," "fake," "fate," "hate," "nate," and "joke" of the world (36–41). The novel lists other similar disturbances that suggest a linguistic and ideological slippage away from the official order of history. Uncle Sam's own language is submitted to a deconstruction that exposes its cynical undermeanings:

> It is our manifest dust-in-yer-eye to overspread the continent allotted by Providence for the free development of our yearly multiplyin' millions, so damn the torpedoes and full steam ahead, fellow ripstavers, we cannot escape history! . . . I tell you, we want *elbow-room*—the continent—the *whole* continent—and nothin' *but* the continent!" (8)

Despite Uncle Sam's successful restoration of the Times Square slogan, the memory of these subversive errancies cannot be wiped out. The more control he exercises over significance, the greater the risk of an entropic breakdown. Marginalized perspectives such as those of the Rosenbergs return to haunt the official discourse, creating a partial transfer of symbolic potential from victimizers to victims. By summing up the forces of alterity under the code of the "ungraspable Phantom" "made of nothing solid" (336), Uncle Sam exposes himself to the haunting of the repressed specter of history that for Jacques Derrida is always disruptive and revolutionary, mixing a "coming back" with a "coming for the first time" (*Specters of Marx* 4, 6–7, 10–11).

The most dangerous disruption for Uncle Sam's triumphalist spectacle ensues during a Times Square blackout, just before the executions. Plunged "into a nighttime far deeper than that from which this morning they awoke" (492), the crowds undergo a "tribal implosion." Though Uncle Sam returns promptly to blaze a "New Enlightenment" on his people, his light reveals a scene of "rampant nihilism, bestiality, liberated freak shows, careless love and cheating hearts, drunkenness, cock-sucking, and other fearsomely unclean abominations [. . .] not exactly Cotton Mather's vision of Theopolis Americana" (495). The fact that the blackout occurs when Nixon calls on Uncle Sam to drop his pants as part of a collective ritual of self-baring indicates that the custodian of the Cold War ideology is himself a version of the Phantom Other he created.

When the "enemy other" is brought onto the execution stage, further disturbances are created. Julius's frailty suggests a Christ figure to some, but Uncle Sam dismisses that impression as a ploy, "the Phantom's last weapon" (508). Ethel's execution disrupts the official script even more profoundly: Her defiance prevents the crowd from commiserating with her, and her unexplainable survival of the first electrocution attempt robs watchers of their catharsis. The doctor's horrified announcement, "This woman is still alive" (513), points to Ethel's irreducibility as the official culture's other.

Of all Coover's characters, Nixon understands best the seductive power of the other. In the novel he tries to negotiate a balance between a conformist career in the service of a "Tyrant Other," Uncle Sam, and moments of "breakaway wildness" that bring him perilously close to a subversive other, Ethel. Nixon begins by patiently exploring the Rosenbergs' lives in order to understand the centrifugal forces that pulled them out of the mainstream. He is intrigued by how close his own destiny as a struggling middle-class young man came to intersecting theirs. As daughter of poor immigrants, Ethel Greenglass hoped to establish an identity by performing as singer, actress, and union speaker. Like Nixon, she was committed to politics and the public dimension of language, speaking always as if "to a vast audience" (408). But Ethel remained trapped in her marginality through lack of opportunity and political craftiness, while Nixon managed to drift to the center of power. In an ironic reversal of stereotypes, the Rosenbergs remind Nixon of Horatio Alger's self-made Americans (129): They are resourceful and committed, even though to the "wrong" ideology. Their entire existence haunts the establishment "with a strange dark power" (352). For Nixon, this is the power that brings "History itself alive—perhaps by the very threat of ending it!" (352).

Himself coveting that power, Nixon seeks a rapprochement with the other. As he moves beyond stereotypal definitions of the Rosenbergs, Nixon is filled with an unexpected longing to touch their lives. In his secret reveries, Nixon pictures himself in the role of the other's champion, sharing stories and feelings with his rediscovered sister-lover, Ethel. Or he loses himself in hallucinatory visions of otherness, exotic ethnicity, and homosexuality. Despite his efforts to contain the damage done by opening the subconscious "gates and flood[ing] the syntax routes" (181), Nixon begins to function as an empathetic historical medium, remembering "things that had never happened to me, places I'd never been, friends and relatives I'd never met who spoke a language I didn't know" (144). His identification with others makes him feel awkward and yet "richer somehow" (145).

In Coover's words, Nixon lives "close to the center, yet not quite in the center, off to the edge a bit, an observer" (interview in LeClair and McCaffery 74–75). As Nixon becomes more tolerant of his own outsider position in relation to the Eastern political elite, he undertakes "Something Truly Dangerous" in chapter 21: a journey into that "strange space between" to "reach" the other. Propelled by a desire to "provoke a truth for the world at large to gape at: namely, that nothing is predictable, anything can happen" (365), Nixon takes a train to Sing Sing, moving against the general flow of the crowds toward Times Square. His contradictory motives for seeing the Rosenbergs reflect his ideological "in-betweenness": He wants to get the Rosenbergs' confessions and stop the executions; to establish "a partnership in iconoclasm" (368) with them, exposing the arbitrariness of the power structures; and to act as a mediator between Eisenhower and Julius Rosenberg, or small-town traditions and city revolution. Inside the prison, he moves deftly through a number of studied roles (as debater, "progressive" Republican, champion of minorities); but then he startles Ethel and himself by putting aside his inhibitions and embracing her, exhilarated by the "delicacy of innocence, the tang of the unexpected, the nutty flavor of playfulness, the subtlety of the first encounter" (437).

Nixon and Ethel seem to gain a momentary sense of togetherness from this interaction. But Nixon's gain is far greater: Through his identification with Ethel, Nixon rediscovers his potential for feeling and adventure, as well as the memory of an America of "warmth and brotherhood I had not known since those mornings we all huddled around the kitchen stove in Yorba Linda" (439). Ethel functions both as an object of desire (Lacan's "petit object a") and a signifier of otherness (Lacan's "grande autre") that expands Nixon's identity, giving him a new understanding of history: "[W]as this what the dialectics of history was all about," he wonders, this "ecstasy" of reunion? (439) Through his reunion with this replenishing/challenging other, Nixon feels a new freedom from both Uncle Sam and the Phantom: He believes he has escaped "outside guarded time" (442), in a nonpolarized posthistory. Ethel, on the other hand, does not feel empowered by their embrace. Nixon's self-congratulatory notion that "he is making history this evening, not for [him]self alone, but for all the ages!" (439) does not benefit Ethel. As the guards approach to take her to her execution, Nixon runs away, abandoning Ethel to her fate. His effort through the remainder of the novel is to atone for his transgression and rescue his political career. Finding himself transposed inexplicably onto the Times Square execution stage, "his pants a tangled puddle at his feet" (469) and the message "I am a Scamp" written by Ethel on his buttocks, he turns his humiliation into a face-saving political speech on the theme of national vulnerability ("*we have ALL been caught with our trousers down!*" [473]) and the sacrifice that all patriotic Americans need to make in order to meet their "responsibilities in the world" (482).

In spite of his retrenchment effort, Nixon remains suspended between the remembrance of Ethel's "life-giving embrace, where everything seemed possible, once more" (475) and his capitulation to the crudest political game. The two final episodes he narrates—his effort to justify himself to his wife, groveling like a dog, and his submission to Uncle Sam's sodomizing embrace—put even his human identity into question. The scene of his violation by Uncle Sam suggests what Ethel must have felt in his own crude embrace. Nixon responds to Uncle Sam's attention with a mixture of terror and love, accepting his "election" to power as his own execution and acknowledging the irreducible ambiguity of the "other." (He begins to suspect that his Sing Sing encounter had not been with the victimized Ethel but with a disguised Uncle Sam.) Nixon himself can be read as a version of the other, at once a victim and victimizer. He is, in Daniel E. Frick's description, Coover's "secret sharer," a "double" that spotlights "the tug-of-war between Coover's designs—his desire to expose the corruption of America's dominant culture—and his artistic self-doubts, his troubling visions of political powerlessness" (83).

By making Nixon the book's main narrator, Coover created not only an authorial alter ego, but also a credible focalizer for the process of historical construction and manipulation. Politics for Nixon is an all-encompassing, fateful performance in which he tries to play multiple roles: as stage manager, assistant director, producer, and even hero in the failed attempt to rescue Ethel from the towers of Sing Sing. In his competition with other resourceful scriptwriters (Uncle Sam, the Poet Laureate *Time*, President Eisenhower), Nixon's first instinct is to fall back on the tested roles of "tragic lover, young author, athlete, host, father, [. . .] businessman" (361), played in school shows. But as his awareness of the personal and national crisis increases, Nixon decides "to step in and change the script" (363), taking risks against the master narrative controlled by Uncle Sam. Through his "rival act of authorship," which challenges the "narrative of America's election as the 'stuff we make up to hold the goddam world together,'" Nixon becomes an agent for Coover's own "subversion of a national mythology [. . .] from the inside" (Frick 84–85).

In his lengthy musings on the "riddle of history" (115), Nixon wavers between a belief in "case history, the unfolding patterns, the rewards and punishments, the directed life" and a suspicion that we live in a "lawless universe" where if "there was a certain power of consistency, there was also power in disruption" (363). The latter hypothesis scares Nixon with its radical polysystemic possibilities: for if "there was no author, no director, and the audience had no memories," being "reinvented every day," then perhaps "there is not even a War between the Sons of Light and the Sons of Darkness! Perhaps we are all pretending!" (362). Intellectually, Nixon can admit that the struggle against an ideological other is a cover-up for "the motive vacuum" (363). Politically, however, he finds this lack of determination unacceptable. Therefore, Nixon the ideologue prevails over Nixon as author-historian. Betraying his polysystemic belief "that all men contain all views, right and left, theistic and atheistic, legalistic and anarchical, monadic and pluralistic; and only an artificial—call it political—commitment to consistency makes them hold steadfast to singular positions" (363), Nixon chooses a "singular position" in the end, allowing himself "to be possessed by Uncle Sam, be used by him, moved by him" (262). In surrendering to Uncle Sam, Nixon acknowledges the more masterful author, who is literally and symbolically a rapist, violating history in order to assert his domination over the private and public domain.

But *The Public Burning* also interrogates this type of dogmatic authorship. The presence of multiple authors (Uncle Sam, Nixon, FBI agents, the Poet Laureate *Time*), representing partly different interests, creates unexpected distortions and revelations in the official narrative, like the "H-polarizer" 3-D glasses carried into Times Square by a forgetful moviegoer. As he staggers out of a horror movie about a

Frankensteinian professor who dips living people into hot wax to make them into historical figures, the unnamed spectator superposes the apocalyptic blaze he saw at the end of the movie over the Times Square pageantry. Though literally a "misreading" produced by the "eye-straining, H-polarizer haze of alcohol" (283), the moviegoer's hallucination reveals deeper truths. By seeing the Times Square "public burning" for what it symbolically is, a "final spectacle, [. . . an] atomic holocaust" (286), the 3-D spectator proves "the only sane person left on the face of the earth" (287). He is also an excellent illustration of the kaleidoscopic vision of the polysystemic writer who "maximizes the effects of ambiguity, indeterminacy, and paradox *with a view of occupying the domain of the excluded middle*" (Maltby, *Dissident Postmodernists* 145–46). Dragged away by the guards when he tries to offer himself as sacrificial substitute in the electric chair, the 3-D spectator leaves a weighty message behind ("BEWARE THE MAD ARTIST" [288]) that simultaneously affirms and warns against the power of the word to rewrite history.

Rewriting remains an ambiguous cultural strategy in *The Public Burning*, promising to create space for the manifestation of the repressed other but failing to sustain it. In spite of its momentary disruptions, the misogynist master discourse reasserts itself periodically: Eisenhower refuses to grant the Rosenbergs a pardon because "in this instance it is the woman who is the strong and recalcitrant character [. . . t]he man is the weak one" [249]; Uncle Sam advises Nixon to keep his "little wife well tilled, willed, I mean" [332]; and Nixon himself exchanges his risky fantasy of identification with subversive femininity for a profitable homosocial bond. Other Coover narratives have challenged the masculinist discourse more directly, though with varying degrees of success. In "The Babysitter," the male characters project their violent libidinal fantasies on the unnamed babysitter, but the girl's private discourse occasionally finds its way into the male fantasies, disrupting them with her own desires. *Gerald's Party* (1985) exposes the polarized roles available to women in a lingering patriarchal world: those of nameless suburban wife or promiscuously loved and victimized showgirl. This gender polarization is enhanced by the genre of the parlor mystery and the voyeuristic conventions of theater that Coover's novel parodies. The rape/murder that Coover's characters, led by Inspector Pardew, attempt to unravel implicates them all in a violent game of male desire.

Echoing Foucault's concern with history's destructive "imprint" upon the body (*Language, Counter-Memory, Practice* 148), Judith Butler has argued that patriarchy is responsible for the abuse and destruction of woman's body ("Subversive Bodily Acts" 129–30). Coover's *Spanking the Maid* (1981) bears this observation out, linking sexual violence to the master's cultural performance. This novella undertakes a polysystemic critique of mastery in all its forms: rational mastery over existence, male mastery over woman's domestic space, and the writer's mastery over the (female) body of the text. These various forms of control are embodied in the story's male figure who enacts a correctional, sadomasochistic fantasy on his maid. Though the master's "disciplinary interventions" (*Spanking the Maid* 25) seem less ominous than those of Kafka's sinister typewriter in "Penal Colony," which imprints the "sentence" directly on the victim's body, they still manage to enforce the simple "lesson of the master" on the body and mind of the maid (57).

The figure of the master recalls Coover's various pattern makers (from J. Henry Waugh to Uncle Sam) who strive to reduce contingencies to an abstract, dogmatic order. What is different in this case is the master's identification of the problem

with femaleness exclusively, seen as hopelessly fluid and unruly. Coover's male protagonist seeks to control female errancy with "method," "habit," and a "correctional"
ideology reinforced by the culture's symbolic forms (philosophy, religion, literature)
that provide the master with the proper "manuals" for educating his female subaltern. But these manuals—parodied after the Victorian "Guides for Domestics" and
the corresponding literary styles of the period—fail to ensure the master's control
over a polysemous corporeal and spiritual reality. The world's "natural confusion and
disorder" (71) strikes back, producing fertile mutations in the master's Puritan ideology. His discourse is befuddled by ironic puns or homonymies (Varsava 134–35)
that undermine its self-definition: The "lecture[r]" is revealed as a "lecher" (29),
whose "civil service" to humanity is a "sibyl service" (32) or service rendered to the
"civil surface" of womanhood (99). As gaps and ambiguities multiply, both master
and maid escape "the closed system of their own psyches, their thoughts condemned
to oscillate aimlessly between two poles" (122). The disintegration of the master-
discourse makes room for the maid's repressed signifiers. Against the master's injunction to keep the bed cover smooth like "a blank sheet of crisp new paper" (36),
she discovers the beauty of a ruffled surface, already inscribed with unruly desires.

A *Night at the Movies* (1987) expands the exploration of the sadomasochistic ideology that subtends gender relations in popular culture. Most pieces parody stock
plots (a raunchy gangster version of *Guys and Dolls*, a pornographic replay of *Casablanca*,
an evil-triumphing-over-good rewrite of *High Noon*, etc.), reinforcing the division
between insecure and abusive males and submissive hypersexed women. But at least
one piece, the short movie "Gilda's Dream," mixes gender identities and genres in a
way that partly empowers the other. Similarly, the Western feature called "Shootout
at Gentry's Junction" allows the Mexican gold-toothed bandit to speak in an elo-
quent mixture of languages, usurping the discourse of the Northern master and
making off with his women. The phallic stereotypes of the Western, which Coover's
fiction beginning with *The Public Burning* has associated with the confrontational culture of the Cold War, are submitted to a parodic reworking also in *Ghost Town*
(1998). Retold from the perspective of the post–Cold War 1990s, the Western plot
appears depleted of meaning, trapping its "lone rider" in a "land of sand, dry rocks,
and dead things" (3). Befuddled by the "vast empty plain, where nothing seems to
have happened yet and yet everything seems already over, done before begun" (4),
the archetypal wanderer pursues the traditional mirages of the frontier: an elusive
town "sitting on the edge like a gateway to the hidden part of the sky" (5) and the
equally intangible figure of the quested woman. What he finds at the end of a perilous journey is a ghost cowtown full of "futility and stupidity and veritable craziness" (12) and a female apparition who promises to put some meaning back into the
"monumental void" (4). She performs for the Western hero the ambivalent role of
object of desire and civilizer, instructing him "by gaze alone on the ways of the universe and the means of quelling the spirits of evil in the human heart" (115).

The problem that Coover's Western hero faces is that of reconciling civilization
and the wilderness, order and lawless desire. In the novel, he plays alternatively sheriff and outlaw, but his misadventures confirm the absurdity of both roles in a wilderness where "[t]hey aint nuthin wuth dyin fer [. . . and] the end of whutall else is
emptiness and the end of adventurin is emptiness too" (40). Coover's rambler finds
periodical relief from his inner and outer desert in the graceful apparition of a
"schoolmarm" glimpsed most often through a curtain or a veil. She awakens in him

chivalrous feelings and respect for cultural conventions (including those of grammar). But her every little victory is forthwith reversed by the seductive interventions of the barroom floozy Belle. Not surprisingly, the Western hero has a difficult time trying to keep his idealized image of civilizing femininity separate from the threatening image of uninhibited femaleness. Belle challenges his obsession with the schoolmarm, but the hero defends his stereotype of femininity (she is "kindly and reefined and pure as an angel" [115]) because his own identity as a "man on a mission" (116) depends on it. On his last rescue trip to town, however, the Western hero is confronted with the folly of the chivalrous tradition he defends. In a series of carnivalesque revelations, the schoolmarm whom he frees from prison, the jealous black mare that threatens to trample him underfoot, and an angry Belle who wants him hanged for jilting her turn out to be versions of the same creature. Unable to deal with the "two-faced card" (145) he draws in the Claims Office, suggestive of the Janus-faced career of the Western hero and of the metamorphic nature of femaleness, the hero has no other option but to return to the "monumental void" (4).

The polarization of female roles into chastising virgin and nourishing vamp was at the center of Coover's earlier *Pinocchio in Venice* (1991), but this novel, which updates Carlo Collodi's *Le aventure di Pinocchio* (1883) by returning its hero to a decadent Venice reminiscent of Thomas Mann's *Death in Venice,* suggests that the inability of male imagination to accommodate metamorphic femaleness is linked to the more general failure of modern civilization to balance order and contingency, self and other. In Coover's rewriting of Collodi's famous bedtime story, the boy-puppet turned Nobel-winning art historian revisits Venice in search of his lost muse-savior-teacher, the protean Blue-Haired Fairy. His pursuit can be read as a version of the search for the *grande-autre*—a "great Other" that is again seductive and monstrous—but also as a broader search for the "soul" of his culture. Pinenut/Pinocchio cannot solve the enigma of his Blue-Haired Fairy without simultaneously pondering modern civilization's problematic divisions between life and art, nature and metaphysics, body and spirit.

Pinenut's return to Venice exacerbates these contradictions: Like his famous predecessor Petrarch, the Professor comes here in search of a "synthesizing metaphor" for his work-in-progress, "a vast autobiographical tapestry in which are woven all the rich, varied strands of his unique personal destiny under the single predominating theme of virtuous love" (14). He finds instead disruption, anarchy, and self-doubt. This *Homo aestheticus,* who seeks everywhere "a true instance of art reflecting the reality beneath the surface" (15), is confronted by a sequence of tragic-grotesque mishaps that reawaken his doubts about his puppet-turned-human body. As Pinenut gradually reverts "back to wood" (37), he feels that "his own nature is fatally betraying him," robbing him of the dignified meanings he has cultivated for nearly a century (59). Pinenut's remnant of dignity is shattered when, in a sequence that repeats Pinocchio's misadventures, he is robbed of all his possessions, chased by a horrifying winged beast, and almost mangled by a police dog. He is spared the final humiliation when the old mastiff, Alidoro, recognizes him as the Pinocchio he befriended in Collodi's novel and takes him to an improvised shelter watched over by his mistress Melampetta.

The metaleptic leap into a world of fabulation, with speaking beasts that remind Pinenut/Pinocchio of his own hybrid nature, has a restorative effect upon him. The philosophic-scatological conversations of Alidoro and his mistress, parodying the metaphysical tradition of the Renaissance, open critical inroads into the Professor's life-denying aestheticism. The two dogs' unabashed sensuality encourages Pinocchio to refocus on the figure of his desire, the "enigmatic creature" who has provided "magic assistance" (66) for his life and work. As retold by Pinenut, the story of the Blue-Haired Fairy begins ambiguously, with her playing the roles of revitalizing sister and playmate on the "Night of Assassins" when, chased by murderers, he bangs on her door for help. After he is waylaid again by his enemies La Volpe and Il Gato, landing in prison, Pinocchio returns to the Fairy's house to find only a tombstone inscription announcing that she died of grief. The earnestness of Pinocchio's penitence makes the Fairy rematerialize, this time as a magic *mamma* who turns the puppet into a flesh-and-blood boy. Their relationship is strongly oedipal: With the father out of the picture, the Fairy can play the role of protectress, muse, and civilizer. She responds to Pinocchio/Pinenut's acts of disobedience with deathlike cataleptic states that blackmail the boy into compliance. Pinenut's entire career as a "famous scholar and exemplary citizen" (75) is built on a guilty notion that goodness is "something that [had to be] re-created from day to day, from moment to moment, by living and dying men" (75).

Pinenut's mongrel friends mock his "logomaniac" notion of a redemptive spiritual life (105), but they cannot offer a better alternative to "life's ceaseless mutations" (115). Faced with the troubling scattering of his body, parts of which begin to fall off, the Professor embraces even more fervently his myth of spiritual salvation. Upon visiting the San Sebastiano Church, he identifies with Bordone's depiction of Jonah and Veronese's St. Sebastian. Like them, he seeks not only delivery from a "decrepit, foul-smelling" material world (112), but also transubstantiation from rudimentary bodily form to the condition of "transfigured spirit" (115). But his utopian vision of a world in which "*Change* was *changeless,* Becoming *was* being" (115) is disturbed by the unexpected encounter with an updated version of the Blue-Haired Fairy and of Nabokov's Lolita. Pinenut's former student Bluebell storms into his metaphysical reveries, offering to nurse his withered genitalia back to life. In spite of Pinenut's effort to write off this encounter with the Other as a mere chance meeting with a "vulgar American coed with a soft blue sweater" (136), he experiences profound disturbances. As his body continues to crumble, assaulted by "galloping wet rot" (181), Pinenut hangs on to his rare "flash[es] of blue" (153), his memories of the Blue-Haired Fairy and his sightings of the blue-sweatered coed.

Like Stencil's infatuation with the polymorphous Victoria/Virginia/Veronica/Vera/Viola in Pynchon's *V.,* Pinenut's obsession with his "vaporous fantasy" in blue is both enabling, giving some coherence to his "ceaseless mutations," and disruptive, throwing him off course and forcing him to painful reevaluations of identity. The Blue-Haired Fairy's interventions have ambivalent effects on him precisely because of Pinenut's inability to imagine her as an integrated female figure. She represents many disparate things—"[s]ister, mother, ghost or goat" (183)—that he loves madly but cannot reconcile. The Professor's main effort has been to control the polymorphous nature of his female muse by spiritualizing her, translating her seduction from a sexual to an intellectual level. Pinenut's failure to encompass Blue-Haired Fairy's polymorphous nature affects his own condition and view of the world. Trained on the "categorical imperatives" of his spiritualized Blue-Haired Fairy, he cannot participate in the "ongoing dialogue between form and history," reason and desire (176) that Venice encourages.

While Pinenut misses the chance to be rejuvenated by the city's hedonistic Carnival, he makes amends on an intellectual level, moving from simple pure thoughts to "pure complex thought[s]" in an effort to embrace an "impure world" (107). As he is carried along by the carnivalesque procession, Pinenut begins to understand the underhanded role that body has always played in his life and work. The Professor's own crowning study, *Mamma,* was conceived as a "homiletic account of his idiosyncratic search for the magic formula by which to elevate his soul from vegetative to human form, as though body, far from being a corrupting adversary, were in itself a kind of ultimate fulfillment" (253). Though this work is left unfinished by Pinenut's incapacity to reconcile his two visions of maternity—one virginal-metaphysical, the other sexual—it marks a change of heart in the Professor. Having previously argued that the "bridge between It-ness [. . .] and I-ness" (250) is "virtuous [spiritual] love" (14), Pinenut now concedes that humans cannot transcend their material condition (their "It-ness"). One needs, therefore, to plunge "into the alluring labyrinth of the magical city [. . .] as a lover might enter the body of his beloved" to experience a "true mystical communion with the Other (266).

Like Nixon in *The Public Burning,* the Professor experiences one such "ecstatic and visionary moment" before the novel is finished. Exactly when he feels most excluded from the general "revelry" (251), Bluebell reappears to free him from his wheelchair and take him on the "most exciting ride of his life," atop the Apocalypse roller coaster. His "reasoned approach abandoned" (267), the Professor embraces Bluebell as "ideal beauty's very image and all he would ever know of the divine" (264). But once again, Pinenut seems better suited for the role of sacrificial clown than that of romantic lover. On the pretense of arranging another meeting between him and Bluebell, Pinenut's former schoolmate and city tycoon Eugenio wraps the Professor in dough, turning him into a donkey-shape pizza for the cannibalistic enjoyment of the crowds. As he peeps through his pizza eyeholes at the "demonically Carnivalized Piazza" (280), Pinenut has the shocking realization that he cannot expect to be saved again by his Blue-Haired Fairy. She now appears to him as his true *assassina:* "[W]hipping him with guilt and the pain of loss, [she] has broken his spirit and bound him lifelong to a crazy dream, this cruel enchantment of human flesh. In effect, liberated from wood, he was imprisoned in metaphor" (288–89).

While delivery is not possible, Pinenut's final experience teaches him the value of merging. In an ironic reversal of his childhood deliverance from the belly of a monster shark, Pinenut is lured by a female voice into the fishlike insides of the Santa Maria dei Miracoli church. Identifying the voice with the Quatrocento Madonna dressed in midnight blue he dreamed of as a child, Pinenut remonstrates her for having robbed him of his innocence and freedom with her "terrifying heartbreaking parade of tombstones and canon" (320). The Madonna responds by taking on various alluring masks, from Bluebell to composite "mamma," whose face displays features of all the other women Pinenut has known. Pinenut's static male imagination proves again incapable of comprehending the polymorphous nature of femaleness and his own hybrid identity, which encourages bonding with an other. (As the blue Madonna explains, she deliberately left a seed of the naughty puppet inside the human Pinocchio in order to help him bond with her [325]) But what Pinenut cannot grasp mentally, he performs instinctively, allowing his "decrepit misshapen creature, neither man not puppet, [to get] entangled in blue hair and lying in an unhinged sprawl in the embrace of a monstrous being [. . .]. Hideous. Beautiful" (329). In an act of "rare creative communion" (250), which both annihilates and reconstructs him, the Professor melds with a polymorphous Blue-Haired Fairy who, in Coover's vision, represents an emanation of "Venice's corrupt and mongrel history"

(176). The novel's narrative structure backs this denouement, moving from the deceptive linearity of its opening chapters to an increasingly mixed and revisionist approach. Narration is carnivalized, its rules of verisimilitude stretched to include continuous character conversion and an irreverent recapitulation of history that valorizes the disruptive energies of human desire, "mongrelizing" plots and identities.

According to Coover's more recent novel, *John's Wife* (1996), contemporary culture is still largely unprepared for such moments of "mystical communion with the Other" (266), lacking the visionary power or willingness to perform the polymorphous bonding that violates conventional ontological and cultural boundaries. Most of the novel's suburbanite characters are disbelievers in love, defining it as "reason's sedative, or else [. . .] as a chemical reaction to certain neural stimuli, sometimes locally pleasurable, generally overrated" (14). A number of male characters allow that "love"—conceived in transpersonal terms as divine "grace" (Otis), artistic design (Gordon and Ellsworth), or "the original pioneers' love of adventure that brought them out here, the settlers' love of the land that caused them to stay and put down roots, the love of the early town planners for order and progress" (John)—may still play a role in providing culture with a structure. But they are themselves incapable of partaking of that feeling. A rudimentary mechanism of desire and violent gratification replaces love as social glue. Even though the web of desire converges mostly on John's wife, it fails to offer a satisfactory structure against the "meaningless frenzied blur of life" (10). John's wife remains nameless, a figure of desire clad in conventional roles: "[e]verybody's favorite Homecoming Queen" (12), "[c]oveted object, elusive mystery, beloved ideal, hated rival, princess, saint, social asset" (73). Though each man sees her as an abiding center, she remains an "unreadable" signifier for all. John's gorgeous wife pops up unexpectedly in the novel but she disappears just as suddenly, leaving behind the "holy vision" of an ethereal womanhood that contrasts sharply with images of real female desire—expanding uncontrollably like Pauline's pregnant body.

This lack of centering, caused by the failure of emotional bonds and the semiotic emptiness of the feminine figure, reflects also at a narrative level. As in *Gerald's Party*, Coover's narrative shifts from one character focalization to another and from past recollection to present fantasy, mimicking an electronically produced "hyperfiction" that generates endless level swaps and associations. As usual, Coover's male pattern-seekers (Reverend Lennox, photographer Gordon, novelist and reporter Ellsworth, entrepreneur and developer John) try to force life's "mind-numbing volume of mazy detail" (70) into some semblance of order. The area over which Coover's men seem to have most control is that of female sexuality. The novel's women tell similar tales of "[r]apes and whippings," "gangbangs" and "incest" (52), in which they function as interchangeable "glory 'oles" (43) for predatory men who violently intrude in their "shapeless li[ves]," configuring them "with narrative thrust and plot and conversion to the future perfect" (40). There are no alternative stories to tell because men have reduced the pioneer spirit that had once animated American culture to a form of scavenging. Marge's solitary "insurrectionist venture" to oppose John's handpicked candidate for mayor ends in humiliation.

Yet even when they fail, women like Marge or the supersize runaway Pauline present a challenge to the dominant male plots. Male narrative imagination falls constantly short in its effort to control the vagaries of desire or the unpredictability of history. Neither Ellsworth, the town's self-appointed chronicler, nor John, the community's master builder, can prevent the "surface excitements" that endanger "history's deeper design" (89). Their Pioneers Day celebration, designed as "a kind of community embrace of the 'unembraceable'" (311), only precipitates disasters. In typical Coover fashion, the novel climaxes in an orgy of violence, as the local policeman Otis leads a posse in pursuit of the "huge womanish thing snorting like a wild animal" (pregnant and hungry Pauline), held responsible for the break-ins and shootings that spread around town on Pioneers Day. The posse sets fire to the Settler's Woods, trapping Pauline (who looks momentarily like a mystical mother figure to Otis) inside a firestorm. These disasters happen outside the "healing presence" of John's wife (402), her prolonged absence in the novel being regarded as a proof that change and disaster, not harmony and stability, are the conditions of the universe. After her reappearance, the male characters try to put "the town's lives back together again" (413), but none of their efforts manage to "shield [them] from the dusky terrors of the flux" (334). Like other Coover novels, *John's Wife* can be read as a satire on man's efforts to contain "the dusky terrors of the flux" through narrative orders, but also as a guarded praise of narrative imagination *in its failures,* because these failures liberate the polymorphous potentialities of life.

While Coover's polysystemic critique is focused primarily on the symbolic forms that construct unidimensional systems, Thomas Pynchon's fiction starts from the economic base and expands concentrically into the psychological and ideological infrastructures of Western narrative imagination. Pynchon's choice metaphor for monologic "forms of capitalist expression" is the parabolic arch of the phallic rocket. This arch is implicated in the "pornographies of love, erotic love, Christian love, boy-and-his-dog, pornographies of sunsets, pornographies of killing, and pornographies of deduction [... that] lull us with [...] that Absolute Comfort. [...] The self-induced orgasm" (*Gravity's Rainbow* 155). This obscene arch stretches also over the (post)colonial enterprise, which treats the cultures of others as "outhouses" for the base desires of metropolitan cultures. Pynchon exposes the "pornography of blueprints" (224) that Western imagination has enforced on the world, looking for alternative human arrangements that cross the assigned ontological and ideological lines.

By comparison to other narrative projects (feminist fiction and film, minority and "borderline" fiction) that challenge our notions of cultural identity from a "subaltern" position that becomes the performative center of an alternative discourse, Pynchon takes the opposite route: He focuses on the hegemonic body of historical knowledge and narrative produced by the West in order to critique its closed-system approach and intrude an "adversarial" energy into it. More specifically, Pynchon's narratives activate the marginalized perspectives of Western culture (paranoids, bums, clairvoyants, suburban women), encouraging them to disrupt the linearity of a given system of "truths" through active (mis)reading and dissonant rearticulation. From this viewpoint, Pynchon's project appears no less "dissident" than those of feminist and borderline writers, a fact only recently recognized by criticism (in addition to Maltby's *Dissident Postmodernists,* see Dugdale, especially chapter 3; also my articles "Systemic Transgression and Cultural Rewriting in Pynchon's Fiction" and "Rethinking Postmodern Liminality). While encouraging our nostalgia for "Holy-Center-Approaching" (*Gravity's Rainbow* 508), for pattern and order in history, Pynchon draws us into the interstitial spaces of contingent variables, back to what Franz Fanon has called the "zone of occult instability where the people dwell" (190).

Pynchon's narratives begin by questioning forms of systematic knowledge derived from history, but then project a plethora of provisional explanations that cannot be integrated. *V.* (1963) constructs endless "versions of history" (209) that converge (at least in Herbert Stencil's imagination) on a "V-structure." But, "[j]ust when the structure looks like a system, a thing, a confluence, a shape, then it lapses back into an arbitrary piece of graphics" (McHoul and Wills 167). The absence of a stable perspective opens a space for alternative narrative propositions. *Gravity's Rainbow* (1973) engages conflicting modes of reading history, organizing the world variously "according to theology, physics, history, myth, popular culture, cybernetics, literature, chemistry, mysticism, biology, politics, economics, psychology. [. . .] Each system of belief or body of knowledge [. . .] offers to explain the nature of reality, culture and of the individual's place within the larger whole, but on condition that all other claims to knowledge are renounced" (Madsen 15). By simply coexisting, these systems challenge each other's limiting criteria for knowing. Pynchon's novel encourages us not only to tolerate conflicting perspectives on history, but also to transform them through imaginative reconstellation.

This process of reconstellation is illustrated also in Pynchon's two Californian novels, *The Crying of Lot 49* (1966) and *Vineland* (1990), whose female characters disrupt traditional (male-centered) semiotic orders, replacing them with paradoxical "nodes" of meaning. The secret postal network discovered (or imagined) by Oedipa Maas, "by which X number of Americans are truly communicating whilst reserving their lies, recitations of routine, arid betrayals of spiritual poverty, for the official government delivery system" (*The Crying of Lot 49* 170), functions as a typical countersystem that lives off the entropy of the dominant order, turning waste into new insidious patterns. The collage of oral, visual, and electronic bits of narrative assembled by the inquisitive Prairie Wheeler in *Vineland* works in like manner, complicating official history with eccentric, interpersonal stories. More recently, *Mason & Dixon* (1997) seeks alternative mappings meant to rescue American (and by extension Western) civilization from the repressive politics of the dividing line traced by Pynchon back to the Enlightenment rationalism. Neither of these projects is successfully completed: Pynchon's novels are packed with characters who can read the boding signs of history, but few of them are capable of acting decisively on their "information." However, even their smallest interventions open the culture's self-regulated narratives up to alternative possibilities.

2. NARRATIVE AS AN "INTERVENTIVE" MODE: FROM SURFICTION TO AVANT-POP

The historical rereadings proposed by Robert Coover and Thomas Pynchon manage to open up the culture's prescriptive "plots," infusing them with new desires and interests. But as long as they remain focused on "mass-produced and institutionally controlled" networks of information (LeClair, *The Art of Excess* 16), in which individuals are mere nodes, they cannot escape the power systems they are about. Coover's more recent work has wavered between an ironic recycling of pop culture (not only *Gerald's Party*, but also *Briar Rose* and *Ghost Town* deserve to be called "the poppest" of Coover's novels [Christgau 7]) and a radical denunciation of "reality" as a "series of overlapping fictions that [cohere] into a convincing semblance of historical continuity and logical truth" (*The Public Burning* 122). Historical continuity and logical truth are what Coover's characters have been struggling with. The town historian Ellsworth keeps "a file cabinet stuffed with bios ready to be plucked for print," but "life's disorderly overabundance" (*John's Wife* 70) defeats his purpose. The stories that Ellsworth tells have recognizable, predetermined plots that are difficult to rewrite. As another character in *John's Wife* muses, "We are born into the stories made by others, we tinker a bit with the details, then we die" (138). Coover's exquisite gift for parody manages to turn "repetitive behavior or occurrences" (156) into great moments of comic-heroic prose. But "tinkering" with the details of the stories available to us is not enough. What we need is a radical "reset[ting of] the basic patterns," "breaking down the boundaries for a moment, producing monsters we secretly know to be more real than the good citizens who eventually subdue them" (224, 225). Traditional stories need to be "reset" in a way that allows for real transgressions. Otherwise, as Coover's recent rewriting of "Sleeping Beauty"—*Briar Rose* (1997)—suggests, characters will remain entangled in their "storied strands," and the promise of polysystemic opening will end in "sequential disenchantments."

Pynchon's own novels disrupt traditional historical constructions, looking for the "untold" behind history's "master plan" (*Gravity's Rainbow* 720), but these polysystemic alternatives are often "not only absorbed by the System, they are understood to sustain it. [. . . Their] critical perspectives are quite consciously derived from cultural practices which are either archaic or marginal; in other words, they cannot affirm any course of action, which [. . .] could deliver humanity from the System" (Maltby, *Dissident Postmodernists* 161). Pynchon's concern with people who fall through the cracks of history (runaways, floaters, outcasts) gives *V.* and *Gravity's Rainbow* a "remarkably scattered" perspective that is both their virtue and their weakness. *The Crying of Lot 49* and *Vineland* valorize these liminal energies more carefully, performing a rearticulation of recent history through marginocentric characters like Oedipa Maas and Prairie Wheeler. Still, the efficacy of this rearticulation remains uncertain: *Vineland* has been alternatively described as a revisitation of the 1960s couched in a "right-leaning aesthetic," whose "reference[s] to ideas and artifacts of mass culture" make this novel "the most self-consciously 'popular' of Pynchon's books" (Tabbi 90, 91), and as a successful break through the discursive screens of printed and electronic culture to the repressed data of experience (Cornis-Pope, "Systemic Transgression" 82–87). Pynchon's more recent *Mason & Dixon* (1997) is an even more interesting case in point. Though concerned with the "crimes of demarcation" and "ortholinearity" at the root of American culture, this novel proceeds on the whole chronologically, mimicking the structure of an eighteenth-century picaresque novel. *Mason & Dixon* offers a major revision of our traditional representations of modernity, but does not include a similarly profound revision of the narrative line. In spite of the frequent interruptions, anachronistic allusions, and interweaving of story strands, Pynchon's novel remains committed to an incremental model of narration, plodding ahead like Mason and Dixon's line "by the customary ten-minute increments,—each installment of the Story finding the Party advanc'd into yet another set of lives, another Difficulty to be resolv'd [. . .]" (706).

A more effective form of systemic disruption can be found in surfiction and the postmodern feminist novel. Both types of fiction make informational and systemic integration more difficult, denouncing their controlling mechanisms. They also seek to expand fiction's repertoire of choices, using narrative innovation and cultural rewriting to "intervene in experience and change it" (Sukenick, "Post Modern Fiction and Oppositional Art" 181). Introduced by Raymond Federman in the early 1970s, the term "surfiction" designates "the kind of fiction that tries to explore the possibilities of fiction; the kind of fiction that challenges the tradition that governs it; the kind of fiction that constantly renews our faith in man's imagination and not in man's distorted vision of reality" ("Surfiction—Four Propositions" 7). Including initially a broad range of experimental forms, the term became subsequently identified with the work of a specific group emerging in the 1970s (Walter Abish, Rudolfo Anaya, Russell Banks, George Chambers, Steve Dixon, Raymond Federman, Kenneth Gangemi, Madeline Gins, Marianne Hauser, Steve Katz, Clarence Major, Paul Metcalf, Ursule Molinaro, Gilbert Sorrentino, and Ronald Sukenick), rallied around alternative outlets such as the Fiction Collective press and *The American Book Review.* According to Federman, what differentiates this group from the "first wave of self-reflexive [. . .] fictioneers" (Burroughs, Barth, Coover, Gass, and Pynchon) "is a more daring, more radical use of language," but also a "rejection of traditional forms of narrative, and especially of mimetic realism and mimetic pretension" (*Critifiction* 31). The disruptive strategies employed in surfiction ("fractal organization, discontinuity, interactivity, ellipse, eclipse, non-sequitur, incompletion, association, chance, coincidence, achronicity, synchronicity, improvisation, intervention, self-contradiction, overlap, mosaic, modularity, graphic composition, sonic formation, rhythmic symmetry, vortextualization and eddyfication, rhizomatic interconnection, hypertextual hopscotch, paradox, wordplay and in conclusion, inconclusion" [Sukenick, *Narralogues* 19]) have broader consequences than the "quantitative deformations of conventions" (LeClair, *The Art of Excess* 21) practiced by the systems novelists. Their role is to denaturalize and disrupt the "perverse logorrhea" of conventional representation (Federman, "Surfiction—Four Propositions" 8). As Sukenick put it in his latest novel, innovative fiction "has become an emblem of holdout against an idolatrous, icon conned, oscar crazed hollywooden world," refusing the "projection of image" or any "counterfeit, including imagination's sleight-of-mind" (*Mosaic Man* 160). But the disruption of conventional representation is only a first step toward a significant rearticulation of fiction's experiential and cultural field. Though critical of the effort to translate life into a "fraudulent [. . .] edifice of words" (Federman, *The Voice in the Closet* 6, 12; my pagination), surfiction does not give up articulation. As Christopher Nash has argued with regard to recent antirealistic fiction, "we sense intuitively that while it rejects verisimilitude as a stable and abiding good, it is nonetheless *veristic* in its aims," "referring to a substantive reality *beyond* the text" (206). Surfiction can best be described as a polemical form of narration that continually reformulates its modes of articulation, "push[ing] out to the edge of culture and of form" in order to "allow more reality into the work" (Sukenick, *In Form* 135). This type of fiction is doubly "interventive": seeking a new narrative contract, a "non-factitious relation" to reality and its readers, while also trying to "intervene in and change reality, experience" (Sukenick, "The Rival Tradition" 4, 6).

This concept of "interventive fiction" challenges the self-reflexive aestheticism of a metafictionist like William Gass, for whom "[f]iction doesn't make a difference" in social reality nor should it try to make one (*Unspeakable Practices* 253). It also oversteps Coover's notion of an iconoclastic fiction that does not discard traditional forms but rather assaults them from "inside," relaxing their ideological strictures in order to "speak the unspeakable" (*Unspeakable Practices* 252, 253–54). What a surfictionist seeks ultimately is a new narrative epistemology capable of mediating between the "generative" word, "[i]conoclastic," "[a]lways beginning," and the analogic language of fiction, "genetic" and "iconic" (Sukenick, *Mosaic Man* 9). Inevitably the "pictures" presented by the language of fiction become "Multiplied. Viral. Simulacra, i.e., Facsimiles. Androids. Manikins. Gelded. Gilded. Scripted. [. . .] Story becomes history" (9), so that the novelist needs to return continually to the generative word that can start the process of articulation afresh. Content neither to parody the "exhausted forms" of realism like earlier metafiction, nor to abandon the world-making function of the word, surfiction is "engaged in the process of inventing reality," "understand[ing] that the story is its own real process" (Major, "Making Up Reality" 154). Clarence Major finds surfiction more liberating politically than "novels of bitter protest" because it responds to the "trauma of social experience" by moving "a new set of eyes into the open" (*The Dark and Feeling* 25):

> The novel *not* deliberately aimed at bringing about human freedom for black people has liberated as many minds as has the propaganda tract, if not more. [. . .] Art [. . .] is to be improvised on and extended, if not broken. [. . .] By "radical black aesthetic" I mean a *new* attitude of freedom of expression, a new approach, from within, to one's experience, *being*. A new way of understanding the interplay between one's conscious and unconscious life—how this activity moves through the world. (28)

The emancipatory potential of surfiction has not been properly understood, being downplayed at times even by its practitioners. In response to an interviewer who noted that his novels interrogate the "realities" that American culture lives by, Major insisted on separating the experiential "reality" of his fiction from social reality: "The kind of novel I'm concerned with writing is one that takes on its own reality and is really independent of anything outside itself. [. . .] In a novel the only thing you really have is words. [. . .] The content exists in our minds" (interview with John O'Brien, in Major, *The Dark and Feeling* 136). Still, a few paragraphs later Major acknowledged the complex web of Judeo-Christian, American, and African American references that his "definitely Southern novel" *No* (1973) evoked. He made clear, however, that this evocation of Southern culture was revisionistic rather than merely metonymic. What Major rejects both in this 1973 interview and in other statements published since is not the referential function of fiction but rather a politically based definition of black aesthetics that privileges social realism and the radical form of protest laid down by Richard Wright. Arguing that "there is no single 'black aesthetic'" with an "inherent set of guidelines," except in the sense that "black writers were always working against a single dominant impulse in American culture: the use to which white America put blackness" (interview in McCaffery, *Some Other Frequency* 247–48), Major has been seeking innovative modes of narrative capable of dealing imaginatively with both individual and cultural experience. Like Ralph Ellison and more recently Ishmael Reed, Major has chosen art over "'black' anger and 'clenched militancy'" (Ellison, *Shadow and Act* 120). But he has also challenged Ellison's attempt "to deflect racial provocation and to master and contain pain" through a Hemingway-inspired concept of autonomous art (*Shadow and Act* 111–12), replacing it with a postmodern understanding of the interplay between self and world, history and art.

We can better grasp the nature of surfictionist revision through Sukenick's idea of "mutiny":

> Mutiny does not need a program, it does not need an ideology. [. . .] It does not proceed from alienation, but is an impulse from the inside to reclaim its own identity. It is an eruption of the spiritual unconscious. Mutiny is a movement of collective conviction and revulsion, a refusal to proceed as usual, a diversion of the channels of power to more constructive ends. A mutiny does not have to win [. . .]. There's nothing to win, there's simply the diffusion of a vision as the agent of change. (*Narralogues* 22)

What both Sukenick and Major emphasize is surfiction's need to remain revisionary, engaging the culture from a position of permanent "mutiny" and trying out alternative modes of articulation. A novel like Gilbert Sorrentino's *Crystal Vision* (1981) "tells many stories, but no *story*. With few exceptions, each of the seventy-eight chapters is a little narrative. But there is no larger plot to which these narratives belong. There are, likewise, many storytellers, but no single narrative voice" (Mackey, "Representation and Reflection" 206). The storytellers dispute models of individual and collective narration, hoping to achieve an ultimate synthesis. (One character is engaged in writing an encyclopedic novel that will incorporate everything that happens in *Crystal Vision*, others rewrite stories from Sorrentino's earlier books.) In the long run, neither the implied author nor any of Sorrentino's surrogates gain control over the narrative process that generates many self-modifying presentations. What surfiction seeks is not unity but a difficult balance between competing visions and styles. Surfiction engages us in an open-ended dynamic on several levels: *temporal,* since this type of process-oriented fiction is provisional and self-corrective as time itself; *epistemological,* since the novel allows no privileged position for evaluating its "reality," requiring an interactive hermeneutics that reformulates both reflecting consciousness and world; and *sociocultural,* engaging its discursive agents in a form of "experiential thinking" that seeks to "undercut official versions of reality in favor of our individual sense experience" (Sukenick, *In Form* 67). For surfiction, like for Derrida, "writing must be in the interval between several styles" (*Spurs* 139), problematizing any single perspective or mode of representation.

We need to take this performative variability into account when we consider the surfictionists as a group. Though criticism has often discussed the work of Raymond Federman and Ronald Sukenick under a common rubric, there are important differences between their approaches. As summed up by Sukenick in a conversation with Federman, they involve a contrast between an experiential approach and a language-centered, textualist one. While Federman has valorized the self-controverting rhetoric of fiction that operates for him like the "liar's paradox" ("The truth of the statement is canceled by the lie of the statement, and vice versa" [Federman and Sukenick, "The New Innovative Fiction" 140]), Sukenick has been interested in the generative, experiential aspect of fiction: "Fiction, like all other forms of culture, is in process of creating our experience" (141). Both writers are acutely aware of literature's crisis of credibility in the post-Holocaust and post-Gulag era, "after all those identities were exxed off the surface of the planet" (Sukenick, *Mosaic Man* 181), but they have responded to it differently. Federman has addressed the crisis of literature from the perspective of a playful/agonistic authorship (see chapter 5), which involves a polemical confrontation between teller and listener, speech and writing. Federman wants "the fiction writer to be present in his writing, present as a voice that manipulates, and controls, and sets in motion. [. . .] Fiction today is more like a rehearsal than a finished performance. [. . . It's] being shaped right there on the space of the paper, on the stage of the writing" (Federman and Sukenick 142–44). The subversive potential of Federman's fiction relies mostly on rhetorical and linguistic surprise, on the notion of the work as "an enormous joke" (147). Surprise is also important in Sukenick's fiction (see chapter 4) but, in addition to the improvisational "disorder of style," it involves unforeseen reconfigurations at the experiential level. Despite Federman's affinity for a "non-pronominal fiction which totally disintegrates character," as in the case of Beckett and the "New New French Novelists" (143), his work remains interested in the "story" level at least in the minimalist sense of "little anecdotes strung together to make a sequence, although the sequence is no longer that of a plot" (144). Sukenick wants to replace plot with "more basic" configurations of "incident, movement, action" (143). Both writers share, on the other hand, a desire to "subvert official morality" and "certain ideas people have about life, about society" (147). But while Federman has viewed the "frantic and homogenized" American culture of the "telematic era" as an enemy to the imaginative writer ("The Last Stand of Literature" 191), proudly protecting his unassimilated "European point of view" (McCaffery, Hartl, and Rice 131), Sukenick has incorporated the maverick spirit of American popular culture, enrolling it in the service of his revisionist, "post-European" poetics ("Post Modern Fiction and Oppositional Art" 180; "Unwriting" 27).

From this brief comparison of two of its main practitioners, it becomes clear that surfiction pursues competing emphases: on one hand, a notion of art as a liberating medium that offers itself as experience or can return us to "a place beyond words where experience first occurs" (Kosinski, The Devil Tree 32); on the other, an insistence on the nonrepresentational, "resisting" nature of innovative art. Surfiction takes note of the "mutually exclusive positions" it generates, but also strives to achieve a "cunning [. . .] balance" between them (Federman, Double or Nothing 9). Federman's success at articulating a "real fictitious discourse" in his books has depended on his willingness to unwrite/rewrite the already extant stories of his life, blocking their "masturbatory recitation" (The Voice in the Closet 17; my pagination), while simultaneously releasing new possibilities within them. Described by Federman as stories that "cancel" themselves as they go, his novels have shunned both extreme disruption and the "fairy tale" of realism, moving tentatively ahead through trial and error, anxious to make some sense of the writer's traumatic survival from the Holocaust and transplantation to the New World.

Similarly, Sukenick's fiction has tried to bridge the "schizoid" division between writing and life, reclaiming segments of experience obfuscated by conventional realism. Like Federman, he has been "fascinated with writing and rewriting [his] autobiography," recognizing that the most "meticulously factual" narratives run into "heartbreaking ambiguities" (Mosaic Man 191, 193). But while his work has emphasized the improvisational freedom of narration, it has also maintained a productive tension between flow and pattern, experience and (re)writing. In Sukenick's novels, a metaphoric-cryptic discourse that raises the expectation of a hidden "message" competes with a notational, "instantaneous" prose that re-creates the feel of a particular experience. One can find similar experiential vignettes in novels written by Federman, Katz, Major, and Sorrentino:

> I swung at it, and I connected for a line drive over second base. A solid shot. I'll never forget how good that felt, connected from earth under my feet, up my legs, turned by

my wrists and hips, up my spine, through my arms, into my wrists and hands on the bat. Smash. I was one piece, moving into the ball. Thick as the Palisades. Fluid as the big Hudson River. (Katz, *Florry of Washington Heights* 32)

In their refusal of a depth hermeneutic, such passages appear "antithetical to what is the most conventional, imitated, standardized feature of modernist fiction [. . .], the epiphanic illumination" (Stevick 34). Still, they do allow "flashes of insight" to appear, until a provisional "fable, a gloss, begin to develop" (Sukenick, *The Death of the Novel* 154). Sukenick's *98.6* (1975) illustrates the rewards and limitations of narrative improvisation. This novel moves from realistic documentary in the first part that describes the violent implosion of the 1960s' culture, to social and literary experimentation in the second section focused on the career of a Californian commune, and finally to utopian fantasy in the last section that offers an alternative world vision in which Arabs have made peace with the Jews, Robert Kennedy has not been assassinated, and "Palestine" is a fertile country of the mind. By radically disrupting our expectations, the last section foregrounds the "fictive quality" of the novel we have been reading while also stressing the generative potential of narrative imagination, its "saving power" in "solving dilemmas" (Sukenick, "The Rival Tradition" 6).

As Sukenick has often argued, a *generative* approach to narration can enhance fiction's sociocultural impact. By replacing the culture's fossilized representations with self-generated "magical connections" (*The Death of the Novel* 154), surfiction replenishes our sense of reality. But this notion of spontaneous rearticulation is not left unquestioned. While continuing to celebrate "invention," "improvisation," and "freewriting," surfictionists have become more aware of the tension between their effort to break conventional rules and their need for new frameworks of articulation. Surfiction often contrasts two versions of language: one "official," bearer of the master codes, the other private and spontaneous, like the "Bjorsq" idiom in Sukenick's *98.6*. The random, body-oriented "Bjorsq" promises to reconnect us with immediate experience. But as Sukenick's more recent novels and essays have suggested, even an intensely private language like Bjorsq "can hardly avoid being representational in some sense" ("The Rival Tradition" 6). The impulse to articulate resides in all language, and it is indistinguishable—especially in the classic oedipal economy of narration—from the desire to formulate and control. In chapter 4, I will examine Sukenick's struggle with this question, highlighting his critique of the "pornographic" economy of realistic representation and effort to develop a generative approach to fiction based on the principle of the "mosaic," which uses "items of the museum of culture to create a new, if indeterminate, configuration" ("Post Modern Fiction and Oppositional Art" 180). Here I will exemplify with some of the more recent work of Marianne Hauser, Clarence Major, and Steve Katz, which suggest both the range of possibilities within surfiction and the extent to which its politics of narrative has matured in response to the polarized culture of the Cold War decades.

As Marianne Hauser's *The Memoirs of the Late Mr. Ashley: An American Comedy* (1986) makes clear, surfiction has become more critical of the myth of experiential and semantic plenitude that informed the performative poetics of the 1960s. At issue in Hauser's novel is the status of "voice" and its success/failure at articulating a coherent narrative world. This problem has been central to Hauser's English novels, which explore the ambivalent role that fantasy—psychotherapeutic in *Dark Dominion* (1947), mythic-archetypal in *Prince Ishmael* (1963), and cultural-linguistic in *The Talk-*

ing Room (1976), *The Memoirs of the Late Mr. Ashley* (1985), and *Me & My Mom* (1993)—play in both reinventing and insulating individual reality. *Prince Ishmael* renders uncertain our notions of agency and selfhood, adopting the figure of Caspar Hauser—a legendary nineteenth-century foundling known as "the Child of Europe"—as its narrator-protagonist. This "tottering spook" from another age disturbs not only nineteenth-century notions of social circumstance (the Nuremberg aristocracy cannot decide whether he is a genuine prince or a charlatan), but also our twentieth-century expectations of voice and identity. The dying (or already dead) first-person narrator at the end of the novel confronts us with the riddle of "origin and death" (Hauser, "About My Life So Far" 126), pointing ahead to the postmortem voice of *The Memoirs of the Late Mr. Ashley* and to Hauser's obsession in that novel with the artist as an "illusionist," "acteur manqué," and a "liar" (interview with McCaffery, *Some Other Frequency* 111). *The Talking Room* (1976) is narrated through another problematic voice: that of a pregnant thirteen-year-old who has a boozy "dropout from suburbia" for "mom and the test tube for pop" (1, 24). B's discourse is crossed and displaced by the voices of her mother and her lesbian friend, "Aunt V.," coming from downstairs. The interplay of these various voices, each caught in its own narrative of desire, transgression, and subjective reinvention, creates a "surfictionist" mix of "dream speech" and reality that looks forward to Hauser's later masterpiece.

The imaginative possibilities of Hauser's metamorphic narrative voices—whose problematic status mimics Hauser's own interstitial identity, born in the interspace between French and German cultures (Alsace-Lorraine), from a family with mixed French German and Jewish Arab heritage, and transplanted successively to the Orient, the American Midwest, and the countercultural New York where she discovers her "androgynous" nature (McCaffery, *Some Other Frequency* 109, 112)—are best illustrated in *The Memoirs of the Late Mr. Ashley*. These "memoirs" are delivered through the first-person voice of a male narrator, a fact that—as the author wryly comments—should not be seen as an "oddity" in women's fiction: "The gender switch is merely one of many switches, transferences, disguises, masks—tricks, if you will—through which a writer travels to create an epiphany" ("Literary Cross-Dressing" 3). And yet, by transgressing every other expectation we have about narrators (Mr. Ashley is not only male, but also gay, alcoholic, misogynous, racist, delusionary, and "dead" at the time of narration), Hauser upsets profoundly the logocentric premise of her novel. Mr. Ashley's postmortem memoirs foreground numerous tensions between confession and invention, past events and references to a present time "after I had been incinerated" (*The Memoirs of the Late Mr. Ashley* 14).

The narratorial voice is "diegetic-retrospective," to use Suzanne Fleischman's terminology, the voice of a "historian" or "memorialist"—rather than that of a "performer" (61–63)—unable to intervene in the course of events. At the same time, this voice is itself uncontrollable in its recriminations of others and sadistic wielding of language. Hauser's narratorial stance is well suited for what Drew Ashley calls the "oxymoronic sixties," torn between belief and cynicism, reality and simulation. It also befits Ashley's cultural complexities as a "fugitive from the Bible belt and the seedy boarding house of a pious mother" (Hauser, "About My Life So Far" 138); as a husband absconding from his wife's "all-American Electric dream" of a kitchen; as adoptive father-teacher-lover for an illiterate all-American gigolo, Richie; as failed gentleman-scholar, author of a never-completed treatise on the "sheen and sham and shamble of borrowed [Southern] culture" (*The Memoirs of the Late Ms. Ashley* 29).

Even more intriguing than the status of the narrative voice is the process of self–re-creation that it triggers. Hauser's novel dramatizes the difficulty of articulating a concept of self and world in an age of mass-produced messages, when "[r]eality, however rotten or rosy, can't compete with the comics, the TV-shows, the all night movies." Mr. Ashley's lifelong travesty as a successful *pater familias* and scholar is summed up in the tape with prerecorded typing noise that the narrator plays whenever he wants others to think he is diligently at work. This tape becomes in time a self-usurper, trapping the narrator in the mechanics of self-plagiarism. When his wife, Gwen, discovers the "Busy Fingers" tape after Drew's death, she wonders whether this was another

> facet of my narcissism. I always was enthralled by the reflections of my multiple selves, she recalls, the many echoes of my voices, my mirror images which I would scrutinize as though to chronicle them for posterity. [. . .] In spite of all my talent, all her money, I have produced only canned sound. (361)

But this "canned sound," the reader comes to realize, is metaphorically speaking Drew's other "work," the fiction of his life. With his "genius for intrigue," the narrator pieces together a life story from many incongruous fragments: "At times I am almost as good a liar as [Richie] is. And isn't every lie a masked confession? [. . .] When I give him what he wants—now clamor, now shoot-outs, I'm story-talking" (17). Like the dreamy monologue of the thirteen-year old in *The Talking Room* (1976), or the wistful discursions of an aging mother and her daughter that run parallel courses in the nursing home world of Hauser's novella *Me & My Mom* (1993), Drew's subjective "story-talking" defies the constraints of factuality, turning a "calendar of memories" dominated by images of defeat into a richly layered "vision of beauty." But his process of imaginative reinvention cannot escape the constraints of culture. His "memoirs" are replete with borrowed voices and styles.

Narrators in the earlier work of Federman, Hauser, Kosinski, Katz, Major, and Sukenick used imaginative reinvention against a representational tradition that had reduced Jews, blacks, women, or gays to "dummies"/"manikins"/"androids"/"cyperpods"/men and women "of parts" (Sukenick, *Mosaic Man* 205). Their improvisational fiction sought to "unwrite" the dominant narratives of "othering," reclaiming their own right to record "everything and anything that comes into the circle of my senses or memory or personal space or all of these at the same time" (Major, "Self Interview" 127). At the same time, these books denounced what Clarence Major called the "lies" of their own "arrangements" (interview in McCaffery, *Some Other Frequency* 255), accepting only provisional configurations. Their recent novels seem more cognizant of what Major has called the "artifice" of their freedom (Major, *The Dark and Feeling* 141). Their narration is inclusive, more tolerant of the conflicting interests and "mythic residues" that inform every act of narration. Drew's narrative is like "a tape erased, reused, erased & used again ad infinitum," haunted not only by his private demons, for "death is no exorcist," but also by many familial and cultural voices. The "impure" poetics of the novel, imitating both high-brow fiction (the confessional style of Proust, the descriptive vignettes of Joyce) and "the vulgar idiom of our teen-age motivated culture" (*The Memoirs of the Late Mr. Ashley* 71, 94), is an ironic comment on "the zeitgeist of an age that has no soul" (to quote Hauser's epigraph from Oscar Wilde). The novel's vaudevillesque scenes expose the male-dominated culture of the 1960s and 1970s that blends a rebellious hedonism with a vestigial Cold War mentality in which racism and sexism are still prevalent.

In Clarence Major's *Such Was the Season* (1987), a similar type of overcoded, media-created hyperreality forms the backdrop of the story. The narrator, a black Indian matriarch from Atlanta, Georgia, has an uncanny way of participating simultaneously in the life of her family and the ersatz reality of TV soaps. But unlike other Major characters who blended into the shows they were watching, using their TV for the "weird vegetable sort of copout security" it afforded (*Reflex and Bone Structure* 36), Aunt Annie Eliza is a "plain down-to-earth common sense person" (*Such Was the Season* 16) who can more successfully disentangle what Rick Altman calls the "TV flow" from the real-world "household flow" (40, 43–44). She is a shrewd observer of the human comedy on screen and in actual life. Drawing on the oral traditions of her double heritage, Annie Eliza's narration proceeds tentatively, restless "like the fall leaves in the wind" (121), following the natural structure of "talk." Her narrative strategy is simple enough: Even when it stalls or dissolves the plot in a mass of circumstantial detail, it connects the story to larger human interests without needless syntactic detours.

On the whole, this novel seems less concerned than Major's earlier fiction with the issue of narrator/character identity and the instability of their narrative vision. Following Henry Miller's example, the black protagonist of Major's first novel, *All-Night Vistors* (1969), tried to rebuild his sense of reality, traumatized by the Vietnam War and racial violence at home, through reckless erotic escapades and torturous narrative retrospection. Major's subsequent novels problematized the process of narrative reconstruction even further, looking for alternative ways of establishing reality such as "visual thinking," which pursues "connections between things more on the basis of visual associations than verbal or logical ones" (interview in McCaffery, *Some Other Frequency* 258). Written in the first person and the present tense, *Reflex and Bone Structure* (1975) reflected on the difficulties of writing, the unreliability of narrative imagination, and the endless mutability of characters. The trio of male protagonists composed of a whimsical narrator and theater director, an actor-revolutionary (Canada Jackson), and an off-Broadway actor, Dale, try to define their identities through their love for black actress Cora Hull—whose name carries the promise of a virginal center—and through storytelling. Failing in both, the male characters remain trapped in the claustrophobic spaces of their impoverished erotic and narrative imagination. All characters in this novel watch movies, plays, or TV shows, allowing their lives to be framed by formulaic plots. Like in Jerzy Kosinski's fiction, television tangles and flattens out everything, "night and day, big and small, tough and brittle, soft and rough, hot and cold, far and near. [. . .] Alas, [. . .] in this country, when we dream of reality, television wakes us" (*Being There* 5, 89). At the same time, by interrupting his characters' phony plots and refocusing attention on the "bone structure" of their lives, the narrator manages to offer—as the novel's inscription page suggests—"an extension of, not a duplication of reality," opening the process of narration to futurity, where "characters and events are happening for the first time." The strenuous process that an African American narrator has to follow in order to convert a postwar America, plagued by racism and gender polarization, into a narrative of new possibilities is illustrated also in Clarence Major's *Emergency Exit* (1979) and in *My Amputations* (1986). The latter novel illustrates—through its clever story of usurpation in which a well-read parolee impersonates a black novelist whom he has taken hostage—the difficulties of establishing an effective identity as an African American writer.

By contrast, *Such Was the Season* finds its sociocultural grounding more quickly, without taking us through elaborate metafictional questionings. But it still shares with Major's other novels an awareness of the versatility of storytelling as "living, moving document" (Klinkowitz, *The Self-Apparent Word* 85, 108). Aunt Annie Eliza's narrative begins with the promise of a "killer-diller week" (1), "full of all kinds of unexpected things happening" (2): the "homecoming" of her nephew Juneboy, a successful doctor from Yale who returns to Atlanta to rediscover his roots; her daughter-in-law Renee's failed campaign for a seat in the state senate; her son Jeremiah's downfall in the tomato industry scandal, a local Watergate of sorts; the mysterious disease of the incumbent state senator; a botched assassination plot on the president; a policeman's ball; and so on. Between these events that reveal a vestigial Cold War world on its way to becoming a more open, multicultural society, the narrator interpolates everyday conversations, sums up TV shows, or drags up "deep memories" (48). Annie Eliza's unhurried reconstruction of the context in which the events of a week unfold helps her cope with the altered world of her sons and nephews and with the "ailments of an old woman" (13). *Such Was the Season* is a novel about adjustment to a changing cultural ethos in the aftermath of the civil rights and women's movements. The narrator, who has owned her house for thirty years, proudly reports on her son's careers and wives, attends a dinner party "for Atlanta's Negro royalty," and confers secretly with prominent black politicians. But she never loses her elemental candor. She knows how to put things in perspective both by evoking the past of her hybrid race and by viewing skeptically its new gains.

Particularly important for Annie Eliza's comparative understanding of the changes undergone by African Americans is her retrieval of a few tales about her ancestors (the Sommers of Atlanta), who "lent [their] flesh and spirit to the continuation of the culture" (60). She incorporates these into her narrative, trying to fill the enormous gaps in the historical chronicle of her race. Annie Eliza's effort is recuperative and articulatory, not disruptive. As the consciousness of her family and race (there are hints that her performance is oriented toward an unidentified younger narratee), she cannot leave out any "membering" (186), however inconsequential. Her narration is also corrective, rejecting facile explanations and mendacious stories. What she finally discovers is not a single "truth," but the individual truths of many stories she helps articulate. Like the narrator in *Reflex and Bone Structure*, she can say, "I'm extending reality, not retelling it" (113).

Major's more recent *Painted Turtle: Woman With Guitar* (1988) cautions us again that self-narration is a difficult process. The reconstruction of Painted Turtle's complex story as a female folksinger of Zuñi ancestry proceeds slowly, mediated by her narrator-lover, the Navajo guitar player Baldwin Saiyataka. Yet this novel also illustrates the positive readjustments that multicultural subjects cause in traditional representations. The energy discharged by the hybrid character of Painted Turtle and the nonlinear approach of Zuñi storytelling she evokes stimulates the narrator's imagination, forcing him continually to revise his biographical reconstruction. The traditions of oral narrative and blues finally prevail also against the uninspiring background of a racially divided Chicago in the late 1940s. After a period of domestic troubles, discrimination in demeaning blue-collar jobs, and a mindless pursuit of "Dirty Bird" (Old Crow whiskey), the itinerant blues singer of *Dirty Bird Blues* (1996) finds his mooring in his music and in dreamlike narrative sequences that purge him of his anxieties. Like Aunt Eliza's reconstructions, the narratives of Baldwin Saiya-taka and Manfred Banks manage to provide an imaginative space for the negotiation of conflicting historical truths and personal perspectives.

Florry of Washington Heights (1987) by Steve Katz is a successful narrative for similar reasons. This novel is concerned with the "possibility" of fiction, refusing to be an "experimentation in failure" (Klinkowitz, *Literary Disruptions* 20) as some of Katz's ear-lier narratives. Its narrator, William Swanson (Swanny), is another fictionalized "Steve Katz" persona, patrolling a comic-surrealistic world, but this time only occa-sionally disturbing the boundary between autobiography and fiction. In Katz's *The Ex-agggerations of Peter Prince* (1968), epistemological uncertainty led to genuine narrative impasses: "[D]on't think this book isn't influenced by the fact that sixty times each second it gets dark in here, making over the period of years it takes to write a book [. . .] an appreciable amount of darkness" (136). It also encouraged a pluralization of narrative hypotheses: "Sometimes it's so hard to tell what has really happened. [. . .] That's why I want to develop multiple possibilities simultaneously" (157). The au-thorial narrator continually begged the question of narrative justification, struggling to maintain some control over a metamorphic character who was "pieced apart," "his past eras[ing] itself like a disappearing wake" (162). The technique of "erasure" worked paradoxically both to disrupt "dull literary conventions" and to create alter-native paths through the novel. A good example is the episode of Peter Prince's Ital-ian encounter with an eccentric Danish sculptor and his family. This incident is put under "erasure," crossed out but still legible; interspersed in it are passages that reveal the autobiographical basis for this episode. Thus the reader is encouraged to partici-pate in a contradictory narrative dynamic that simultaneously fictionalizes and prob-lematizes a personal experience, rendering the distinction between fiction and autobiography uncertain.

Florry of Washington Heights still begins with a recognition of narratorial limitations:

> Whoever tells you he knows everything about his own neighborhood you can be sure is fooling himself. Something else always goes on in the schoolyard while you sit on the park wall, or vice versa. When I was a kid, I learned that. So when I grew up I never expected to understand all the world, not even with all the education I could get [. . .] as a lawyer. [. . .] To have the illusion that you can understand the world you've got to come from a small town." (7)

Nonetheless, what emerges is a successful retrospective of childhood in "Washing-ton Heights" told by an articulate and obliging narrator, Swanny. This Manhattan neighborhood is "not unlike a small town" (7), giving its inhabitants the illusion of control over their world and stories. It is true, however, that narrative mastery is not taken for granted: Life's events continually surprise Swanny with their potential for unpredictable associations. In order to set free this dormant energy of reconfigura-tion, Swanny lingers on anecdotal incidents and conversations. His narrative is made up of self-contained vignettes that suggest, beyond their experiential verve, a larger puzzle that needs slow deciphering.

Especially two of the experiences recalled (the love story between Fred Sugar-man, star of the Bullets A & S, and Florry O'Neill, former "steady girl" of the Fan-wood gang leader Jack Ryan, who is subsequently raped and strangled; and the final "rumble" between the mostly Jewish gang of the Bullets and the Irish Fanwoods) open onto larger existential and social "mysteries." One wonders at times how much reality this conflict, reminiscent of *West Side Story*, has outside the narrator's need to play out the two sides of his ethnic heritage: "Was I making it up? Was this only my fight? Is this going on in my Irish-Jewish head?" (119) This type of self-questioning is not irrelevant here: *Florry of Washington Heights* shares with other of Katz's narratives a knack for retelling, borrowing certain events and characters from previous books (especially *Wier & Pouce*, 1984), narrating them in alternative versions that render "the 'truth' into all its dimensions, never ruining a good story" (*Wier & Pouce* 29). The effect of this imaginative rewriting is not only to pluralize truth, but also to submit "reality" to a transformative consciousness that enhances the narrator's cultural knowledge. By way of his revisionist narrative, Swanny moves from the adolescent world of pop heroes and simple resolutions to the pragmatic vision of adulthood. Applying narratorial hindsight, he relates the raw and insecure world of a childhood spent during the Great Depression to the treacherous culture emerging after the Korean conflict in which he is a lawyer, "spreading the cheeks of the law" (71) in order to survive. As several scenes featuring a caricatured Henry Kissinger suggest, the latter world is the political fruit of the former. The Cold War culture bears the unredeemed burden of past racial and social violence; in turn it lends its own con-flict-ridden imagery to the narrator's boyhood memories (Broadway divides Swanny's childhood world into a "North and South Korea" [104]; choosing a club name and uniforms exposes Swanny's friends to absurd retaliations from the rival gang: "It was like having to go to Korea or something" [180]). This process of cul-tural clarification is continued in *Swanny's Way* (1995), conceived as the last volume in a loose trilogy that includes *Wier & Pouce* and *Florry of Washington Heights*. Swanny shares here the task of storytelling with his former enemy, Jackson Ryan, whom he entrusts with commenting upon a series of "manuscripts" containing his memories and fan-tasies. Revisiting their adolescence from the perspective of the 1970s, the two nar-rators manage to transcend some polarizations of their youth, imagining a world in which interracial cohabitation and homosexuality are no longer tabooed.

The first person narrators of Hauser's, Major's, and Katz's more recent fiction perform their tasks successfully, putting together a narrative in spite of all the uncer-tainties encountered. But their diegetic and cultural status is unsettling in various ways: The very choice of voice—male in Marianne Hauser's *The Memoirs of the Late Mr. Ashley*, female in Major's *Such Was the Season*, fictionalized-autobiographical in Katz—is partly polemical, reversing expectations. The narratives they produce are thus self-problematized, making us aware of many "empty spaces" and of the novelist's effort to fill them "with luminous motion, and things. [. . .] How a novel can fill with them like a barrel with sponges. They rise like pieces from a sunken ship and lie noiselessly on the tide" (Katz, *The Exagggerations of Peter Prince* 165). From this point of view, recent surfiction falls into Klinkowitz's category of "'experimental realism' in which the sim-ple act of vision becomes not just an integral work of art but an interpretation of our cultural act of seeing as well" (*The Self-Apparent Word* 121). This type of realism revisits the available modes of narrative articulation, interplaying them imaginatively.

The recent fiction of Federman and Sukenick has moved in a similar direction, emphasizing the task of rewriting over that of deconstruction. Recent surfiction seems less concerned with the impossibility of narrating "lives," more eager to de-velop its own alternative story against the dominant cultural narratives. Though within the Cold War era that Federman's *Smiles on Washington Square* (1985) evokes, the most logical narrative hypothesis is that Moinous (a "foreigner" returned from active service in Korea) and Sucette (a well-to-do American woman with leftist sympathies) will be granted no further opportunity to come together beyond their chance encounter "across a smile" at an anti-McCarthy rally, this novel mobilizes a whole array of strategies to turn a *nonstory* into a "love story of sorts." But with the emphasis on fiction's role in "reinventing" reality come increased responsibilities. Federman's *To Whom It May Concern* (1990) shares the task of rethinking postwar history with its readers—real or implied. The same commitment to re-creative narration is also present in Federman's collection of sixty-two "microfictions" and prose poems, *Loose Shoes: A Life Story of Sorts* (2000). While exploiting the formal versatility and easy availability of on-line publication, these hybrid, "non-direction" narratives maintain their experimental ability, employing the electronic format to break out of conventional literary and discursive modes. Federman's recent work makes thus clear that surfiction's new resolve to carry out a successful program of narrative revisioning should not come at the expense of submitting too "easily to **the possible**. On the contrary, I know that literature, today as always, faces the impossible, faces the inadequation of language and of thought to apprehend or even comprehend reality, and yet, always in quest of new forms, literature will succeed in giving life once again to **the impossible**" (*Critifiction* 129). Nor should innovative fiction allow itself to be recuperated by the marketplace or popular culture. What we need is not "an easy, facile, [. . .] high-tech prose," but the kind of writing that will "systematically erode and dissipate the setting of the Spectacle," creating "a space of resistance to the alienated devotion to images" ("Avant-Pop: You're Kidding!" 176–77).

By contrast, Sukenick has accepted the challenge of integrating plots and techniques from pop culture into his recent "interventive" fiction. His Hollywood novel, *Blown Away* (1986), and the experimental prose poem *The Endless Short Story* (1986) revealed Sukenick's awareness of competing narrative modes in contemporary culture—verbal and cinematic, language-oriented and image-oriented. These works demonstrate surfiction's ability to renew itself syntactically and epistemologically, but they also suggest that innovation risks constantly to be co-opted by mass culture. *Doggy Bag: Hyperfictions* (1994), on the other hand, is entirely immersed in the "white noise" of contemporary culture, welcoming both peripheral experiences that "you ordinarily screen out when you focus on something" (Sukenick, "The Rival Tradition" 10) and the cacophony of styles that define our postmodern condition. While sifting through the rubble of Europe's decadent high culture and America's consumerism, Sukenick's "hyperfictions" recycle both canonic and popular modes of narration, using their differential "noise" to combat the cultural entropy of the postindustrial West. From this point of view, they invite comparison with "Avant-Pop" fiction characterized, according to McCaffery, by

> its appropriation of mass cultural imagery and idioms, its information density, its recognition of the mythic [. . .] dimensions of pop cultural materials, and its use of collage and other non-linear, multidimensional formal structures as a means of suggesting the pluralism of worlds existing within the Mediascope; and by its emphasis on collaborative, interactive approaches to the composition process whereby pop cultural references, character types, plot lines, and other motifs [. . .] are reconfigured into new contents and meanings

which illuminate the original's biases and limitations [...]. ("Reconfiguring the Logic of Hyperconsumer Capitalism" I)

Sukenick's theoretical interest in "avant-pop" dates back to the period when he researched the post-beat and punk scene for his book-length essay *Down and In: Life in the Underground* (1987), a re-creation of the Greenwich Bohemia from the late 1940s to the 1970s. But his own fiction, exploring—as early as *Out* (1973) and *98.6* (1975)—the forms of violence embedded in the language and plots of popular culture, can be regarded as a forerunner of this trend, together with the work of Woody Allen, Kathy Acker, Donald Barthelme, Robert Coover, Tom Robbins, Kurt Vonnegut, and Steve Katz, whose *The Exagggerations of Peter Prince* has been described by Lance Olsen and Mark Amerika as "*the* Avant-Pop novel par excellence" (2). Sukenick's fictions have focused increasingly on the "blizzard of white noise, random codes, and competing realities" (McCaffery, "Avant-Pop" xiv), while also trying to rewrite the plots of contemporary culture in ways that would free them from the "negative hallucinations" of the Cold War era. Larry McCaffery is justified, therefore, in juxtaposing Sukenick's work with that of a new generation of avant-pop writers (Mark Amerika, Rikki Ducornet, Eurudice, Harold Jaffe, Mark Leyner, Lynne Tillman, Curtis White, and Stephen Wright) who have tried to define a place for reflexive fiction in an age of information density and electronic reproduction. But some distinctions remain. While this new breed of artists has replaced the "avant-garde's conception of artistic radicalism in terms of an *evolutionary paradigm* (an 'us' engaged in a deadly [...] struggle with 'them') with a more flexible model based on *co-evolution*" or interaction with mass culture (McCaffery, "Reconfiguring the Logic of Hyperconsumer Capitalism" 12), Sukenick is an "interventionist" rewriter even in his "avant-pop" fiction. While accepting his younger colleagues' challenge to "synthesize modernist intelligence and avant aesthetics with pop/punk/postmodern popular resistance" (Curtis White, "An Essay-Simulacrum on Avant-Pop" 80), Sukenick has pursued this by interplaying critically different worlds and projects, opening them up to new possibilities.

Sukenick's most recent novel, *Mosaic Man* (1999), is a fine illustration of this type of revisionist synthesis. This novel reconfigures not only Sukenick's previous work, rewriting motifs from *Up, 98.6, Blown Away,* and *Doggy Bag,* but also the poetics of post-Holocaust/post–Cold War fiction. *Mosaic Man* attempts a comprehensive recovery of an "other-worldly" realm of lost or ignored possibilities. As the authorial narrator Ron puts it, "in a world that can't honor" the needs of imagination, "The other world is all you have!" (*Mosaic Man* 180). This other world of untapped potentiality is sought first of all at a personal level: in the story of a post-Holocaust Jewish writer who suffers from "an advanced case of being human, [when] being human may be a terminal disease that's run its course" (16); or in the "cross-fiction" (192) of Sukenick and his literary friends who want to reinvent fiction as an interventionist art. The "other-worldly" is sought also at the level of cultural history through an approach that uses improvisation and parody to rewrite our master narratives (from the five books of Moses to highbrow modernist texts and lowbrow popular plots), flushing out the oppressive virus of polarization. Sukenick finds this virus not only in Nazified wartime Europe, but also in McCarthyist and hyperconsumerist America, in the hedonistic machismo of the 1960s, in the Cold War geopolitical partitions and their subtle reinvention after the fall of the Berlin War, and even in the "Mosaic" vs. the "Aronian"

(Rabbi vs. Mogul) division of his own Jewish heritage (248). Threatened with being "xxx-ed" by the religious, political, and literary "writing on the wall" (182), the post-Holocaust/post–Cold War writer finds his home in the re-creative "mosaic" of his fiction that honors the visionary "unworldly" as much as the "worldly-wise" (180).

3. REVISIONISTIC NARRATIVES FROM THE INTERSTICES OF THE COLD WAR: POSTMODERN FEMINIST FICTION

Like surfiction, the innovative feminist novel has been engaged in a critique of the dualistic, exclusionary mode of thinking embedded in the master narratives of modern history. But while surfiction has emphasized the performative dynamics of fiction, creating imaginative ruptures in established reality, the feminist novel has sought to liberate women from the stereotyping imagination of male-authored fiction, articulating an alternative narrative epistemology in response to both traditional "objectivism" and postmodern "relativism." Feminist theory and practice have helped elucidate what Betty Friedan's pioneering study, *The Feminine Mystique* (1963), called "the problem that has no name": women's "bored, diffuse, feeling of purpose-lessness, non-existence, non-involvement with the world" (191) during the period corresponding roughly to the Cold War. This identity crisis has been explained in terms of the "inescapable either/or: motherhood or individuation" alternative forced upon women in the decades following World War II (Rich, *Of Woman Born* 154), and of the lingering patriarchal tradition that operated through "romance narratives" to obfuscate women's interests and imaginations. The Cold War narratives of "containment" found reinforcement at home in the

> cult of domesticity as a form of political and social containment for the sexual energies of the post–World War II [. . .]. [T]he responsibility for this containment in the post-war era fell on women, whose role was to resist and channel the "natural" sexual energies of men. Female sexuality thus had the burden of supporting the monolithic goals of cold war America through the practice of duplicity: the woman had to attract and stimulate male sexual drives but not gratify them. (Nadel 117)

Reflecting this duplicity at a narrative level, the roles available to women in the 1950s were still those of "Devourer/Bitches or Maiden/Victims" of some fatherly "Super-Male" (Russ 84, 96). Midcentury women writers could either produce new versions of the seduction plot that represented for women "success, failure, education, and the only adventure possible" (112), or abandon female protagonists altogether. A third option added at the beginning of the 1960s was to "write about heroines whose main action was to go mad—but How She Went Crazy [. . .] also los[t] its charm in time" (85).

A better alternative—identified by Joanna Russ with the feminist utopias of the 1970s of which her own *The Female Man* (1975) was a good example—was "to pick up the novel-as-it-was," think "I can't use this," and create "a new field" (*To Write Like a Woman* 121). Russ's 1971 essay "What Can a Heroine Do? or Why Women Can't Write," proposed two new narrative "fields" for women's creativity: lyrical-associative fiction and explorative science fiction. Reminiscent of Dorothy Richardson's pioneering effort to create a specifically female narrative style that replaced the "modernist compositional symmetries" with a streamlike rendering of women's subjective lives (Brian Richardson, "Remapping the Present" 295), Russ's postmodern lyric mode replaced chronology and causation with an "associative" principle, organizing "*discrete elements* (images, events, scenes, passages, words, what-have-you) *around an un-spoken thematic or emotional center*" (*To Write Like a Woman* 87). The other narrative "field" advocated by Russ, explorative science fiction, had an even stronger rearticulative potential, revising our experience of the social and offering a "positive subjunctivity" (16). The feminist utopias written in the 1970s by Ursula Le Guin, Marge Piercy, Sally Gearhart, and Joanna Russ herself reconsidered the technoscientific narratives of modernity from the perspective of those who do not control them, "women, non-whites, the poor" (37). Concerned with the "woman's estate and the problems of social structure" (135), explorative science fiction was "demystifying about biology, emphatic about the necessity for female bonding, [. . .] and serious about the emotional and physical consequences of violence" (58), seeking alternatives to the gothic-misogynistic plots of the 1950s and 1960s. The role of the new fiction envisioned by Russ was to "re-perceive experience, not because our experience is complex or subtle or hard to understand (though it is sometimes all three) but because so much of what's presented to us as 'the real world' or 'the way it is' is so obviously untrue" (vi).

Though interconnected in Russ's 1970s essays, thematic and poetic revision were pursued somewhat disjointedly by other feminist writers of the first wave. As Larry McCaffery also argues:

> the first wave of feminist fiction dealt primarily with the "content" of feminist insights within fairly traditional structures. [. . . B]ut it quickly became obvious to a number of important female authors that the basic assumptions and conventions underlying realistic fiction—its reliance on reason and causality, [. . .] its requirements for a dramatic action in which conflicts could be resolved, its implications about what constituted "heroism" and "significant" action—were inherently male-defined and hence in many ways inadequate to convey the most salient features of women's lives. ("The Fictions of the Present" 1170–71)

As a consequence, the earlier content accommodation was followed beginning in the mid-1970s by more radical forms of innovation (from surreal deconstruction to feminist political utopias) that mixed cultural and narrative rewriting. In this newer version, feminist fiction has made consistent use of "differential" gender thinking to disrupt the system of "phallogocentric" representations, creating discursive structures more responsive to women's experiences. But a certain reticence toward postmodern experimentation has lingered on in feminist theory and practice:

> Feminist fiction seldom is as self-conscious and artificial as are male fictions, and experimentation usually serves the ultimate end of realism. The conflict between representationalism and experimentalism is resolved in this manner: it is women's real lives that defy the laws of the text; as women write their selves, so do they destroy the laws. (Zimmerman, "Feminist Fiction and the Postmodern Challenge" 177)

Derived from the "anti-intellectualism based on bad Marxism that plagued the women's movement of the early seventies" and "the identity-based politics of feminism in the eighties" (Robinson 1), this continued resistance to the avant-garde betrays an unqualified faith in the uniqueness of women's experience and the ability of conventional language to render it. The narrative recuperation of "women's real lives" has undoubtedly radicalized the content of the novel, complicating cultural expectations and recasting traditional gender roles. But this thematic reorientation has not automatically radicalized the narrative practice itself, maintaining women's fiction within the orbit of traditional realism. As long as they endorse the "phallocentric symbolic order" of realism, women writers may "feel guilty about [their] desire to obtain mastery over language," resisting or "fantasiz[ing] away responsibility for such an unspeakable wish" (Moi 118). To put it simply, our sense of "lived reality" depends directly on the models through which we see and represent it. Formal revision remains, therefore, a "powerful tool of feminist critique, for to draw attention to the structures of fiction is also to draw attention to the conventionality of the codes that govern human behavior, to reveal how such codes have been constructed and how they can therefore be changed" (Greene 2).

Narrative experimentation has been present in women's fiction from Gertrude Stein and Djuna Barnes to Doris Lessing and Kathy Acker, but has sometimes passed unnoticed even though "it helped make a social revolution, playing a major role in the resurgence of feminism in the sixties and seventies" (Greene 2). Ellen G. Friedman and Miriam Fuchs have identified three cycles of female experimentation—before 1930, between 1930 and 1960, and after 1960—suggesting that female narrative innovation has had a notable history, antedating through Gertrude Stein's *Three Lives* (1909) the beginning of male modernism in Proust (introduction to *Breaking the Sequence*). Women's experimental writing has also challenged and refined the work of mainstream feminists. Thus, while feminists in the 1980s focused on "defining, constructing, or recovering a genuine self out of the schizophrenic condition of being female in this culture," the work of experimental women writers "bl[ew] apart conventional notions of a unified self" and invented new "discourses to convey *their experience of being female* because they [did] not find conventional discourse up to it" (Robinson 11, 12). Women's experimental fiction cannot be understood properly without reference to feminism and postmodernism, but its departures from these two paradigms are just as important as its similarities.

The relationship with feminist literary and cultural theory is not difficult to ascertain. Though innovative writers like Margaret Atwood, Gail Godwin, Doris Lessing, Alice Walker, and Toni Morrison have periodically distanced themselves from programmatic feminism, their work shares the feminist "analysis of gender as socially constructed and its sense that what has been constructed may be reconstructed" as well as the feminist "understanding that change is possible and that narrative plays a part in it" (Greene 2). The label is useful also in distinguishing this project from forms of male innovation. Instead of simply transplanting postmodern male strategies, feminist fiction has sought specific solutions to analogous questions. For example, the "experimental fictions by women seem to share the decentering and disseminating strategies of postmodernist narratives, but they also seem to arrive at these strategies by an entirely different route, which involves emphasizing conventionally marginal characters and themes, in this way *re*-centering the value structure of the narrative" (Hite, *The Other Side of the Story* 2). While affirming "woman's side" of the story, feminism fiction has carefully dissociated it from the "masculinist definition of what this 'side' consists of" (Greene 11).

The problem with both classic realism and male postmodernism is that they have often locked women in predictable plots, constraining them to subordinate cultural positions. Therefore, in addition to recovering the "other side of the story" (marginalized characters, suppressed truths), women's experimental fiction has had to define "another *kind* of story" (Hite 11), an alternative narrative practice. The maternal semiotic of sound-play and somatic rhythm theorized by Julia Kristeva (see "The Subject in Signifying Practice" 22, 24–25; *Revolution in Poetic Language,* chapter 1) or the *écriture féminine* advocated by Hélène Cixous (245–64) and Luce Irigaray (*This Sex Which Is Not One,* especially chapters 4 and 11) offer such an alternative, pitting a fluid, metamorphic type of narrative discourse against the rigid truths of androcentric realism. Moving away from the empirical approach of first-wave feminism, the French poststructuralist feminists have tried to put "woman" (multiple, "never being simply one"—Irigaray, *This Sex Which Is Not One* 31) back into discourse, giving "a new language to those other spaces" (body, nature, woman) repressed by the symbolic discourse of the West (Jardine 33, 72–73). For example, Luce Irigaray's *Speculum of the Other Woman* (1974) begins with a comprehensive critique of the monopoly held by a unified masculine subject in the Western tradition, but then sets out to define a new female subjectivity and mode of thinking/speaking. For Hélène Cixous, there is no better way for displacing the patriarchal language and cultural order than the avant-garde "texte féminine": "A feminine text cannot fail to be more than subversive. It is volcanic; as it is written it brings about an upheaval of the old property crust, carrier of masculine investments" (258).

But this project has obvious limitations. As Mary Jacobus cautioned more than two decades ago, the call "for a special language for women" should not "mean a refusal of language itself; nor a return to the specifically feminine linguistic domain which in fact marks the place of women's oppression and confinement" (12). Cixous's "new insurgent writing" seems to do both, seeking to "wreck partitions, classes, rhetorics, regulations and codes" (250, 256) while also essentializing female discourse, deriving its model of authentic language from the idealized features and functions of the female body. Irigaray is caught in a similar contradiction when she emphasizes the need for dialogic interaction between gendered subjects and discourses but admits that there is an "irreducible mystery between man and woman" predicated on an irreducible "ontological difference" (conversation with Luce Irigaray, in Olson and Hirsh 160, 161). The practice of "écriture féminine" is equally ambivalent: It both recalls features of the male avant-garde (Lautréamont, Joyce, Genet, Artaud, Bataille, Sollers) and exposes masculinist representations by "miming" deconstructively women's assigned roles in the phallic discourse (Irigaray, *This Sex Which Is Not One* 76). This subversive "miming" risks falling back "into the language and sexual ideology of the status quo" or becoming entirely absorbed "into the world of female 'difference'" (Frye 23). Both outcomes raise the question of the feasibility of articulating a specifically female perspective in a male-dominated language.

The issue of language remains central to the feminist debate and practice. As bell hooks puts it:

> We are wedded in language, have our being in words. Language is also a place of struggle. Dare I speak to oppressed and oppressor in the same voice? Dare I speak to you in a language that will move beyond the boundaries of domination—a language that will not bind you, fence you, hold you? Language is also a place of struggle. The oppressed struggle in language to recover ourselves, to reconcile, to reunite, to renew. (*Yearning* 146)

Some feminist theorists have gone so far as to argue that until the whole cultural base changes, women should "stop reading novels" if they do not want to remain stuck in the old prescriptive plots. As Margaret Homans summarizes, "Oedipal, phallic narrative and narrative histories misrepresent women's experience; for that, new forms are needed that break narrative conventions and that perhaps even depart from narration altogether" (6–7). Feminist novelists have taken a more pragmatic view on the issue of language but their work returns to it continually. "I want to be better with words"—Marge Piercy wrote in *Small Changes* (1973)—"I want to be able to answer them back. But I don't believe that's how you do anything. I only want to use words as weapons because I'm tired of being beaten with them" (267). Feeling "betrayed" by the prevailing novels that "lead her to want what she could not approach" (29), Piercy's heroine feels tempted to bypass the entire novelistic tradition. But this proves not only impractical but also impossible since that tradition is part of her acculturation. Kathy Acker's female Quixote faces the same quandary: "Being born into and part of a male world, she had no speech of her own. All she could do was read male texts which weren't hers" (*Don Quixote* 39). But her response is more radical: In contrast to her seventeenth-century male counterpart who remained trapped in a world of chivalrous deceptions, Acker's punk heroine follows the example of the medieval Arab scholars and their twentieth-century poststructuralist heirs, "cutting chunks out of all-ready written texts and in other ways defacing traditions: changing important names into silly ones, making dirty jokes out of matters that should be of utmost importance to us" (25). What emerges from this textual intercutting is a scathing critique of male and female fictions, from the narratives of the Vietnam era to the new feminist mythologies. Abhor, the part robot–part black heroine of *Empire of the Senseless* (1988), proposes an even more scandalous response to the dominant culture, "taking layers of my epidermis [. . .] and tearing each one of them off so more and more of my blood shoots in your face. This is what writing is to me a woman" (210).

Not all feminist writers have followed Acker's extreme course, "throwing [their] body against language, throwing language at the body of [their] readers, desperately and bravely trying to narrow the gap [. . .] between blood and word" (Landon 7). While calling into question the patriarchal traditions of fiction, most female writers have retained the novel as a useful medium for recovering the historical experience of women. Recent feminist narratology has also reevaluated narrative, arguing that, in spite of its long history of privileging the male point of view and experience, narrative is "potentially polyvocal" (Susan Stanford Friedman, "Lyric Subversion" 180).

Innovative women writers need the novel, but in a form that is "internally dialogized" (Bakhtin, *The Dialogic Imagination* 324), radicalized against itself. The procedures proposed for this task range from the "'pointless' and 'plotless' narratives stuffed with strange minutiae" and obsessed with "the unspeakable, undramatizable, unembodiable action," theorized and illustrated by Joanna Russ (*To Write Like a Woman* 89, 90) to the rewritings discussed by Joanne Frye (6), "in which women are able to develop a capacity for complex selfhood in interaction with contemporary realities, resisting the old stories and telling lives in new ways" (8); and from the plots of lesbian awakening studied by Bonnie Zimmerman ("Exiting from Patriarchy") to the fiction valorizing "marginal" body experiences and making "space for the unthought" theorized and practiced by Nicole Brossard (81). The narrative relay can also be dialogized as in Louise Erdrich's novel *Tracks* (1988), which reconstructs the struggle for identity of

a Native American woman through alternating male and female voices. By simultaneously addressing a female narratee (the heroine's daughter) and interpellating the
patriarchal tradition that threatens to reappropriate female experience, Erdrich's
novel is "able to 'space-shift' [its] language and thus [its] perspectives [. . .]. Through
Fleur's struggles against [. . .] white male domination, and through the dialogism
present between the author, her narrators, and her audience, Erdrich ex/changes her
narrative strategy with female experience" (Nelson-Born 9–10).

The task of replacing traditional plots and narrative categories with new, more
responsive practices for representing female experience brings women's fiction
closer to postmodernism. In spite of their mutual weariness, feminism and postmodernism have reasons to cooperate. As "two of the most important political-cul-
tural currents of the last decade" (Fraser and Nicholson 19), feminism and
postmodernism need each other to correct and enhance their insights. Feminism
can start by calling into question the excessive fragmentation of the postmodern
sociocultural vision and the marginalization of female issues in the male avant-
gardes. On the surface, "None of the metaphors for the postmodern (writing, the
sublime, conversation, or aesthetic practices) seems congruent with the concerns of
feminist discourses or practices" (Flax 210). Yet it is clear even for Lyotard that
postmodernism can learn from women writers "a certain way of coming to what is
unknown, a sort of patience with the necessity to answer, rejecting the necessity to
have results immediately or as soon as possible" (conversation with Jean-François
Lyotard, in Olson and Hirsh 188). Revisionist postmodern writing is in this sense
"feminine," working against "the temptation to grasp, to master," and opening
"blank fields, blank spaces [. . .] in order to let events happen" (Olson and Hirsh
184). In turn, the postmodern perspective can illuminate the "tug of war" between
those who have encouraged and those who have undermined totalizing tendencies
in feminist literature. The relativizing perspective of postmodernism can be used to
interrogate feminism's bias for realistic narrative forms or essentialist categories
like "sexuality, mothering, reproduction, and sex-affective production [that] group
together phenomena which are not necessarily conjoined in all societies" (Fraser
and Nicholson 31). Feminism's liberating "politics of experience" must not be compromised through narratives that make use of outdated notions of selfhood and
conventional strategies for constructing that precious object called "shared experience" (conversation with Donna Haraway, in Olson and Hirsh 60, 61).

One question that postmodernism has helped women writers rethink is that of
female narrative agency in the Cold War era and after. Simply saying "I" does not
empower women or other marginalized subjects. According to Luce Irigaray, in
"speaking as a woman" it is important "to say not only 'I' but to say 'I-she' (Je-elle) —
that is to live that 'I' and define it not only as a simple subjectivity that expresses itself, but in terms of a dialectic between subjectivity and objectivity. [. . . This]
permits me to make visible that the subject is two [. . .] and to pose all sorts of dialogic questions" (conversation with Luce Irigaray, in Olson and Hirsh 153). The
paradoxical interactions between two aspects of the female subject and between
these and patriarchal culture complicate the female speaker's position:

> By virtue of speaking as a woman, any female narrator-protagonist evokes some aware-
> ness of the disjunction between internal and external definitions and some recognition
> of her agency in self-narration. To speak directly in a personal voice is to deny the ex-

clusive right of male author-ity implicit in a public voice and to escape the expression
of dominant ideologies upon which an omniscient narrator depends. (Frye 51)

Postmodernism problematizes the notions of autonomous consciousness and authentic individual self on which male narrative authority has relied; but for similar reasons,
it subverts counterclaims to a stable female identity. As Michel Foucault warned, there
are "two meanings of the word 'subject': subject to someone else by control and de-
pendence; and tied to his own identity by a conscience or self-knowledge. Both meanings suggest a form of power which subjugates and makes subject to" ("The Subject
and Power" 781). Therefore, in reclaiming woman as agent feminism should seek new
forms of subjectivity that refuse the "simultaneous individualization and totalization
of modern power structures" (785). Understanding power and agency as relational,
feminism can engage other free "individual and collective subjects" in a narrative "field
of possibilities in which several ways of behaving, several reactions and diverse com-
portments, may be realized" (789).

On both sides of the former Cold War divide, women's fiction has had to struggle with questions of narrative agency, looking for ways to reinscribe the female subject in fiction. The narrating "I" in the fiction of Doris Lessing, Maxine Hong
Kingston, Louise Erdrich, Christa Wolf, Gabriela Adameşteanu, and Tatyana Tolstaya manifests not only self-fragmentation and nonlinear multiplicity, but also a dialogic "capacity to engage the normative and simultaneously elude and critique it, to
evoke realities at the same time that it interrogates our ways of defining them" (Frye
55). The narrating "I" is engaged in a reconstruction of both female experience and
of the categories through which this experience is being reflected and defined (female identity, character and plot, memory and representation). Nothing is taken for
granted: Innovative feminist fiction challenges traditional assumptions that identity
is destiny, that character features are stable, that female memory works just like male
memory, or that historical "reality" is a matter of transcendental truths rather than
an interpretive construct. Women's novels often make visible the struggle to reem-
plot the past and keep the narrative open to alternative futures. From this point of
view they emulate the postmodern critique of plots and selves but resist its stronger
deconstructive claims, preferring to redefine rather than dismantle narration. The
best work by women balances a "double allegiance: to the formal experiments and
some of the cultural aspirations of the historical male avant-gardes; and to the feminist critique of dominant sexual ideologies, including the sexual ideology of the
same avant-gardes" (Suleiman, *Subversive Intent* xvii).

In negotiating this double allegiance, feminist writers have most often chosen to
develop their own innovative narrative strategies rather than follow the solutions offered by male experimentalism. Therefore, Julia Kristeva has proposed the term
"postmodern feminism" ("Women's Time" 211–12) for the more recent trend concerned with rethinking the categories and strategies of both postmodernism and
feminism. By comparison with alternative terms like "feminist postmodernism,"
"women's metafiction," and so on, this term has the advantage of emphasizing the
feminist agenda, suggesting that progressive feminist thinking must not only incorporate the postmodern solutions but also move beyond them. Against the fashionable postmodern aversion toward all integrative theories, innovative feminism has
remained committed to larger explanatory narratives of social identity, constructed
according to a theory that is "explicitly historical, attuned to the cultural specificity

of different societies and periods and to that of different groups within societies and periods" (Fraser and Nicholson 34). From the combination of a "feminist standpoint epistemology" (conversation with Sandra Harding, in Olson and Hirsh 18), engaged in elucidating the sources of gender inequalities, and a revisionist narratology has resulted a stronger version of postmodernism that uses postmodern insights "to get more useful understandings of history and knowledge, relationships of theories of representation, [and . . .] the problems of totalizing theories" (3).

Responding to the emphasis on the production of meaning in various branches of contemporary thought, postmodern feminism has offered its representations as overtly political and revisionistic. Examples are easy to find, from the innovative women's fiction that has raised important questions regarding the male-oriented economy of traditional narration, to feminist visual and performative art that has interrupted the dominant male modes of production in modernist culture. Through its willingness "to focus culturally not only on inward relationships but on the broader outward relationships that control them," feminism has managed according to Lucy Lippard to "change the character of art" (151). Other evaluations have been more circumspect. For Linda Hutcheon, feminist art represents a "potentially positive oppositional and contestatory" mode within postmodernism, which has given us an awareness of the cultural "systems of power [that] authorize some representations while suppressing others" (*The Politics of Postmodernism* 143) but not always the tools to transform them. Hutcheon praises Angela Carter, Gayl Jones, and Christa Wolf for their innovative uses of oral and written narrative to re-present the ignored history of women, but she also reminds us of the problems involved in confronting the "complexities of [historical] representation" in a language that is still inherently sexist. Similarly, the feminist photography of Cindy Sherman exposes the stereotyping of female identity by the male gaze but does not define more gratifying roles for women.

Such works reconfirm Hutcheon's view that postmodernism is only an incipient critique, providing feminism with "strategies of parodic inscription and subversion in order to initiate the deconstructive first step" (168) but no genuinely reconstructive tools. And yet, contemporary feminist fiction, poetry, and film offer other examples of innovative art that take us beyond "feminist postmodernist parody." One such example, mentioned by Hutcheon in passing, is "body art" (142) that responds to the dematerializing effects of modernist culture by valorizing body both as "matter" and as discourse. The first emphasis, inspired by the more recent work of Judith Butler (*Bodies that Matter*), Elizabeth Grosz (*Volatile Bodies*), and queer theorists like Leo Bersany, Teresa de Lauretis, Eve Kosofsky Sedgwick, Gayle Rubin, and Jeffrey Weeks, valorizes the materiality of the female body—conceived in Deleuzian terms as a heterogeneous structure of "intensities and flows" (Grosz 163)—as a form of resistance against "male knowledges and paradigms" (19). The second emphasis, articulated theoretically in the work of Foucauldian feminists, regards the (female) body as a discursive site that modern technological society both defines and represses continually (Inman 222). The role of art is to challenge the "disciplining" effects of the power-knowledge regimes on the body, by releasing sexual, mental, and spiritual energies in the "field of language." But, as Nicole Brossard cautions, the field of language both enables and constrains poetic energy, trying to appropriate it for certain networks of meaning and rhetorics of the body and soul (75). Therefore, postmodern "body art" must practice a "politics of awareness" (79), relating woman's body to the body politic and the body of language.

Other examples of liberating postmodern feminism include the first-person narratives whose protagonists claim "personal agency" in order to examine "specifically female experiences" and to "engage the narrative process itself in rejecting fixed plot and [...] a teleological structure for [...] lived experience" (Frye 9); "feminist metafiction" that enlists "realism while also deploying self-conscious devices that interrogate the assumptions of realism" (Greene 22); and what Teresa Ebert calls "ludic feminism" (ix), which attempts to bridge materialist feminism with feminist textuality (Ebert's chief exemplification is Donna Haraway's "cyborg" theory—see below). Molly Hite finds feminist innovation on the whole more radical than the "dominant modes of fictional experiment [...] precisely inasmuch as the context for innovation is a critique of a culture and literary tradition apprehended as profoundly masculinist" (*The Other Side of the Story* 2). Ebert also praises the "performative power of language" (26) and the destabilizing role of female "desire" (83) in the examples she looks at, but she faults women's experimentation for abandoning "revolutionary knowledges" and "transformative practices" in favor of ludic-parodic strategies (x, 4).

The limitations of mainstream "ludic" feminism have been offset by more "ex-centric" varieties of feminist innovation associated with "blacks, gays, Native Peoples" (Hutcheon, *The Politics of Postmodernism* 17) or with Third World postcolonial writers who have taken the postmodern project beyond its initial white middle-class heterosexual emphasis. The work of previously marginalized groups of women writers has been effective in rethinking the priorities of feminism and of fiction-writing in general. In the very act of positing a "black self" in the "Western languages in which blackness itself is a figure of absence, a negation," African American writers have challenged the logocentrism and ethnocentrism embedded in Western discourse (Gates, "Criticism in the Jungle" 7), developing their own brand of "theorizing [...] in narrative forms" (Christian, "The Race for Theory" 52). Therefore, as W. Lawrence Hogue has argued recently, novels like Toni Morrison's *Song of Solomon*, Maxine Hong Kingston's *The Woman Warrior*, or Leslie Marmon Silko's *Ceremony* should be read "with reference, not to race or racial tradition exclusively, but to the European-American philosophical debate about the shift from modernity to postmodernity" (x). Their inclusion alongside the "classically postmodern texts" (92) of Pynchon, Barthelme, and Coover should be done not only in the name of "aesthetic criteria," as Harold Bloom has suggested in his introduction to the *Modern Critical Views* volume on Toni Morrison (1), but also with an understanding of the transformative effect that these works have had on narrative ideologies—those of African American and recent postmodern fiction included.

The entire field of postwar feminist experimentation deserves a careful rereading capable of grasping the ideological implications of feminism's "theorizing [...]" in narrative forms." Building on the emerging feminist critique of patriarchal discourses as well as on the polemical "female aesthetic" proposed several decades earlier by Virginia Woolf, Dorothy Richardson, Katherine Mansfield, and Djuna Barnes, the new women writers of the 1960s and 1970s engaged the culture "with a heady sense that they might change their lives and be part of the making of history" (Greene 36). For these writers "re-vision" was "more than a chapter in cultural history: it [was] an act of survival" (Rich, *On Lies, Secrets, and Silence* 35). The feminist novel had to challenge not only the oppressive ideology of the Cold War era that confined women to "sexual passivity, male domination, and nurturing maternal love" (Friedan 43), but also the perpetuation of gender stereotypes in the emerging male postmodernism. In spite of its liberating narrative agenda, early male experimentation created the impression of misogynism bias with its insistent focus on "unenlightened" women and sadistic males who overcame their feeling of social inadequacy by "club[bing their] way with [their] cock" (Sukenick, *Up* 104).

As Erica Jong's protagonist in *How to Save Your Own Life* (1977) noted, women owed their oppression not only to identifiable social forces but also to narrative traditions:

> [A]ll of the greatest fiction of the modern age showed women falling for vile seducers and dying as a result. They died under breaking waves, under the wheels of trains, in childbirth. *Someone* had to break the curse, *someone* had to wake Sleeping Beauty.[. . .] *Someone* had to shout once and for all: Fly and live to tell the tale. (236)

In order to change the historical destiny of women, feminist fiction needed to free them from such crippling narrative conventions and from a masculinist point of view that had traditionally discarded women's stories as self-indulgent or inconsequential. The "mad housewife" novels (Greene 54)—from Sue Kaufman's *Diary of a Mad Housewife* (1967), to Sylvia Plath's *The Bell Jar* (1971), Johanna Davis's *Life Signs* (1972), Marge Piercy's *Small Changes* (1973), and Alison Lurie's *The War between the Tates* (1975)—took an important step toward recovering the untold stories of wives trapped in their bodies, houses, and marriages. But while they successfully deflated idyllic notions of married life and exposed the wife's complicity in her domestic self-effacement, few of these novels offered real alternatives beyond, say, a repetitive adultery plot. Only rarely was the recovery of the "mad housewife story" accompanied by a reinterpretation of the narrative tradition that had suppressed it. The best early example of this type of narrative revision can be found in Jean Rhys's *Wide Sargasso Sea* (1966). By refocusing attention on the Rochester's Creole wife Antoinette Mason, maddened and dehumanized by imperialist patriarchy, Jean Rhys's rewriting of *Jane Eyre* managed to decenter "an inherited narrative structure and undermine the values informing [it]" (Hite, *The Other Side of the Story* 25).

The divided, claustrophobic voices of these early novels testified to the difficulty of developing a female consciousness and speaking subject. As that consciousness strengthened through the 1970s, characters like Jong's Isadora Wing began to understand that their salvation would come not from male analysts and writers "who have *always* defined femininity as a means of keeping women in line" (*Fear of Flying* 18), but rather from listening to other women and using writing as "the submarine or spaceship that takes me to the unknown worlds within my head [. . .], a new vehicle, designed to delve a little deeper (or fly a little higher)" (210). Challenging women's inhibitions, Jong's replaces the romantic quest "for the impossible man" (101) with unfettered sexuality and iconoclastic thinking. Nevertheless, her bold, licentious plot, which Jong explains elsewhere as the fantasy of the "Zipless Fuck" ("Blood and Guts" 178), remains steeped in a sexual vocabulary and an economy of desire that continues to objectify woman. While Jong may have intended a parodic "reversal of roles *and* of [the] language" employed by "tough guy narrator/heroes of Miller and Mailer" (Suleiman, *The Female Body in Western Culture* 9–10), her fantasy of a "Zipless Fuck," exemplified in *Fear of Flying* with the surreptitious sexual game played by a soldier with a passive female traveler in a train compartment, does not take female liberation too far. Like in Margaret Drabble's *The Waterfall* (1969), "that other story" (70) of romance undermines the heroine's effort to assert her own voice and plot. Drabble's rewriting of George Eliot's *The Mill on the Floss* maintains a third-person narrative approach as long as the female protagonist remains immersed in an "image of romantic, almost thirteenth-century love" (*The Waterfall* 46). First-person narration is used only when the protagonist begins consciously to look for "a system that would excuse me, to construct a new meaning, having kicked the old one out" (47).

The lure and limitations of "the old story" of romance, in which a girl "threw everything out the window to get her man" (Godwin, *The Odd Woman* 192) and where the only choices were "a disastrous ending with a Villain; a satisfactory ending with a good Man" (29), are dramatized also in Gail Godwin's *The Odd Woman* (1974). The novel's "odd woman"—daughter of a weekend writer of love stories, herself a teacher of Victorian fiction—uses literature to reevaluate her own interests as a woman, debate the limits of traditional plots, and oppose her own "oddness" to the predictable destiny of her mother's romance heroines. But Jane Clifford's notion of narratology, emphasizing a predetermined progression "from possibility to probability to necessity" (151), is too Aristotelian to give her revisionist, self-defining impulse much of a chance. Godwin's subsequent novel, *Violet Clay* (1978), challenged more radically the "teleological grip of narrative expectation" through alternative strategies such as the use of a woman-artist's first-person narrative "toward self-definition and causal analysis" (Frye 111, 113). After losing her job as cover designer for Gothic novels, the thirty-three year old Violet Clay reassesses her life from the time of her graduation to her escape from marriage and the South and "belated emergence" (347) as an artist in New York. Violet's venture in "time travel" (347) helps her derive a stronger sense of self from the "shadows of [her] own potential" (127). But Violet's re-created narrative cannot avoid falling periodically back upon the "Book of Old Plots" (57), to find there "the concept of Violet Clay as victim rather than as victoress" (73). Gayle Greene is justified to conclude that, while Jong and Godwin offer significant critiques of romantic fiction, neither manages to produce radically "new plots and endings of their own" (87).

Feminist fiction came into its own when it acknowledged that the narrative articulation of a "completely new type of woman" (Lessing 4) had to start by dismantling the patriarchal narrative tradition that drastically limited the roles available to women, adapting a masculine genre—the quest romance—to women characters but in a way that stereotyped them. The heroic romance usually translates into a species of female Bildungsroman that takes the heroine through a series of psychological ordeals, "which function as the quest element," to the "resulting marriage [. . .] which serves as the heroine's reward and the end point of the narrative" (Benson 105). While even a traditional genre can admittedly "contain—literally and metaphorically—disruptive elements" (Benson 106), the romance remains fundamentally disabling for women, separating the male and female spheres and restricting women to an activity of "learning how to read male behavior" (Radway 151). As long as they stay within the confines of romance, female characters are condemned to remain readers of other people's versions of reality, rather than become writers of their own tales. Doris Lessing's own pioneering feminist novel, *The Golden Notebook* (1962), wavered between a vision of "Free Women" (as Anna and her roommate Molly are called in the third-person narrative sections), committed to their friendship and important political causes, and an acknowledgment in the "notebooks" that these women remain dependent on a "relationship with a man. One man" (*The Golden Notebook* 314). This complex novel struggles to "name" female reality, while maintaining a skepticism about the restructuring capacity of fiction. This conflicting agenda reflects in the novel's contrapuntal structure that interplays Anna Wulf's omniscient third-person narrative with multiple "notebook" sections written from the perspective of a fragmented, self-questioning "I." The divisions of the novel map a multiplex female sensibility that uses different styles to come to terms with her personal relations, writing goals, and political commitments. Even though the novel's variously colored notebooks overlap in many ways, the reader understands that the existential and narrative breakthrough recorded in the golden notebook would not have been possible without Anna's explorations of her personal experiences in the blue notebook, her reassessment of her chances for significant political work in the red notebook, her previous novelistic effort ("The Frontiers of War") in the black notebook, and her new fictional projects in the yellow notebook. While neither the novel-within-novel entitled "Free Women" nor the notebooks fully represent Anna's aspiration to creative agency, their interaction set up a fruitful model for later feminist fiction. In addition to promoting multiple selves and narrativizations in response to Cold War polarities ("Men. Women. Bound. Free. Good. Bad. Yes. No. Capitalism. Socialism. Sex. Love" [*The Golden Notebook* 44]), this model emphasized "self-questioning, the writer built into the center of the work, the question at the center of the writer, the discourse doubling, retelling the same, differently" (DuPlessis and members of Workshop 9, "For the Etruscans" 279).

The feminist novels that followed Lessing's model often focused on an intellectual heroine who shares with other thinkers "the responsibility to name, refine, revitalize those values [. . .] which [. . .] transcend the narrowness of self-interest [. . .]; the responsibility to contribute to the process of political dreaming [. . .]; the responsibility of truth-telling, [. . .] of finding vocabularies and creating perspectives through which hidden or unspoken things can be voiced" (Mepham 18). Modeled after Joyce's Stephen Dedalus but also Elizabeth Barrett-Browning's Aurora Leigh, the intellectual heroine applied this "passion for thinking" to questions of identity, exploring the conditions under which a woman could become a successful speaking and acting subject. She also trained her critical attention on the gender-coded narratives of her culture, challenging their resolutions and seeking to gain "access to language" by other means than the "'masculine' systems of representation" (Irigaray, *The Sex Which Is Not One* 76, 85). Self-reflection became thus an indispensable tool for feminist fiction, which progressed slowly through endless deliberations, engaging conflicting agendas and concepts of genre. Traditional modes of fiction were not simply discarded, but recycled and transformed. For example, Atwood's *Lady Oracle* (1976) exemplified through Joan Foster—an "escape artist" (367) hooked on "Costume Gothics" (175)—both the limitations and the subversive possibilities of the Gothic in the hands of women writers. As Ellen Moers has argued in *Literary Women,* Gothic fiction enables women to explore their real and imaginary fears engendered by patriarchy; but as soon as it is set in motion, their psychosocial exploration becomes incompatible with the narrow confines of the genre. This is what Atwood's own protagonist discovers, so the character decides to botch the ending of her most recent Gothic narrative, "Stalked by Love." What Joan cannot do is "construct a different [story]" (157) that is not "a leftover from the nineteenth century" (266), but rather a "real novel, about someone who worked in an office and had tawdry, unsatisfactory affairs" (352). Atwood's own book makes that alternative plot possible, dramatizing for us a woman's search for a dynamic identity, freed from traditional concepts of gender and genre.

The revisionistic potential of self-reflexive feminist fiction is predicated on what Teresa de Lauretis has called a "view from 'elsewhere'" (*Technologies of Gender* 25)—the view of the outsider, the dispossessed, the self-exiled. Lessing's Anna Wulf was herself an outsider: Brought up in East Africa and living as an émigré in London, she abandons the Communist party, falls silent in the middle of her lecture on art history, and repeatedly disrupts her novelistic efforts accusing the genre (whose history includes her namesake Virginia Woolf) of falsifying experience. In feminist science fiction and political dystopia (Russ's *The Female Man*, 1975, LeGuin's *The Left Hand of Darkness*, 1976, Atwood's *The Handmaid's Tale*, 1985), the confrontation with the culture's plots is even sharper, involving "an imagined elsewhere, unacknowledged alternatives, other stories waiting silently to be told" (Anderson vii-viii). Feminist utopias/dystopias inscribe their nomadic "subjects-in-process" (Kristeva, "The System and the Speaking Subject" 29) in the interstices of traditional narratives, in a *zone* freed from the constraints of chronology, stable characterization, or realistic consistency. (We could exemplify with Offred's "Night" sections in *A Handmaid's Tale* the uneven fragments that shift voice, person, mode, plot in *The Female Man*, or the parallel worlds separated in time and space in LeGuin's novels.) Typologically, feminist science fiction is invested in the figure of the stranger, bastardized alien, or the cyborg—that "monstrous and illegitimate [. . .] hybrid of machine and organism" that takes "pleasure in the confusion of boundaries" (Haraway 149–54). While Haraway's cyborg seems comfortable with its "partiality, irony, intimacy, and perversity [. . .] wary of holism, but needy for connection" (151), other figurations of liminal femaleness are marked by anxiety toward their heterogeneous condition. The stories re-created by these eccentric female subjects are vulnerable to manipulation by the dominant culture. For example, Offred's oral narrative is censored/distorted not only by Gilead's patriarchal establishment, but also by the post-Gilead scholars.

Conceived in response to the "fundamental changes in the nature of class, race, and gender in an emerging [techno-scientific] world order" (149, 161), Haraway's *cyborg writing* promises to resist more successfully absorption by the "technologies that write the world" (175). Feminist cyborg fiction upsets the "informatics of domination" (161) that prevails in postindustrial societies by valorizing the "difference and contradictions crucial to women's cyborg identities" (170). The emphasis on the "unassimilable, radical difference" (159) of the female cyborg is coupled with the "struggle for language and the struggle against perfect communication, against the one code that translates all meaning perfectly, the central dogma of phallogocentrism" (176). Cyborg writing challenges logocentric views, advocating an "irreducible immersion" in writing's "materiality of difference," which is also an "irreducible immersion in worldliness" (conversation with Donna Haraway, in Olson and Hirsh 49, 59). Cyborg writing also tries to give voice to the "Sister Outsider" (Haraway, *Simians* 174)—that is, to all those disenfranchised by the economic and technological New World Order. The goal of cyborg writing is not to discard the global narratives of technoscience but to submit them to an "acid consciousness" (Olson and Hirsh 49) that will foreground "heterogeneities, cracks, counterintuitive moves" in them (69), allowing marginal groups to participate in the "production of knowledge and systems of action" (54).

Haraway's new female symbolic is a successful example of a postmodern poetics and politics of "transgressed boundaries" (*Simians* 154). In contrast to the militaristic cyborg that encloses a "fortified masculinist self," Haraway's cyborg reflects an ontological/cultural hybridity that cuts across gender, race, and body/machine divisions. Other postmodern feminist projects have sought to deconstruct gender oppositions by calling into question the male-oriented economy of traditional fiction. The task of postmodern feminists has been to "work with and against narrative in order to represent not just a female desire [. . .] but [also] to represent the duplicity of the Oedipal scenario [in traditional narration] and the specific contradiction of the female subject in it" (de Lauretis, "Strategies of Coherence" 186). The focus on female desire has allowed innovative women's fiction to represent a more varied gallery of characters (not only middle-class intellectuals, but also working class, minority, and lesbian women) in their struggle against the traditional heterosexual plots of seduction, marriage, or sexual downfall (DuPlessis, *Writing beyond the Ending* 15). To take one example discussed in detail in chapter 6, the rebellious character of Sula becomes possible precisely because of the ruptures and revisions performed by Toni Morrison in the androcentric traditions of fiction. After a false start (the novel begins with a male character and a historicist framework that suggests a chronicle novel or a Bildungsroman), *Sula* (1973) intensifies its focus on a number of unorthodox attitudes and relationships among women. Described by Morrison herself as a novel about a black woman's intimate relationship with "the other I," her confidante and companion, *Sula* endeavors to "repossess, re-name, re-own" (interview with Sandi Russell 45, 46) a space for women's experience within the male-dominated tradition. Room for this type of focus is created through a subtle process of elimination/restructuring. The few male figures that matter in the novel are bodily or mentally "amputated," accidentally drowned, upstaged by sexually liberated women who enjoy their newly gained freedoms. In addition to achieving a "figural breakthrough" (Hortense Spillers 181), *Sula* undertakes a significant narrative restructuring, dispensing with traditional modes such as the encyclopedic narrative, which systematizes information about a particular culture around male interests, and the novel of growth that systematizes the evolution of a character along male patterns of growth.

Alice Walker's violations of the narrative tradition in *Meridian* (1976) and *The Color Purple* (1982) play a similar role, freeing black experience from the constraints of Anglo-American realism and relocating it in other, emotionally more engaging traditions such as oral storytelling and a restorative African pastoral. Conventional realism can offer Walker's women little besides a one-track self-sacrificial plot: "She dreamt she was a character in a novel and that her existence presented an insoluble problem, one that would be solved only by her death at the end" (*Meridian* 117). Therefore Meridian Hill moves backward from her dissatisfying present role as a self-denying social worker, traversing several layers of personal and familial history in search of a more fulfilling personal narrative. The articulation of a more rewarding self depends on how successfully she can renegotiate her cultural heritage (the Christian South of her childhood and the pre-Christian culture of her great-grandmother, Feather Mae), shedding the trappings of modern middle-class life and relocating her roots "in ultimately pre-American, pre-Christian contexts" (Nadel 257).

Celie undergoes a similarly profound transformation in *Color Purple*, from a passive victim of parental rape and spousal abuse to an audacious woman who enjoys a lesbian relationship, writes vernacular letters to God and her missionary sister in Africa, and builds a business around her talent for designing costumes. Through Shug Avery, a flamboyant female singer, Celie learns to value her creativity and challenge conventional boundaries. Through sister Nettie, she learns to appreciate a precapitalist ethos and a pantheistic spirituality as alternatives to crass materialism. We should note, however, that Walker does not espouse a simplistic substitution of older conventions for those of Anglo-American realism. Such either/or choices impoverish culture, flattening its history. As her name suggests, Meridian is "a point of power" in the first novel because she is geographically, psychologically, and culturally a mediator, "a woman in the process of changing her mind" (*Meridian* 25). Like Meridian, Celie develops a hybrid vision, assimilating conflicting cultural and narrative traditions. Her world view is more radical than either the African pastoral, the agrarian Black South, or modern capitalism would have allowed it to be, replacing like *Sula* the male center (designated by Celie's brutish husband, "Mr.——") with a self-sustaining relationship between two marginalized black women. By letting "the margins *speak*" (Hite, *The Other Side of the Story* 122), Walker destabilizes the patriarchal heterosexual ideology of both the dominant white culture and of traditional black male writing.

Even though not all innovative women novelists have taken their revision of the phallogocentric economy of traditional realism this far, postmodern feminism has managed on the whole to restore narration its polyvocality, recovering marginalized voices and stories. On a structural level, innovative feminist fiction has replaced the rhetoric of integration with a ceaseless "circling back over material which allows repetition with revision" (Greene 14). This revisionistic circling frees women's fiction not only from "time as project, teleology, linear and prospecting unfolding" (Kristeva, "Women's Time" 16–17), but also from the circular plot of the traditional novel of development, which initiated women "into rituals of human relationships, so that they may replicate the lives of their mothers" (Ferguson 228). This paradoxical economy of revision that transforms a "closed, 'vicious' circle" into "liberatory cycles" (Greene 16) reflects also at the level of language that performs radical acts of "renaming" and syntactic rearrangement, breaking out of traditional circles.

As Maxine Hong Kingston's *The Woman Warrior* (1976) suggests, the acts of retelling, renaming, and rewriting are intertwined in postmodern feminist fiction. Kingston's authorial narrator describes herself as "an outlaw knot-maker" who "twists" her own history "into [new] designs" (190), reconstructing herself as a "woman warrior" against both the traditional Chinese culture assimilated at home and mainstream American culture absorbed through education. But her task is not easy: As a first-generation Chinese American, "born among ghosts, [. . .] taught by ghosts, and [herself] ghostlike" (6), the narrator experiences the tenuousness of historical referents and of her own identity. Breaking with family traditions, the narrator enrolls at Berkeley where she learns to "see the world logically" (237). But she must do more than pit one culture against another. Taking advantage of her "paradoxical doubled position" (Hogue 113), the Kingston narrator needs to denaturalize the narratives of both her Chinese heritage (mythical, communal, circular) and her American enculturation (rationalistic, individualistic and teleologic), seeking a creative synthesis between them. She begins by reclaiming a forerunner in her father's sister, the nameless "wild woman" who was "free with sex" (9) and who committed "spite suicide," jumping with her baby into the family well after villagers destroyed her family compound. In retelling her story, the narrator connects "No Name" woman to the legendary warrior woman Fa Mu Lan and to her own mother, Brave Orchid, who managed to emancipate herself through education. She tries to make these models relevant for her own life in the "ghost country" of America, rewriting them in ways that emphasize heroic female histories. The female subjecthood she articulates goes beyond the fashionable "free-spirited, postmodern subject who has many subjective positions, who takes an ironic view of the world, [and] who decon- structs all social spaces he encounters" (Hogue 166). Kingston's female narrator functions as a rallying node rather than as a decentered subject, transforming the past into a usable present.

Most postmodern feminist revisions work in similar ways, going beyond a simple refashioning of narrative precursors and traditions. Direct "rewrites" are not lacking (Valerie Martin's recasting of *Dr. Jekyll and Hyde* in *Mary Reilly,* Bharati Mukherjee's science fiction transcoding of *The Scarlet Letter* in *The Holder of the World,* Lauren Fair- banks's updating of *Sister Carrie*), but what predominate are subtler rewritings of the dominant culture's intertextual web, opening it up to new configurations. There are also parodic rewrites informed by an "outlaw" imagination such as Kathy Acker's, which subverts the master plots of the patriarchal tradition (Cervantes's *Don Quixote,* Hawthorne's *The Scarlet Letter,* Dickens's *Great Expectations,* Stevenson's *Treasure Island*), revealing their vestigial relevance in the world of Cold War politics and nuclear threats (*Don Quixote,* 1986), urban sexual angst (*Blood and Guts in High School,* 1984), sadomasochistic violence in the present and near future (*Empire of the Senseless,* 1988), and pop culture manipulation (*My Mother: Demonology,* 1993). Through their "abusive" appropriation of traditional typologies and linguistic "rape of master texts" (Travis 55), Acker's irreverent books explode fiction's patriarchal contexts. By contrast, re- cent multicultural and postcolonial writers have used ironic appropriation not only to explode but also to regenerate the themes, typologies, and language of autobiog- raphy (Audre Lorde's *Zami: A New Spelling of My Name,* 1982; Gloria Anzaldúa's *Border- lands/La frontera: The New Mestiza,* 1987) and fiction (Maxine Hong Kingston's *The Woman Warrior;* Leslie Marmon Silko's *Almanac of the Dead,* 1991, Bharati Mukherjee's *Jasmine,* 1987). Starting from the self-contradictions and exclusions they have suf- fered as African American lesbian women (Audre Lorde), Chinese American women (Maxine Hong Kingston), South Asian American women (Bharati Mukher- jee), or Native American women (Leslie Marmon Silko and Paula Gunn Allen), these authors resort to "oppositional autobiographical [and fictional] self-creation" to transform "multiple jeopardies into multiple strengths" (Susan Stanford Fried- man, *Mappings* 21).

Not all postmodern feminist work has been equally successful in "effect[ing] changes in the way people perceive themselves and others" (Travis 59). Postmodern feminism has run into new difficulties in the "remasculinized" 1980s (Jeffords) and "postfeminist" 1990s, characterized by a resurgence of traditional polarizations and a more general suspicion of narrative innovation. Among the few forms of "rewrit- ing" encouraged by the mainstream culture is that carried out by postfeminism itself as "an emerging culture and ideology that simultaneously incorporates, revises, and depoliticizes many of the fundamental issues advanced by Second Wave feminism" (Rosenfelt and Stacey 341). Though "postfeminism" is not reducible to either an- tifeminism or the "New Age backlash retrofeminism" (as represented by Camille Paglia, among others) that replaces women's social struggle with self-indulgent "sex- ual-empowerment" and anti–Cold War activism with post–Cold War acquiescence (Ebert 254, 256, 265), the cautionary plots it has inspired return women to the tra- ditional either/or choice between careers and motherhood. Gayle Greene finds striking the fact that the most widely read women novelists (Ann Beatie, Ellen Gilchrist, Mary Gordon, Alice Hoffman, Anne Lamont, Joyce Carol Oates, and Ann Tyler) have in common "the privatization and depoliticization of their concerns, the sentimentalization of the family, the resignation to things as they are. Even the feisty feminist writers of the seventies [. . .] participate in this retreat" (200). Lessing's *The Diaries of Jane Somers* (1984), Piercy's *Summer People* (1989), and Atwood's *The Robber Bride* (1993) show a new "skepticism about social systems, social relations" (Greene 205), and the capacity of feminist art to cause change. The retreat of women's fiction from cultural and narrative innovation to angst-ridden confession is well illustrated also in Helen Fielding's *Bridget Jones's Diary* (1998), a book that predictably became an end-of-the-century best-seller. The record of a year in the life of a London junior publisher is filled with "postfeminist" concerns: worries about her fluctuating weight; an obsession with counting alcohol units, cigarettes, calories, and answering-machine messages; and embroilment in hopeless relationships. Even though Bridget tries valiantly to make it as a "singleton," the pressures associated with her choice give her a "freakish," self-deprecating notion of femininity, not very different from the commodified "post-female" bodies that Rita Felski (230) associates with our "rhetoric of fin de millennium." These post-female bodies "serve once again to elide the particularity of women" (236), commodifying all aspects of consciousness and containing their claim to self-determination.

It would be misleading, however, to consider the "third wave" of (post)feminism as a mere retrenchment. The skeptical and diversified feminisms of the last two decades have contributed to a rethinking of postmodern innovation, sharpening its grasp of political and narrative issues. Joan Didion's *Democracy* (1984) illustrates the magnitude and difficulty of this rethinking in an increasingly globalized world. Framed by the American struggle to maintain dominance over the Pacific rim countries, the story of Inez Christian begins in the spring of 1975, one week before sister Janet and her lover, the Hawaiian-born Congressman Wendell Omura, are killed by Inez's demented father, and one month before the fall of Saigon. Refusing chronological treatment, the narrative moves back and forth spotlighting Inez's estrangement from her husband, Democratic Senator Harry Victor; her journey to Vietnam and Malaysia accompanied by her longtime admirer, the CIA operative Jack Lovett; and Inez's return to Kuala Lumpur after Lovett's sudden death, to help with the flood of Vietnamese refugees. The complexity of this story of personal struggle against the patriarchal "codex [of] dreamy axioms" (24) inherited by Inez and against an equally oppressive Cold War political order (significantly, there is an overlap between important moments on Inez's time line and key years in the history of the Cold War: 1952, 1955, 1959, 1960, 1964, 1966, 1968, 1969, 1972, 1975) demands the continuous intervention of "Joan Didion" as implied author and dramatized narrator-reporter. The early chapters are concerned with the difficulty of claiming an authorial voice. Simply repeating the nineteenth-century self-establishing procedure—"Call me author"—will not do since, as the narrator confesses, she started thinking about her characters "at a point in my life when I lacked certainty, lacked even that minimum level of ego which all writers recognize as essential to the writing of novels, lacked conviction, lacked patience with the past and interest in memory; lacked faith even in my own technique" (17). And yet the fact that she misses the confident narrative attitude of a Victorian novelist enables the "Didion" to place the events she retells in a different "field of vision" (17).

Part 1 is mostly about a novel "Didion" can no longer write. The first chapter uses briefly Jack Lovett's first-person voice to reminisce about his participation in secret nuclear tests in Asia and his romantic attraction to Inez. Inez is present only as a form of address ("Listen, Inez" [14]) or as a frustrated interjection ("Oh shit, Inez [...] Harry Victor's wife" [15]). The next two chapters sketch the history of the Christian clan, which is coextensive with the history of American expansion in the Pacific. Neither Jack's secret-agent narrative nor the family chronicle is followed through. By "jettison[ing] [...] those very stories with which most people I know in those islands" and political operatives like Lovett "confirm their place in the larger scheme" (*Democracy*, 19), Didion "destabilizes the authority not only for her own text but the authority, as well, for the text of American hegemony globally since World War II under the name of *democracy*" (Nadel 283).

Didion's self-inclusion in the novel submits her own authorial identity to a revisionist critique. (The novel quotes somewhat ironically Didion's teacherly statements on the techniques of writing fiction.) In the end, the novel she produces is "not the novel I set out to write nor am I exactly the person that set out to write it" (*Democracy* 232). The deconstruction/rewriting of the genre (the abandonment of the family chronicle and colonial epic revolving around the Christians) creates room for a stronger focus on Inez Christian's personal story. Inez's patriarchal childhood in Hawaii and marriage to a self-absorbed, womanizing politician robs her not only of her identity but also of her narrative memory. "Didion" has to step in to record the climactic events in Inez's life (the murder of her sister Janet, the drug addiction and disappearance of her daughter Jesse, the loss of Jack Lovett) but also to free her story from the conventional roles handed down to her by her patriarchal education and the Cold War establishment. Cooperating with her character, the narrator disrupts the controlling narratives enforced by family patriarchs, political handlers, and the mass media, piecing together a complex personal story from both oral and written sources. This act of narrative recovery gives Inez Victor back her narrative agency while also restoring the writer's faith in narration.

Author and character interact thematically but also narratologically in the "autobiographical third person" (Katherine Usher Henderson 72) that the novel develops after rejecting the first-person narration of either a male character (Lovett) or a self-absorbed female narrator. This form of collaborative autobiography resists formulaic plots. For example, part 2 avoids a romanticized presentation of Inez's affair with Jack Lovett, raising instead questions about Jack's shifty identity Cold War operative and emphasizing Inez's increasing assertiveness in her relationships with men. Inez courageously shakes off her role as self-effaced supporting wife to an ambitious Democratic politician, whose human-rights mission in the Pacific is a disguised enforcement of American interests on "USAID Recipient Nations" (92). In part 4, the writer and her female friend are reunited in Kuala Lumpur where Inez is engaged in an effort to save Southeast Asian refugees. Her independent course of action challenges not only the Cold War ideologies but also "Didion's" own expectations of how Inez's story should end. Renouncing "whatever stake in the story she might have had," Inez achieves "serenity" (*Democracy* 216) in her own idea of significant action. The novel ends with a recognition of the unpredictable and quietly heroic character of Inez, whose future can hold further surprises for the writer, but also of the writer's own skills as an "aerialist" (108), dangerously balanced "on the wire in this novel of fitful glimpses" (232). Pitting the role of journalist-confidante who "knows" only what she observes or is told against that of a novelist who inter-prets/reinvents Inez's story, the authorial narrator produces a hybrid work that cri-tiques and expands imaginatively the narrative forms available in the Cold War era (espionage novel, murder mystery, domestic fiction, colonial epic, travelogue). She also clears some ground for a post–Cold War vision that no longer confuses democ-racy with a narrative of Western expansion.

Anticipations of a post–Cold War epistemology that challenges separations be-tween self and other, male and female, global and local, "specify[ing] a liminal space in between, the interstitial sight of interaction, interconnection, and exchange" (Susan Stanford Friedman, *Mappings* 3), can be found also in the recent work of mul-ticultural and/or lesbian writers like Gloria Anzaldúa, Louise Erdrich, Maxine Hong Kingston, Audre Lorde, Toni Morrison, Bharati Mukherjee, Leslie Marmon Silko, and Alice Walker. Based on their example, Susan Friedman has proposed a bold remapping of the field of woman's studies, replacing the emphasis on sexual differ-ence and identity politics with a focus on what she calls "cultural narratives of rela-tional positionality" (47). These narratives redefine identity as a "positionality, a location, a standpoint, a terrain, an intersection, a network, a crossroads of multiply situated knowledges" (19). Although identity "is in fact unthinkable without some sort of imagined or literary boundary" between self and other (3, 19), we can bene-fit from approaching our increasingly fluid "global ethnoscape" (Appadurai 191–92) with nonbinary notions of identification. We must learn to be "crossroads," as Glo-ria Anzaldúa advises in *Borderlands* (195), without giving up the different heritages that constitute our multiple identities. Postmodern feminist theory and literature need not become "postfeminist" or antiexperimental to reach the kind of synthesis that Susan Friedman advocates. On the contrary, post–Cold War fiction can con-tinue to make good use of both retrieval and rewriting methods, combining "the positivist epistemology with the outer-directed [. . .] desire to 'make history' and the subjectivist epistemology with the inner-directed, self-reflexive problematizing of the feminist history writing" (*Mappings* 202–3).

Kathy Acker's last published novel, *Pussy, King of the Pirates* (1996), and Toni Mor-rison's *Paradise* (1998) suggest some of the challenges and opportunities that post–Cold War feminist fiction faces. The challenges are numerous: Acker's retired Moroccan Jewish prostitute O seeks a treasure buried on Pirate Island, but what she is really trying to retrieve is her female self-awareness robbed by patriarchal culture over several centuries and on different continents. She hires Pussycat's band of women pirates to help her with her search, but finds their underground world to be as plagued by greed and violence as the dominant male culture. Acker's witty rewrit-ing of Stevenson's *Treasure Island* and Pauline Reage's pornographic *Story of O* ques-tions both the patriarchal heritage and accepted feminist solutions, proposing an outlaw narrative imagination that crosses boundaries between historical periods, gender roles, and literary plots. Morrison's *Paradise* also concerns a group of (self-) outlawed women hiding in an abandoned Convent, who are attacked by a posse of black patriarchs because they "don't need men and they don't need God" (276). The males who control decision by force of their "deep level" racial purity (193) and po-sition as founders of Ruby, Oklahoma, justify their act of deadly violence by regard-ing the Convent women as a major threat to their patriarchal all-black community. But not everybody shares their representation of events. Patricia Cato, the amateur historian of the community, sums up the plot differently: "nine 8-rocks murdered five harmless women" because the women were "impure (not 8-rock)," "unholy," and "because they *could*" (297). She and other open-minded members of the community question the racist and sexist bias of the official Ruby history, seeking a more inclusive cultural future.

The patchwork structure of Morrison's novel, which interrupts the main plot, teasing out the overlooked stories of the runaway women who become victims of the raid, contributes to the strengthening of this alternative perspective. The chapters are identified by female names, establishing a symbolic web of female nodes that runs parallel to and challenges the patriarchal plot line with which the novel starts. In typical Morrison fashion the denouement is left open-ended: The fate of the women assaulted in the first chapter remains uncertain. Not only do their bodies disappear mysteriously, but there are "sightings" of several of them in the concluding sections and even the expectation of a "miraculous" return in block. These haunting "threshold people" (Turner, *The Ritual Process* 95), crossing from one world into another, call into question the exclusionary narrative of Ruby, exposing its complicity with the polarizing Cold War ideology. (All the Ruby males are veterans of various wars, beginning with World War II, and their appetite for violence is fed, indirectly at least, by the "long, inintelligible war" in Vietnam [160].) Through "keen imagination and the persistence of a mind uncomfortable with [official] stories" (188), the implied author and her various aids open "crevices or questions" in the town's history. In the interstices thus created, a younger generation of Ruby blacks can experiment with a reintegrative vision that moves "past your great-great-grandparents, past theirs, and theirs, past the whole of Western history, past the beginning of organized knowledge, past pyramids and poisoned bows," back to a redemptive planetary home "when rain was new, before plants forgot they could sing and birds thought they were fish" (213).

CHAPTER 3

"Chain of Links" or "Disorderly Tangle of Lines"?

ALTERNATIVE CARTOGRAPHIES OF MODERNITY IN THOMAS PYNCHON'S FICTION

History is not Chronology, for that is left to lawyers,—nor is it Remembrance, for Remembrance belongs to the people. [. . .] [H]er Practitioners, to survive, must soon learn the arts of the quidnunc, [. . .] [producing] not a Chain of single Links, for one broken Link could lose us All,—rather, a great disorderly Tangle of Lines, long and short, weak and strong, vanishing into Mnemonic Deep, with only their Destination in common.

—Thomas Pynchon, Mason & Dixon (349)

[In a non-totalizing, perspectivist view] a given landscape opens itself to potentially limitless topographical mappings. [. . .] Each mapping is the "cry of its occasion" [. . .] infinitely variable, always open to revision. [The divers mappings] can be thought as superimposed on one another and on the landscape, like different navigations through a hypertext.

—J. Hillis Miller, Topographies (281)

---

1. "WRECK[ING] THE ELEGANT ROOMS OF HISTORY": PYNCHON'S POLYSYSTEMIC REVISIONS OF HISTORY FROM THE ENLIGHTENMENT TO THE COLD WAR

The generation of interwar modernists, Pynchon suggests in Gravity's Rainbow, has "wrecked" the orderly rooms of traditional history, threatening the "idea of cause and effect itself" (56). Old narrative orders have been dismantled, but has anything been created in their place? "Will Postwar be nothing but 'events,' newly created one moment to the next? No links? Is it the end of history?" (56). What will our choices be in the age of "posthistory": the way of "love" and "connection" or the way of "death" and entropy? These are questions that Pynchon's fiction has continually raised without ever settling them. Few of his characters are ready to embrace the "symbols of randomness and fright" as readily as the wise statistician Roger Mexico, whose job in Gravity's Rainbow is to predict the rockets' hits on World War II London. When faced with disasters, most of Pynchon's characters fall back on reassuring narratives and "preserving routines" as protection "against what outside none of them can bear—the War, the absolute rule of chance, their own pitiable contingency" (96).

Pynchon shares Karl R. Popper's intuition that history cannot be predicted rationally, a conclusion that the distinguished philosopher of science developed during the historical period (the 1940s) explored by *Gravity's Rainbow*. In *The Poverty of Historicism* and elsewhere Popper refuted the predictive pretenses of "historicism," exposing its fallacious systematizations. Pynchon's approach is more ambivalent, suggesting an insidious teleology behind historical events ("Random [is just] another fairy-tale word" [*Gravity's Rainbow* 395]) but defining it as rationally unknowable. Therefore, the model chosen by Pynchon to sum up history is neither Marxist determinism nor Popperian analytic pragmatism, but postmodern gothic paradox. Pynchon reminds us continually of sinister plots behind history; at the same time, he promises a form of dark illumination that will "change us forever to the very forgotten roots of who we are" (134). Unfortunately, the truth of history always finds us too late.

For the chronicler of the twentieth century who has witnessed a "shrinking world" with its "democratic vistas [ending] in barbed wire" (Orwell, "Inside the Whale" 526), the task of historical interpretation is formidable. The ominous field of forces that entraps Pynchon's characters can only be hinted at in gothic metaphors such as the protean V-puzzle in *V.*, the Tristero counterorganization in *The Crying of Lot 49*, the "Rocket-State" in *Gravity's Rainbow*, the tangle of "tubal" stories in *Vineland*, or the ubiquitous "crimes of demarcation" in *Mason & Dixon*. These figurations suggest a global conspiratorial order described alternatively as politico-economic and metaphysical in nature. The metaphysical descriptions predominate, Pynchon drawing his explanatory metaphors from Kabalistic interpretation, enigmatic zodiacs, angelology, vampirology, geomancy, or tarot cards. Other metaphors are obtained through metaphysical enhancement/distortion of scientific phenomena such as the "delta-t" effect or the "eerie, Messianic, extrasensory force of Gravity (590). However, Pynchon is equally concerned with the modern politico-economic conspiracy that reduces—in Nietzsche's words—life's "confusing multiplicity to a purposive and manageable scheme" (*The Will to Power* 315).

Against a metaphysical-deterministic concept of history that sacrifices contingencies, Pynchon has mobilized a polysystemic vision that emphasizes history's unexhausted potentiality. His fiction pendulates between these two understandings of history, pitting grand metaphysical plots against the unpredictable dynamics of a centerless web. The struggle between the two visions has generated ex-centric, skeptical narratives that (like the St. Veronica cathedral-hospital in *Gravity's Rainbow*) "arise not from any need to climb through the fashioning of suitable confusions toward any apical God, but more in a derangement of aim, a doubt as to the God's actual locus (or, in some, as to its very existence)" (46). Pynchon's novels challenge monologic views of history through their endless multiplication of "limited, contingent, overlapping systems which coexist and form relations without achieving abstract intellectual closure" (Hite, *Ideas of Order* 21). An alternative strategy he has used is *preterition*, "the figure of conspicuous omission. Omission by mention, or mention by omission" (Mackey, "Paranoia, Pynchon, and Preterition" 20). For example, *Gravity's Rainbow* offers a curiously unconventional picture of World War II that passes over some of its climaxes (key confrontations on the Western and Eastern fronts, the Normandy invasion, Hiroshima). But a simple inventory of Pynchon's omissions takes attention away from the equally important recuperative impulse of his fiction. His novels fore-ground a thousand details repressed or overlooked by conventional history.

The novelist's concern is with figures that fall through the cracks of history—floaters, outcasts, victims of "power and indifference" (*Gravity's Rainbow* 209), even lab animals or aging light bulbs. They are, in Pynchon's apt phrase, the "preterite": defenseless humans and nonhumans expended by murderous political and meta-physical bureaucracies. The resulting perspective may appear skewed, but it articu-lates a powerful rereading from the margin, a microtextual history that tells us more about who we are than our grand narratives. Pynchon has the unique ability to evoke essential features of modern culture through his eccentric metaphors. Nazi—and by extension Western—decadence is epitomized in Blicero's "oven game" with its oth-erworldly and Auschwitz suggestions; also in the mythography of the Rocket. The playful "Story of Byron the Bulb" in part 4 of *Gravity's Rainbow* turns an ordinary bulb into a modern picaro and champion for the rights of "discards" on the corporate "Grid" (652). Another story-within-story suggests that the novelist's role is compa-rable to that of Pensiero, the army hair cutter whose scissors work subtle "modula-tions" (643) in his clients' lives. In much the same way, the novelist "cuts" history down to size, reorganizing it around the potentially endless modulations of his mar-ginal subjects. By accommodating contingency and illustrating the effects of "major" conflicts on lowly beings, Pynchon narrows the gap imputed by Nietzsche between "history" (as a retrospective rationalization) and "life" ("Vom Nutzen und Nachteil der Historie für das Leben" 232–33, 243).

Pynchon's rereadings of modern history create paradoxical narrative structures that can be best understood in terms of Even-Zohar's concept of "dynamic stratifica-tion": A provisional order emerges gradually through a process of "polymerization" that takes advantage of the connective possibilities of motifs and characters that be-have like a "supermolecule with many bonds available" (*Gravity's Rainbow* 346). The availability of multiple bonds turns a "serial" form of narration into a hypertextual web of plot coincidences, character associations, and symbolic interfaces that "must be apprehended all at once, together, in parallel" (753). The cohesion of this hyper-text depends not on the strength of any single threads, but on their "woven" structure resulting "from the partial overlapping of many different strands of connectedness" (Spiro et al. 193). Pynchon's hypertextual narrative field behaves like the "complex content domains" described by Rand J. Spiro and his associates: Instead of solidify-ing into "a relatively *closed* system," it remains "open to context-dependent variability," highlighting "multifacedness" and demanding "multiple entry routes" (187–88).

Pynchon's first published novel, *V.* (1963; the William Faulkner First Novel Award), already points to this project, trying to recover (through its foolhardy ex-plorers and cultural bums) some of the traditionally hidden or excluded areas of ex-perience. As Fausto Maijstral's diary in chapter 11 suggests, the reinterpretation of history must begin with the effort to grasp the web of historical events from the off-centered perspective of one of its nodes. Fausto's "confessions" are written in 1956, on what was once a center of Mediterranean culture, Malta. In the enclosed cham-ber of his memory, Fausto tries to make sense of the "freak show" of history (287), struggling to connect his liminal narratorial identity (as a "dual man" in Anglo-Mal-tese culture) to an equally marginalized narratee (Maijstral's daughter, Paola). Fausto's fragmented, self-contradictory journal plays the role of a *mise en abyme* for the larger novel, pitting an insular vision of history, in which knightly combatants defend an archetypal mother-island (Malta, Western civilization) against an entropic vision that sees history as an erratic, patternless surface. While the historical sections (covering the period between 1898 and 1943 around the Mediterranean basin) reinforce a metaphysical-centric reading that assumes that all events fall into "ominous patterns" (452), the "contemporary" sections (the years 1955 to 1956 in Washington D.C., New York, and Valletta, Malta) follow the centrifugal pull of an open system. The episodes recounting the exploits of British Foreign Office agents in the first half of the twentieth century converge on several symbolic locations (Cairo, Alexandria, Florence, Paris, and Valletta) that epitomize modernity's characteristic tensions: imperialist confrontations, international espionage, nationalistic uprisings. By contrast, the behavior associated with the American 1950s is one of narcissistic drifting, as illustrated by Benny Profane and his "Whole Sick Crew." The difference is one between a "profane" view of history that consumes the events or preserves them in a superficial structure of monuments and a metaphysical immersion in history: "[I]n Malta [. . .] all *history* seemed simultaneously present, [. . .] all streets were strait with ghosts. [. . .] In London [. . .] [h]istory was the record of an evolution. [. . .] Monuments, buildings, plaques were remembrances only; but in Valletta remembrances seemed almost to live" (452).

The two views of history also inspire opposing narrative approaches, suggesting alternatively the divergent and the convergent ends of a V pattern. The "contemporary" sections record Benny Profane's picaresque escapades in a nonhermeneutic horizontal sprawl; the historical sections and those contemporary episodes that deal with Herbert Stencil's quest for the vanished figures in his family history (Lady V., father Sidney Stencil) seek a deep structure of meaning and a resolution to their hermeneutic plot. V. tries to negotiate some middle ground between these two modes of historical thinking and their respective definitions of entropy: the drifting disorder of "posthistory" and the deadening burden of history's grand narratives. But Pynchon complicates the idea of cultural entropy borrowed from Henry Adams (as *The Education of Henry Adams* argued, "Chaos was the law of nature; Order was the dream of man" [451]), using it revisionistically to question the tightness of Adams's plot of modern decadence and to foreground pockets of human creativity that resist entropy, like a "resurgence of humanity in the automaton, health in the decadent" (V. 316). This novel represents Pynchon's first attempt at rereading modern history based on a polysystemic type of integration that highlights important coincidences between disparate narrative moments (for example, von Trotha's extermination campaign in Southwest Africa is read as a rehearsal for the Nazi Holocaust: "Allowing for natural cause during these unnatural years, von Trotha [. . .] is reckoned to have done away with about 60,000 people. This is only 1 percent of six million, but still pretty good" [227]).

The Crying of Lot 49 (1966; the National Institute of Arts and Letters Award) also unfolds on two different narrative levels, one syntagmatic-contingent, the other paradigmatic-historical, but the difference between them is negotiated more boldly by Oedipa Maas beyond either Profane's horizontal drifting or Stencil's self-absorbed hermeneutics. The first level involves Oedipa, a suburban housewife and English major from Cornell, in a plot line designed to make sense of the variegated holdings of a California real estate mogul, Pierce Inverarity. The second level engages her in an in-depth cultural exploration that redefines Inverarity's capitalist patrimony as collective (the "legacy was America" [178]), and elevates Oedipa from the position of testamentary executrix to that of self-conscious projector-interpreter of worlds.

The plot line takes Oedipa from her suburbanite home in Kinneret-among-the-Pines to the Echo Courts Motel in San Narciso, the housing estate at Fangoso Lagoons, the aerospace company Yoyodyne, the Berkeley campus, and the Greek Way bar in San Francisco, to learn about the possible connections between Inverarity's investments and the conniving history of an alternative postal and political organization called the Tristero. Oedipa's quest is undercut even on this topographic level: The proliferation of contradictory "clues" makes the task of adjudicating between a symbolic reading of the "hieroglyphic streets" and an immanent one that finds only "earth" beneath the surface (181) very difficult. Oedipa's interpretations are caught in a structure of mutually exclusive propositions. Not only is the order envisioned by Oedipa doubly constituted, pitting the established system against its anarchic counterpart, but each side has conflicting definitions: Inverarity's "legacy" suggests both a reactionary structure of properties and a random clustering of "unreal estates." Likewise, the Tristero/Trystero conspiracy discovered by Oedipa is a mystifying counterorder whose meaning and even spelling changes. In response to an increasingly complex polysystem, Oedipa learns not to expect but rather to make meaning by establishing her own significant "circuitry" in the world's "circuit card" (24). Identifying with the disinherited communicators who "spent the night up some pole in a lineman's tent like caterpillars, hung among a web of telephone wires, living in the very copper rigging and secular miracle of communication" (180), Oedipa discovers the subversive potential of her marginality. Her interpretive strategies become more complex, including projection and "dissonant" rearticulation.

Dissonance (linguistic and cultural) is an important principle of articulation in Pynchon's fiction. For example, Oedipa Maas's name distorts/retranslates a male tradition of questing into a female one. Likewise, Oedipa's disc jockey husband, Mucho Maas, welcomes instrumental dissonance but reduces it to mechanical discord: "You pick your zero point anywhere you want, that way you can shuffle each person's time line sideways till they all collide" (142). Pynchon is well aware of this negative definition of dissonance, providing abundant examples of fractured or distorted communication—some violent (attacks on mail coaches, assassinations of couriers), others merely parasitical (the right-wing Peter Pinguid Society co-opts the Yoyodine office mail system to circulate trivial messages). Against the entropic effects of contemporary culture, Pynchon mobilizes a form of creative dissonance. Like the "muted horn" emblem of the Tristero, which produces both a smothering and a subversive modulation of the trumpet's voice, Pynchon's fiction simultaneously illustrates and transforms the discordant quality of contemporary culture. Dissonance functions both as *repetition with a difference,* counteracting the trivial repetition of mundane existence (Mendelson, "The Sacred, the Profane, and *The Crying of Lot 49*" 131), and as subversive *rearticulation* that, like the forged stamps auctioned at the end of the novel, transpose traditional icons, introducing "noise" in the culture's carefully orchestrated narratives.

The novel's chief example of an alternative order created through subversive rearticulation is the secret Tristero organization. Like the Counter-Force in *Gravity's Rainbow,* the Tristero operates as a partially successful "anti-system" (Nohrnberg 154), "an anarchist miracle" (Sklar 94–95; Dugdale 175–81) that recycles the waste from other cultural sectors. But even an antisystem has its closed, metaphysical aspect that allows only slavish repetition. Oedipa becomes aware of both facets of the Tristero: The metaphysical aspect encourages her paranoia; the open-ended aspect, her creativity. Oedipa's increasing unhappiness with the plots of her culture convinces her that a countersystem like the Tristero can offer "a real alternative to the exitlessness, to the absence of surprise to life" (170). The story of the Tristero suggests to her that dissonant rearticulation in the form of forgeries, garbled names, or apocryphal insertions in classical texts (as in the case of *The Courier's Tragedy,* a seventeenth-century revenge play performed with later interpolations that refocus the play on the "disinherited" Prince Trystero) can play a subversive role, enabling the marginalized to write themselves back into history.

By comparison to *V., The Crying of Lot 49* is more boldly revisionistic, offering a speculative-metaphoric version of a polysystem. But Oedipa's vision of a social and communicational system open to alternative voices cannot fully offset the paranoid concept of a "closed-circuit" universe that still haunts her imagination. *Gravity's Rainbow* (1973; a National Book Award co-winner and Pulitzer Prize finalist) attacks this concept at its root. This 760-page-long novel undermines traditional models of narrative configuration (causality, character motivation, logical emplotment), overwhelming the reader with its virtually inexhaustible web of narrative paths that have required two different reader's guides to explain them. *Gravity's Rainbow* pursues two conflicting hermeneutic projects: The ostensible proposition is that the book's systematic rereading of twentieth-century history will result in a plot of plots, a grand narrative explaining every aspect of modern life. But this totalistic vision is continually subverted, replaced either with a version of chaos theory or with a polysystemic perspective that seeks meaning in the interstices between systems. *Gravity's Rainbow* focuses more successfully than the two previous novels on the hinges and interfaces of history. The "otherworldly" functions as one of the huge interfaces recuperated by Pynchon's fiction, expanding our conventional ontological and cultural map. The other interface (the temporarily stateless "Zone" at the end of World War II that knows "no serial time" [549, 556]) promises to free human actions from the constraints of causality and cultural regulation. Ultimately, Pynchon's entire novel works like an intertextual "Zone" that replaces linear historical development with a "disorderly Tangle of lines" (*Mason & Dixon* 349).

The hypertextual complexity of Pynchon's novel demands continuous "rereading with an insistence that denies it is really used up when one has read the last page" (Stonehill 154). Rereading has always played an important role in Pynchon's fiction. By actively rereading, "'bringing something of herself'—even if it was just her presence" (*The Crying of Lot 49* 90) to the world of male interests and texts, Oedipa managed to alter them partly. In *Gravity's Rainbow* rereading plays a more crucial role, encouraging characters and readers to reexamine the explanatory models that have framed their understanding of history. As Brian McHale has argued in a seminal essay, this novel compels us "to become metareaders, readers of our own (and others') readings—and, more to the point, of our inevitable *mis*readings" of history ("You Used to Know What These Words Mean" 113). One such misreading that the novel denounces is the understanding of history as a unbroken causal narration:

> Most people's lives have ups and downs that are relatively gradual, a sinuous curve with first derivatives at every point. They're the ones who never get struck by lightning. No real idea of cataclysm at all. But the ones who do get hit experience a singular point, a discontinuity in the curve of life—do you know what the time rate of change *is* at a cusp? *Infinity,* that's what! (664)

Pynchon's fiction focuses on points of crisis that "nab" characters away from their familiar environment, projecting them into places that "will *look* like the world you left, but it'll be different" (664). These "cusps" of history, forking out into conflicting possibilities, cannot be reduced to a causal plot.

Causality, Paul de Man argued, represents history "as a generative process [. . .] that resembles a parental structure in which the past is like an ancestor begetting, in a moment of unmediated presence, a future capable of repeating in its turn the same generative process" ("Literary History and Literary Modernity" 164). Against this paternalistic model of history, Pynchon mobilizes a host of "alienating" procedures that interrupt the narrative progression. In the midst of an ordinary conversation, Pökler chants "Victim in Vacuum" for "all you masochists out there, specially those of you don't have a partner tonight" (415); Pointsman explains his Pavlovian philosophy in doggerel; lab animals with Russian names sing of their fate to the accompaniment of a tropical band; and, to make the Brechtian inspiration unmistakable, film director Gerhardt von Göll hums a medley from *The Three-Penny Opera*. The conflict of styles and genres (historical narrative, journal, therapy notes, song, dream, cartoon) creates a composite structure difficult to assimilate under a realistic poetics. The story level is equally conflicting. This novel that begins in misty London where a "million bureaucrats are diligently plotting death" (17) and subsequently moves to the infamous Rocket fields in Nazi Germany is both overplotted and amorphous. Some sections have no discernible plot, being concerned with scientific arguments and psychological speculations. Other episodes parody the heavy-handed plots of spy novels, pornographic fiction, or action movies. Most often evoked is the quest plot. Pynchon's major players are all involved in some version of questing: Slothrop searches for a new identity in the German War Zone; Pointsman pursues a scientific explanation for Slothrop's "rocket-dousing" erections; Major Weissmann-Blicero and his sex toy Gottfried seek "transcendence" inside the Rocket; Enzian tries to ensure a meaningful survival for the black race in postwar Europe; and Tchitcherine chases his half brother, Enzian, and the secrets of German rocketry. Doomed "always to be held at the edges of revelations" (566), none of these quests is resolved satisfactorily.

Each plot or thematic strand is riddled with contradiction. *Gravity's Rainbow* proliferates countless antithetical images: "gravity" and "rainbow," angels of God and angels of war, custodians of the established order and "the Counterforce," "a good Rocket to take us to the stars, an evil Rocket for the World's suicide, the two perpetually in struggle" (727). These conflicting strands provide some continuity within the novel: "*Gravity's Rainbow* is really two giant molecule-chains of paired opposites. There are more than a dozen recurrent binary contrasts spiraling like DNA chains through the novel's 73 scenes" (Fowler 47). Pynchon pits Northern/Western civilization (industrialized, rational, bureaucratic, repressed) against Southern/Oriental/non-Western cultures (pagan, colored, organic, sexual). But more engaging than this vertical (hierarchized) polarization of values is their lateral networking. The horizontal bonding of images and motifs creates unpredictable strings that reveal the deeper interconnectedness of modern culture. For example, by focusing on Slothrop's papermaking ancestors in New England, Pynchon develops an intertextual definition of our Western culture that connects paper to the colonial spoliation of virgin nature and the abstract commerce with symbols. This definition integrates also the idea of paternal bureaucracies that rule the world of sons by edict, spreading their "virus of Death" (723) through writing technologies. On the side of life,

Pynchon enrolls the sensual music of Rossini (frequently hummed in the novel) and the inexhaustible libido of Slothrop who can derail "machineries of greed, pettiness, and the abuse of power" (440) with his erratic coupling.

Pynchon's novel emulates its protagonist's insatiable capacity for bonding, generating multiple associations that transgress polarities. For example, the pivotal image of the Rocket contains its own opposite ("two Rockets, good and evil [called] the Primal Twins [726]), integrating ascent and descent, flight and bondage, rainbow and gravity. The novel's binary logic of Zeros and Ones is likewise violated by paradoxical crossovers such as Blicero's firing of his potent Rocket into the empty Northern spaces of 000 longitude and latitude, or Mexico's search for "the domain *between* zero and one—the middle Pointsman has excluded from his persuasion" (55). These examples indicate not only the problematic nature of the binary matrix, but also a way out of it. We are urged to move from a digitalized world of either/or choices into a world of coexisting possibilities—a polysystem.

Even with its polysystemic revisions, *Gravity's Rainbow* maintains a certain hermeneutic-metaphysical bias, being closer to the male speculative perspective of *V.* than to the "sensitive" female perspective of *The Crying of Lot 49*. *Gravity's Rainbow* shares with Pynchon's first novel not only a few characters, such as Seaman Bodine and Kurt Mondaugen, but also a more general interest in the cultural and literary history of the modernist half of the twentieth century. By contrast, *Vineland* (1990) returns us to the dissonant female hermeneutic of *The Crying of Lot 49* and a stronger focus on postmodern history. A passing reference to a character from *The Crying of Lot 49*, Oedipa's husband Mucho Maas who, "after a divorce remarkable even in that more innocent time" (309), has become the all-powerful executive of "Indolent Records," measures the distance traveled by American culture between the two novels, from bohemian innocence to corporate consumerism. Applying Baudrillard's conceptual distinctions, we can argue that *Vineland* "marks the end of metaphysics, and signals the beginning of the era of hyperreality: that which was previously mentally projected, which was lived as metaphor in the terrestrial habitat is from now on projected, entirely without metaphor, into the absolute space of simulation" (*The Ecstasy of Communication* 16). Despite its hypertextual appearance, *Gravity's Rainbow* was still organized around "scenes" played out on the grand operatic stage of history. *Vineland* converts the "hot universe" of conflicts into "a cold universe" of simulations (*The Ecstasy of Communication* 26), as it focuses on the hyperreal world of 1984, "cut into pieces" by the ubiquitous "Tubelight" (*Vineland* 38, 71). The hyperinformational age evoked in this novel abolishes the distinction between private and public space, flattening both on the screens of computer terminals and television sets.

At first glance, *Vineland* reinforces a paranoid suspicion (shared with postmodern theorists like Baudrillard, Jameson, and Lyotard) of the new hypermedia. Two discursive orders vie in *Vineland*: One is flat, soporific, controlled by digital technology and mass reproduction; the other is personal, interrogative, a form of re-creative storytelling. As Jonathan Rosenbaum explained in an early review, "implicit throughout is the notion that thanks to the dominance, ideology, and druglike powers of the Tube (as Pynchon calls it), disseminating [. . .] 'the ruling ideas of epoch,' the recovery of even recent history has to be carried out through willful and sustained archaeological research into buried documents and testimonies" ("Pynchon's Prayer" 29). Truth must be sought and relayed through "a kind of grapevine, word-of-mouth story-telling model that is explicitly contrasted with the corruptions and dissociations of various TV shows" (29). And yet the task of historical reconstruction is too complex to rely on a single model of narration. Verbal accounts are themselves haunted by misremembering and misconstrual. What is needed is a critical cooperation between oral and electronic narrative, with each mode supplementing and correcting the other.

This cooperation is achieved through Prairie Wheeler, the novel's data-collecting and interpretive node. By combining a rediscovered interest in her family's oral narratives with a natural flair for reading images, this younger counterpart of Oedipa Maas manages to recover some of the ignored interpersonal aspects of postwar culture. Prairie's rereading of recent history, from the revolutionary 1960s to the paranoid and benumbed 1980s, inevitably centers on her missing mother, Frenesi. In her effort to piece together the incongruous aspects of her mother's biography (daughter of a labor activist and a Hollywood "revolutionary," herself a member of the radicalist "24fps" film collective converted into a government informant in the Nixon years), Prairie retrieves an archive of visual images—scrap books, film reels, computer files. She also triggers a process of narrative recollection in the characters associated with Frenesi: Zoyd Wheeler, her "hippie psychopath" ex-husband, former members of the 24fps collective, or inhabitants of the North Californian community of Vineland. The web of images and stories that Prairie weaves together creates inroads into the "Tube-maddened" world of 1984, disturbing its manipulative diegesis. The role of this multimedia reconstruction is twofold: It frees Frenesi's generation of some of its guilt and misconceptions; it also gives Prairie, the disinherited progeny of the Tubal Age, a sense of familial and cultural belonging.

The structure produced by this process of narrative cooperation is nonconventional, "developing its own logic of connectedness through permutations, comparisons, inversions and variations" (Strehle 101). Its "comparative readability" in relation to Pynchon's earlier texts is deceptive. Vineland follows the looping structure typical of Pynchon's fiction, creating new entanglements with every effort to decipher and resolve. The plot is broken up into many mini-narratives that mix historical events (student unrest at the College of the Surf, the establishment of federal rehabilitation camps) with pop fictions (an alien takeover of Zoyd's airplane, a demonstration of the fighting capabilities of the Ninja Vibrating Palm). The reader is encouraged to seek connections across periods and cultures. For example, the novel relates the mythology of the Yurok Indians, who believed in "Woge creatures like humans only smaller" inhabiting features of landscape (186), to the "Tubal" fantasies of the Vineland residents, suggesting that the latter could learn from their mythic forerunners how to turn "dispossession and exile" into an art of "cross-references" (114). The final point is that this kind of cross-reading of past and present, collective and private history, myth and triviality, is essential to our cultural survival.

Though Pynchon's most recent novel, Mason & Dixon (1997), is concerned with the work of demarcation performed by the British surveyors Charles Mason (1730–1786) and Jeremiah Dixon (1733–1779), as recounted in the first decade after the American Revolution by Reverend Wicks Cherrycoke, its implications extend to our own (post–)Cold War World. This novel can be read as an extension and reconceptualization of the quest, started in Gravity's Rainbow and continued in Vineland, for an imaginative "cross-reading" that would scramble the "borderline[s . . .] between worlds" (Vineland 105). Pynchon's novel bears out Edward Soja's warning "that it is space, more than time, that now hides consequences for us"

("Inside Exopolis" 122). Spacial division, we have been reminded in the aftermath of the 1989 geopolitical restructuring, can become reified and contentious, creating the sort of crises we have witnessed recently in Bosnia, Kosovo, and the Middle East. Therefore, literary theory and practice need to bring "a new animating polemic on the theoretical and political agenda, one which rings with significant different ways of seeing time and space together, the interplay of history and geography, the 'vertical' and 'horizontal' dimensions of being in the world freed from the imposition of inherent categorical privilege" (Soja, *Postmodern Geographies* 11).

Pynchon's contribution to this project is twofold. On one hand, *Mason & Dixon* exposes the Western model of cultural self-identification through the violent assertion of difference from others, tracing it back to the polarized philosophies of the Enlightenment. Pynchon's exploration of the eighteenth-century geopolitical world confirms the fact that the success of the Western cultural project depended on a politics of binary separations. As privileged instruments of colonial mastery, cartography and land surveying embodied in the most literal sense the practice of partitioning, encouraging "mutually contesting, starkly conflicting, [local] narratives" (Harlow 107). But while critiquing the geometric imagination that has generated divisive orders from the time Mason and Dixon drew their 244-mile-long line between Pennsylvania and the Southern states to the time of our reading, in the post–Cold War "new world order," Pynchon's novel "contrives [. . .] a map [. . .] that never was" but could be (242): a map of "borderlands" (349), cultural and ontological interfaces, and "fluid Identit[ies]" (469).

Mason and Dixon themselves feature in this novel not only as mercenaries of an ideology of division but also as boundary-crossers. Their assignments in Cape Town, St. Helena, the North American colonies, and North Cape occasion "intercultural narratives" in Susan Stanford Friedman's sense of the word (*Mappings* 137), confronting Mason and Dixon with contact zones between cultures. The two astronomer-surveyors are compelled to reassess periodically their own positionality, dissociating themselves from the Dutch colonialists in Cape Town, from the British imperialists in America, from the metropolitan culture in the Wilderness, and from the control of the Royal Society and the East India Company most everywhere. They become provisional hybrids, visitors in intercultural or otherworldly spaces. Their movement across three continents (South Africa, North America, and Europe) and different ontological realms (dream, superstition, myth) enriches the scope of the narrative, superposing a polysystemic structure over regulating borderlines. Pynchon's multiple spaces and identities upset simultaneously the narrative of European imperialism that prevails in Mason and Dixon's time, the narrative of nationalism that remaps the freed American colonies during the time of narration, and the narrative of U.S. economic and cultural hegemony in the time of our reading. We are encouraged by the novel's many anachronistic allusions to our period ("If you must use [Indian hemp], don't inhale" [10]) to do a historical "cross-reading," moving back and forth between three temporal levels.

While envisioning alternative mappings of the world, Pynchon's maverick imagination never forgets the "crushing homogeneity" (McLeod 90) of the dominant orders, old or new. *Mason & Dixon* challenges them both in style and in content. The novel plays the arcane vocabularies of the 1700s against the high-tech and multicultural discourse of the 1990s and the diaristic approach of eighteenth-century prose against the self-reflexiveness of postmodern fiction. This jumble of styles breaks the linearity of the plot, suggesting that history is a palimpsest in which past, present, and future rewrite each other. The story level is complicated further by delayed revelations and missing links. For example, Eliza Fields's captivity narrative is inserted without warning in the middle of Cherrycoke's narration about Mason and Dixon's West Line. Though Cherrycoke manages to tie Eliza's tale to Mason and Dixon's after the fugitive Eliza joins the surveyors' party, the reader remains aware of a tension between this transgressive female narrative and the rest of the novel in which the rationalistic androcentric perspective holds a privileged, though not unquestioned, place.

Mason & Dixon exibits similar disruptions at the level of historical content. The presence of the "cherubick pest, Cherrycoke" (434) as the book's narrator gives a somewhat whimsical appearance to the recapitulation of history, typical of Pynchon for whom "All History must converge to Opera in the Italian style" (706). The novel is peppered as usually with arias sung not only by humans, but also by a talking Norfolk terrier and an enamoured mechanical duck. The "operatic" is illustrated also through the irreverent portrayal of historical figures. Franklin invites Mason and Dixon to an opium den to demonstrate his new invention, a glass harmonica. Colonel Washington presents his guests with a bowl of new-cured hemp and sings about his dream to follow the Goddess of Love (the planet Venus) in its transit at the end of the world. Mrs. Washington makes her entrance with "Oh, la, call me Martha" (280), and Washington's African manservant, Gershom, tells jokes and sings "Havah Nagilah" like a Jewish comedian. Cherrycoke uses a similarly idiosyncratic perspective to introduce aspects of the Revolution. As we follow Mason and Dixon through the taverns of Philadelphia, Annapolis, and New York, we get a sense that the "endless cups of coffee, sweets too ready at hand, and clouds of tobacco smoke found in coffeehouses all have contributed chemically to the aggressive, disputatious, and ultimately revolutionary spirit" of the American colonies (Hagen 789).

As Dr. Johnson advises Mason: "[T]hink [. . .] how much shapely Expression, from the titl'd Gambler, the Barmaid's Suitor, the offended Fopling, the gratified Toss-Pot, is simply fading away upon the Air, out under the Door, into the Evening and Silence beyond. [. . .] Why not pluck a few words from the multitudes rushing toward the Void of Forgetfulness" (747). *Mason & Dixon* does just that, rescuing words and deeds—the quirky together with the weighty—from the oblivion of History.

2. FROM PATERNALISTIC ORDERS TO DISSONANT WORLDS OF SONS AND DAUGHTERS: V. AND THE CRYING OF LOT 49

Pynchon's fiction has focused from the beginning on the erosion of the paternalistic, closed-universe vision embedded in modernity. This vision has been gradually replaced with a variety of postmodern (dis)orders, from versions of chaos to the excentric, dissonant worlds projected by the sons and daughters of the old Establishment. In *V.* the ethos of the sons is represented alternatively by Benny Profane's disorderly drifting and Herbert Stencil's compulsive quest for the lost Center. The Profane sections are enmeshed in contingency, recording the structureless exploits of "The Whole Sick Crew." Composed of Benny's former navy buddies and a New York cohort, "The Whole Sick Crew" shuttles on the margins of culture, in pursuit of the haphazard energy of the streets "fused into a single abstracted Street [. . .] of the twentieth century" (2, 323). But even a "profane" mode of experiencing can occasionally stumble over a larger pattern. Like a yo-yo operated by an invisible hand (26), Profane drifts through New York's "mercury-lit streets" in search of the fabled "golden screw" that would reconnect conflicting facets of city life (order and disorder, "real-time" and "mirror-time" [40–41]).

If Benny Profane gropes for some coherence at the level of the body, in the interstices of human desire and the liminal spaces of the city, Herbert Stencil seeks order in the metaphysics of the mind and the semiotic density of history. Like Callisto, the artist-scientist in Pynchon's early story "Entropy" (1960), he imagines an orderly superstructure in the lofty "hothouse" of his intellect and tries to project it on Profane's street-level drifting. True to his name, Stencil resorts to a re-creative hermeneutics that fills in the gaps with "stencilized" designs (211), reading new meaning into history. His feverish imagination converts the "surface accidents of history" into a "grand Gothic pile of inferences" (141) that all converge on the mysterious figure of V. His plot line recapitulates history as a progressive breakup of the world of fathers under the pressure of subversive, feminized forces. Stencil's paranoid hermeneutics fits the besieged mentality of the Cold War period. As another character puts it, "In a world such as you inhabit, Mr. Stencil, any cluster of phenomena can be a conspiracy" (140). This "civil-servant without rating, architect-by-necessity of intrigues and breathings-together" (209) needs grand narratives to overcome his sense of marginality "at the bottom of a [historical] fold (140). And yet V. does not grant Stencil's metaphysical quest more credence than to Profane's existential drifting.

The two narrative approaches—one riding the surface of the "present," the other probing the infrastructure of modern Western history—maintain a creative tension throughout the novel. Profane's existential surfing challenges Stencil's depth hermeneutic, forcing it to branch out into conflicting areas: physics and metaphysics, history and "yarn." Stencil's quest for V. contaminates partly Profane's listless wandering. Profane's drifting is governed by sickly mercury lights "receding in an asymmetric V to the east where it's dark and there are no more bars" (2). He and his buddies haunt the V-note bar where they listen to the free jazz of McClintic Sphere. For Stencil, V. is a female "grail" as enticing "as spread thighs are to a libertine, [and] flights of migratory birds to an ornithologist" (50). Her immediate incarnation is the seductive Victoria/Veronica, who had puzzled also his father Sidney. As Herbert reads in his father's diary, "There is more behind and inside V. than any of us had suspected. Not who, but what: what is she" (43).

Herbert Stencil's quest for the slippery figure of V. is thus inevitably connected to an exploration of his father's political world that she challenges. As a Foreign Officer and staunch defender of the moribund Victorian Empire, Sidney Stencil plays a minor role in a series of colonial confrontations, from the 1898 "Fashoda Incident," which pitted the British against the French in the Egyptian Sudan, to the June 1919 anti-British Disturbances in Malta. Sidney's assignment is to debrief informers and interrogate suspects, leaving the interpretive work to his overseers. But he oversteps his assignment when he engages in speculation about the role that unruly passions play in history, creating open-ended "Situations" that threaten British stability in the world. Intent on defending the patriarchal world order not only for Whitehall, but also for "the white halls in his own brain" (42), Stencil senior proposes to monitor human behavior, penetrating nerve networks and "anatomizing each soul" (443) His proto-fascist philosophy enhances racial and cultural divisions, pitting "northern/Protestant/intellectual" values against "Mediterranean/Roman Catholic/irrational" ones (174).

Sidney's Eurocentric point of view is successfully challenged not only by a shifty anticolonial force that mixes reformism with terrorism, but also by representatives of his own culture, like world traveler Sir Hugh Godolphin and his son Evan. Sir Hugh's 1884 Antarctic expedition allows him a glimpse of a secret structure of interconnected passages under the skin of the earth, which calls into question the cultural divisions enforced by British imperialism around the world. Sidney's own Manichaean vision gets blurry with age. As he watches the events around "the Great War" (141) move toward increasing randomness, he begins to doubt the possibility of human control over history. He blames the emergence of a new Age of Decadence on the erosion of the fatherly rule by the "Son, genius of the liberal love-feast which had produced 1848 and lately the overthrow of the Czars. What next?" (444). The patriarchal order is also challenged by resourceful female figures like V., whose seductive incarnations Sidney himself has tried unsuccessfully to resist. Sidney's initial reading of V. is entirely negative, associating her with the entropic forces that promote a universe of "cross-purposes" (455). However, Stencil's last mission to Malta, that "Mediterranean womb" of civilization (446), allows him to arrive at a more complex definition of femaleness, at once subversive and counterentropic. Briefly reunited with a version of V., Stencil loses interest in his reactionary mission to control the "treacherous pasture [of] this island" (460), dismisses his informers and waits out the June anti-British Disturbances that end as precipitously as they began. The imponderables that deflect the revolution also cause Stencil's mysterious disappearance on June 10, 1919, swallowed by a waterspout while watched by an apparition resembling V.

If "Old Soft-Shoe Sidney" inhabits a darkening, Machiavellian world of fathers, Profane shuttles through a degenerate world of sons. Profane's wanderings begin on Christmas Eve 1955 with a trip to Norfolk, where he visits the drunken soldiery of the new world superpower. He finds a cast of seedy heroes with comic-folkloric names (Pig Bodine, Dewey Gland, Pappy Hod, Fat Clyde, Tiger Youngblood) engaged in the kind of zany experimentation that looks forward to Pynchon's Californian novels (like *Vineland*, V. has a "potential berserk [who is] studying the best technique for jumping through a plate glass window" [2]). Sickened by a sense of "Great Betrayal" (276), this motley crew later expanded with an assortment of New York bohemians dabbles in proto-postmodern interests (existential philosophy, abstract expressionism, and free jazz). While their creativity does not appear to be more than a form of "Catatonic Expressionism" (277) or a faddish "rearrangement" of quotes from Wittgenstein, Sartre, Ionesco, de Kooning, and Riesman, it still offers a qualified response to modernist entropy. As Eigenvalue, the novel's "soul-dentist" observes, "This sort of arranging and rearranging was Decadence, but the exhaustion of all possible permutations and combinations was death" (277).

Though Profane's friends have characteristics that anticipate the subversive liminal characters of Pynchon's subsequent novels, their subworld looks more like a splinter of the encroaching pop culture than a true counterculture. Cold War cultural and gender polarizations continue to haunt it. As an executive for Outlandish Records, "Roony" Winsome draws on his Southern chauvinism in his pet project, which is a new recording of Tchaikovsky's 1812 Overture against the background of bombs dropped on Moscow. His wife, Mafia, develops a faithful "sisterhood of consumers" (V., 113) with her sado-racist novels of "Heroic Love." Even more telling is the case of seaman Pig Bodine, a modern-day lecher who runs a lending library of "stag-movie fantasies" on USS *Scaffold*. First featured in the "Low-lands" story (1960), Bodine siphons, in Pynchon's own words, "an unacceptable level of racist, sexist and proto-Fascist talk," typical of the juvenile male culture of the early 1960s when "John Kennedy's role model James Bond was about to make his name by kicking third-world people around" (introduction to *Slow Learner* 11). Other characters challenge polarizations but in ways that only fake "cultural harmony" (91). The plastic surgeon Schoenmaker makes use of "diametric opposite[s]" in his work, giving his Jewish friend, Esther, an Irish retroussé nose that is a "misfit nose" in reverse.

Benny Profane epitomizes the popular *Zeitgeist*, playing simultaneously the role of clown and subversive folk hero. A stranger both in mainstream America and in its ethnic subcultures (like Leopold Bloom, he is the son of a Jewish mother and a Catholic [Italian] father), Benny adopts the street as his makeshift home after he is discharged from the navy in 1954. A self-described schlemiel, he "yo-yoes" through Norfolk, Washington, New York, and Valetta, absorbing experiences. But Benny's erratic wandering puts him in contact with other no-name drifters (bums, seasonal workers, refugees from extermination camps), allowing him to discover a submerged underlayer of human history. While working on Zeitsuss's underground patrol, which hunts pet alligators escaped into the New York sewer system, Benny comes across the rat chapel erected by mad Father Fairing (a Maltese informer-priest transferred to America in 1918) in a remote section of the sewer. Convinced of the imminent end of humanity during the Depression, Father Fairing diligently converted rats (voluptuous Veronica among them) to ensure the survival of Christianity. The priest's insane diary speaks of a "little enclave of light in the howling Dark Age of ignorance and barbarity" (107). As a "Depression kid" himself (335), Profane understands the priest's need to project a mythic order in his underworld. But his practical sense allows him to resist the seduction of the "chalkwritten walls of legend" (109). Unlike Sir Hugh Godolphin, Fausto Maijstral, Sidney and Herbert Stencil, who probe the depths of a symbolic underworld for clues about the course of modern history, Profane returns to his street-level drifting.

And yet Profane cannot ignore history. As he resurfaces, he discovers that the world has pushed toward increased reification (everything has "run afoul of the inanimate" [270]) and cultural violence. The "werewolf season" of August 1956 brings about mindless killings in New York (272) as well as major political confrontations in the Middle East, Hungary, and Poland. Trying to find his own bearing in this unstable Cold War world, Benny accompanies Herbert Stencil and Paola Maijstral on a trip to Valletta, the island port with answers to the V. puzzle. They find Malta reeling in the crosscurrents of international soldiery. But Profane's descent into this cradle of civilization has little effect: Benny becomes only mildly interested in the Malta connections recalling Father Fearing's cryptic messages in the New York sewer system. In the last chapter before the epilogue, Benny runs off with a "WASP" toward the dark edges of the Mediterranean. The 1956 world remains for him a jumble of "screwless" fragments that "go nowhere and can con [them]selves into believing it to be somewhere" (426).

A stronger symbolic order emerges in young Stencil's compulsive quest for V. His identity is predicated on this quest: "[H]e was quite purely He Who Looks for V. (and whatever impersonations that might involve)" (210). In one of these impersonations, V. is possibly Herbert Stencil's own mother, vanished mysteriously in 1944; his father disappeared under equally strange circumstances at the end of World War I. The inheritance of Herbert Stencil (born in 1901) is thus shaped by the early decades of the twentieth century. His adult career is likewise intertwined with the dramatic events of mid-twentieth century, from 1939 when he volunteers for the Foreign Service to the 1950s when he devotes his sleepless energy to the pursuit of the historical-symbolic links in the "V.-jigsaw" (44). Indebted to Robert Graves's description of an archetypal anima, "Mother of All Living, the ancient power of fright and lust" (*The White Goddess* 24), Stencil's pursuit of V. promises both a "forbidden form of sexual delight" and "an adventure of the mind in the tradition of *The Golden Bough* or *The White Goddess*" (*V.* 50). Stencil Jr. enters remote environments and times like a "quick-change artist" (50), identifying with off-centered perspectives that allow him to eavesdrop on history. Enriched by his "imaginative anxiety," these revisitations reveal meanings overlooked by the original players. Around each "seed" in the V. plot, Stencil "develop[s] a mass of nacreous inference, poetic license, forcible dislocation of personality" (50), expanding the connective possibilities of the event.

One of the far-reaching connections that Stencil's quest foregrounds is that between the decadent Zeitgeist of the early twentieth century and the disoriented world of the 1950s. Past and present are connected through a narrative of division and dispersal that runs from the hot imperialistic conflicts of the early decades to the Cold War and from modernist decadence to postmodern drifting. Stencil's shuttling between the two worlds foregrounds this master plot but also activates a counternarrative of reclamation that seeks to retrieve excluded margins. Stencil's "impersonations" in the historical world of his father (as the "amateur libertine" waiter P. Aïeul; the "peregrine and penniless" Maxwell Rowley-Bugge; the train conductor Waldetar who drifts away from his fellow Sephardic Jews; the Moslem desert guide Gebrail who understands the Koran as a product of "twenty-three years of listening to the desert") foreshadow Profane's world of schlemiels and cultural bums. Stencil's impersonations of cultural others (the Jew, the Moslem, the libertine, etc.) wander through a world of ancient deserts in search of lost histories (67). But they also stumble over a thick crowd of European secret agents and Baedekered tourists whose presence in these ancient lands suggests an insidious expansionist plan.

Chapter 9, which reconstructs (presumably through Stencil's mediumistic participation) the story of Mondaugen, makes clear that the "grand vaudeville" (60) of Europe's presence in the Third World has sinister implications, mixing a "tourist's view" with violence against the other. As a young engineering graduate from Munich, Mondaugen carries out complicated scientific experiments in the 1922 German Protectorate of Southwest Africa, scanning atmospheric radio disturbances for a meaningful "code." What he finally captures with his listening equipment and "Mondaugen" (moon-eyes") is the massacre of the natives by a well-trained colonial army cheered on by a conclave of Europeans (German, Dutch, English, Polish) gathered on Herr Foppl's farm. In Foppl's fortified plantation house, he also "remembers" (or dreams up) the criminal activities of a German officer during the 1904 extermination campaign ordered by General von Trotha against the rebellious Hereros and Hottentots. Despite his lame denial ("I'm an engineer, [...] politics isn't my line" [224]), the violence of colonial politics encroaches on Mondaugen's serendipitous voyeurism. Shocked by his own guilty identification with the victimizers, Mondaugen retreats temporarily into the bush. He leaves behind a decadent collection of Europeans whose self-serving philosophies are summed up by no other than Lieutenant Weissmann—the sinister Dominus Blicero of *Gravity's Rainbow*.

Not surprisingly, at the center of Foppl's world is another version of V.: Vera Meroving, Lieutenant Weissmann's companion from Munich. She adds to the mood of moral abdication and irresponsible voyeurism her "eye-clock" (219), a surrealistic reminder of her continuing reification. As a Foucauldian reminder of the destructive effects of history on the body, V. has undergone a significant transformation from the time we first met her as Victoria Wren, an eighteen-year-old "girl in heat. Blithe and so green" (61). The convivial, idealized image of Victorian womanhood has given way to the android figure of the 1920s, at once victim and victimizer in a culture of increasing destructiveness. This metamorphosis complicates V.'s narrative function, allowing her to play disturbing roles in the world of fathers. At nineteen, after a quarrel with her widowed father, V. vows to become a "citizen of the world" (151), moving to Alexandria where she entices several Foreign Office agents, including Stencil senior, in a complicated seduction game (the year is 1898); at twenty in Florence, she seduces Sir Hugh Godolphin into sharing his discovery of a secret underground world called "Vheissu"; at thirty-three, she falls passionately in love with fifteen-year-old Mélanie, defying even the loose mores of anarchistic Paris; and the list goes on. A master of disguises, V. is not only ubiquitous (she materializes in England, France, Holland, Spain, Crete, Corfu, Malta, Asia Minor, and the sewer system of New York) and ontologically inexhaustible (a human being, an android, a rat, a place), but also ageless, spanning the entire history of the twentieth century. Her various names and national identities (Victoria Wren, Virginia, Veronica Manganese, Hedwig Vogelsang, Vera Meroving, Mme Viola) suggest that she is a cross-section of Western Europe in its expansionism in the Third World. To Herbert Stencil, V. appears as a nerve center in the general chaos, a secret imperialist plot that traverses the twentieth century. As a proof of V.'s continued relevance to the world of the sons, bits of her disassembled body reappear in the "contemporary" sections. But as Stencil Jr. correctly intuits, V. is not just the embodiment of a "colonial doll's world" (62), but also its defiance. As a figure of divergence in the weakening Establishment, V. takes on many embodiments (from "beast of venery" to spy/plotter/priest), each increasingly more challenging to the androcentric Western worldview.

In her most subversive translation, V. is the countersystem of Vheissu: a secret underground structure that connects East and West, North and South, nature and culture; a perfect example of that alternative female horizon that valorizes new "modes of thinking, writing and speaking" (Jardine 25; Jardine translates "Vheissu" as "V-is-u" or "'you,' that pronoun of alterity" [249]). The challenge of this alternative order is deemed too dangerous to be allowed to survive. Western governments plot the co-option of the Vheissu underground structures to bolster the established order. V. herself is reduced to a "fetish-construction" (386) of the Western world, forced to serve an androcentric vision of history in which the interests of fathers ("hero[es] of the Empire" like Sir Godolphin, who conceals the existence of Vheissu from the public [142]) and sons (who see a female "cabal great and mysterious" everywhere [143]) coalesce. By contrast to these compromising fathers and sons, daughters like Paola Maijstral play potentially redeeming roles that bring the powers of affect to bear on the Western world (which is why she inherits V.'s mythic five-figure comb). Though neither Paula nor V. rise to the self-conscious level of questing exemplified by Oedipa Maas in *The Crying of Lot 49*, they do share some of the latter's capacity to retard the entropy of the world through their re-creative powers. This liminal female energy is re-asserted in the end of the novel as Stencil takes off to chase another piece of V.'s scattered body, her glass eye mentioned by oneiromancer Mme Viola. Refusing to believe that V. is dead, Stencil adopts V.'s "remarkably scattered" (363) perspective, reorganizing history around eccentric new associations that expand his understanding of his own cultural heritage.

The Crying of Lot 49 rewrites the masculine quest for truth in a "feminine" key, enhancing the counterentropic possibilities of female narration only hinted at in the previous novel. In her various roles as executrix of a patriarch's will and explorer of America's counterculture of sons and daughters, Oedipa mixes questing and drifting that were mostly separate in V. From a narratological point of view, Oedipa's challenge is to understand whether she is merely an "affected" entity in her culture's plots or a controlling "agent," the one "with whom lies the power [. . .] to do, to cause, and to affect" (Toolan 245). The conversion to an active role is not easy: As a socialized suburban wife, Oedipa plays initially the "curious Rapunzel-like role of a pensive girl [. . .], prisoner among the pines and salt fogs of Kinneret, looking for somebody to say hey, let down your hair" (20). Oedipa's transformation begins with her "chick trip" to Inverarity's headquarters, the city of San Narciso that gives her a "hieroglyphic sense of concealed meaning, of an intent to communicate" (24). Pierce's enigmatic legacy stirs her curiosity about history but also her desire to become "the dark machine in the center of the planetarium, [trying] to bring the estate into pulsing stelliferous Meaning" (82).

By choosing to participate in meaning making, Oedipa "redeem[s] herself a little from inertia" (82) and oversteps the role willed to her by Inverarity. Oedipa's initial position in Pierce's patriarchal world is hinted at through the description of Maria de Los Remedio Varo's painting, "Borrando el Manto Terrestre" ("Weaving the Terrestrial Blanket"). The central panel of Varo's triptych represents six "frail girls" imprisoned in a circular tower. As Oedipa reflects, there is little one can do against this gendered form of imprisonment: You "may fall back on superstition, or take up a useful hobby like embroidery, or go mad, or marry a disk jockey" (22). Of these alternatives, only embroidery seems to hold some promise. The tapestry woven by Varo's captive girls "spilled out the slit windows and into a void seeking hopelessly to fill the void: for all the other buildings and creatures, all the waves, ships and forests of the earth were contained in this tapestry, and the tapestry was the world" (21). Oedipa's narrative "embroidery" functions in similar ways, taking over the world it purports to represent and generating liberating "constellations" (90) in it. Oedipa approaches this task gradually, shedding (as in her "strip Botticelli" game with Metzger) layer after layer of her suburbanite acculturation. Renouncing her Rapunzel-like dependency on men, she relearns to "breathe in a vacuum" (171), reinventing herself as protagonist. She becomes the "latest incarnation of Oedipus" imagined by Joseph Campbell, standing "this afternoon at the corner of Forty-second Street and Fifth Avenue, waiting for the traffic light to change" (4); but also a rewriter of that traditional male role, who unites questing and inventing, femaleness and maleness in a provisional synthesis.

As soon as she casts off her inhibitions, Oedipa picks up the first signals of an alternative history unfolding around her. She registers the postmaster misprint on husband Mucho's letter, the muted horn symbol drawn in the ladies' restroom, and hears Fallopian's account of the Peter Pinguid Society. Oedipa's self-actualization as a narrative agent is thus directly dependent on the retraining of her cultural imagination.

Oedipa's political awareness develops in the course of her efforts to piece together the historical puzzle of the Tristero, but also in her wanderings through the streets of San Francisco that apprise her of her own cultural isolation. This discovery incites Oedipa's desire to connect to and take responsibility for "those nameless drifters" like herself (181). Learning from stage director Randolph Driblette that history is an endless reenactment of certain basic scripts, she draws bold parallels between the ancient assassination plot recounted in *The Courier's Tragedy* and the pile of bones that underlie Inverarity's estate. Her approach may appear paranoid, but it is perhaps "the only sane response" to a postindustrial culture paralyzed by a glut of information: "[P]aranoids sift the endless number of signs bombarding us, and use them to create structures of meaning. They re-interpret empty data into complete, coherent systems. Paranoids are therefore not only creative, they are the true heroes of our time" (Schwarzbach 59). But Oedipa is not content to weave a paranoid tapestry in the isolation of her tower. Such isolationism, she realizes after visiting Winthrop Tremaine's surplus store with its assortment of Nazi mementos, only encourages the neofascist tendencies of her culture. Abandoning her genteel reserve, Oedipa confronts the strongholds of male power: the Yoyodyne aerospace company, the stockholders' meeting, and the mysterious stamp auction at the end of the novel.

As she walks through the gate of the Yoyodyne giant, guarded by two sixty-foot missiles, Oedipa discovers signs of breakdown at the heart of the male capitalist enterprise. In her subsequent conversations with engineer Stanley Koteks who indicts corporations for stifling creativity, with the elderly Mr. Thoth who recounts Tristero's subversive efforts to displace the Pony Express, and with Ghengis Cohen who informs her of Pierce's eight stamp forgeries that distort honored cultural icons, Oedipa gathers further evidence of cultural and political dissonance inside the American Establishment. She associates these energies of subversion with a variety of "run-away" subgroups (disaffected corporate workers, gays, the Inamorati Anonymous, Negroes on the graveyard shift, leftist opponents of Inverarity's capitalism) and with the Tristero symbol, the coiled muted horn dating back to the Renaissance. Reorienting her quest, Oedipa seeks out their histories of disinheritance and betrayal, but also the stories of their efforts to establish countercommunities in the margin of hegemonic power.

Oedipa's contribution to this alternative social order is limited politically: Her enlightened but impractical intention of spreading Inverarity's legacy among the disinherited of the world would have been promptly blocked by the system. Oedipa's more significant contribution is at a psychonarrative level where she plays the role of mediator-enhancer of alternative stories. Oedipa's fascination with the parallel postal and political organization of the Tristero suggests that she is ready to become a facilitator of "another mode of meaning [and communication] behind the obvious" (182), a "tryster" who establishes mysterious encounters between people and things. Tristero/Trystero evokes tantalizing phonetic associations not only with "tryst" (illicit rendezvous, conspiratorial encounter) and "tryster," but also with the "three-star" card in the Tarot system from which Pynchon has drawn several of his symbolic references. The "Stars" card depicts "an allegorical image of a naked girl kneeling down beside a pool, as, from a golden jar, she pours a life-giving liquid into the still waters" (Cirlot 310). Above this figure are a bright star and seven lesser ones. As Juan Eduardo Cirlot explains, the "ultimate meaning of this symbol seems to be expressive of intercommunication of the different worlds, or of the vitalization by the celestial luminaries of [. . .] the purely material Elements, of earth and water" (310). In her self-defining fantasies, Oedipa hankers after the mythic ability to create "stelliferous Meaning" or to "flash some arrow on the dome to skitter among constellations" (81). By way of this reconstellating power she hopes to wrest "some promise of hierophany" (31) from the contingent. But her mythic project is no less problematic than her political one: Oedipa Maas's name is one or two letters away not only from Oedipus but also from Maat, the Egyptian goddess of righteousness who admits into the Afterlife only those who speak persuasively, whether in truth or in lie. The references to Thoth further complicate Oedipa's mythic heritage. Oedipa encounters a modern version of the ancient scribe whose job was to record the results of judgment in the Hall of the Two Truths, where each deceased heart was weighed against a feather. For the modern Thoth, the feather is not only a symbol of the lightness of truth but also the tool of its spurious rewriting. Thoth's account of the false black-feathered Indians who attack the Pony Express and massacre real Indians confirms for Oedipa the idea that history has been highjacked by forces that practice dissemblance. Like her mythic patrons, Oedipa remains an ambiguous searcher for truth: simultaneously self-blinded and shrewdly re-creative.

The Tristero countersystem is plagued by similar ambiguities, beginning with the name of its founder—Hernando Joaqúin Tristero di Tristero y Calavera—which suggests both a distinguished knightly tradition and its parody. As one of its many incarnations (D.E.A.T.H.—"Don't Ever Antagonize the Horn") makes clear, the Tristero can take terroristic forms. Oedipa finds, for example, that the seventeenth-century sect of the Scurvhamites identified the Tristero with a "blind, soulless," annihilating anti-God (155). But her explorations into contemporary culture also reveal a weaker, "secular Tristero" behind the demonic one. Unlike the historical Tristero, which created ominous dissonance in the dominant culture, this secular counterpart seems content to circulate trite messages in a closed communicational system. Oedipa divides her attention between a primary metaphysical order for which the "post horn [is] a living and immediate symbol" and a degraded secondary universe that has lost its redeeming emblem, the horn sign appearing now exclusively in stamps and books (Mendelson, "The Sacred, the Profane, and The Crying of Lot 49" 132–33). The difference between the two systems collapses at the end of the novel as an impending revelation is suggested but deferred indefinitely. The auction of Pierce's forged stamps takes place on a Sunday, in a room packed with men wearing "black mohair" and "pale cruel faces." The auctioneer spreads his arms in a priestly gesture, preparing to cry "the lot 49." We suspect that the message has to do with Tristero as the "spiritual terrorist [. . .] the angel of calamitous annunciation, the angel of death, and the angel who will blow the last trumpet—or posthorn" (Nohrnberg 148–49). But the final message never comes.

Though Oedipa is receptive to the metaphysical promise of her "zany Paraclete" (The Crying of Lot 49 68), she does not act as a carrier for his deferred cosmic messages. She is more interested in the humble human messages (such as the old sailor's last letter to his wife) that solicit the "secular miracle of communication" (180). On this level, she performs acts of imagination that take her into "all manner of [half-understood] revelations" (24). Her revelations are cultural and linguistic rather than metaphysical, reorganizing the past around complex "act[s] of metaphor" (129). Following Pierce's advice to "[k]eep it bouncing" (178), Oedipa incorporates everything into her narrative—document, fiction, myth, dream. She listens to oral stories, studies "corrupted" versions of Wharfinger's seventeenth-century play, and explores the "infected" streets of San Francisco to understand the metamorphoses of history. To her modern eye, history unfolds as a sequence of violent confrontations: Wharfinger's post-Renaissance tragedy is about "the abyss of civil war"; the right-wing Peter Pinguid Society memorializes a Confederate ship commander who got into a scuffle with a Russian ship; Metzger's movie, *Cashiered,* is set in the Dardanelles during World War I; Lake Inverarity conceals the smuggled bones of American soldiers killed in a World War II battle and more recent vestiges of the Cold War. These widely scattered historical periods are brought into common focus by such recurrent metaphors as the smuggled pile of bones or the auctioned "lot 49." According to John Dugdale (9), the "lot 49" can be read as a comprehensive metaphor for continental United States, with its 49 states bounded by the 49th parallel and its history haunted by an uncanny recurrence of the same figure (Franklin's experiments of 1749, the Gold Rush of 1849, the start of the Cold War in 1949).

Oedipa's own unifying metaphor (the omnipresent Tristero conjuration) rereads Western history as an ongoing confrontation between two different modes of political and semiotic organization. Tristero makes its first appearance in the sixteenth century as an adversary of the Thurn and Taxis family that ran Holland's mail monopoly in alliance with the Holy Roman Empire. Its subsequent associations are with the seventeenth-century Scurvhamite heretics, the French Revolution, and—after the downfall of the old European establishment—the American Civil War and the ensuing anarchist and right-wing organizations. The emigration of the splintered Tristero to America, where it finds another empire in the making, underscores Tristero's ambivalent role as demolisher/rebuilder of the imperialistic order. The distinction between system and countersystem risks collapsing during the Cold War that translates the ideology of deterrence into a cynical notion of shared power over the world. Emory Bortz imagines someone called Konrad saying: "Together our two systems could be invincible [. . .] we, who have so long been disinherited, could be the heirs of Europe!" (113).

Not surprisingly, Inverarity's capitalist empire overlaps with the Tristero organization. Pierce owns the shopping center that houses the main places with information about the Tristero (Zapf's Used Books, Tremaine's surplus store, and the Tank Theater). However, by allowing Oedipa to come by that information, Pierce facilitates her discovery of a differential energy within his monopolistic establishment that recycles squandered cultural possibilities in the loops of the "W.A.S.T.E." countersystem. Made up of the disenfranchised "squatters" of corporate capitalism, this "secret, second America [. . .] in many ways [. . .] preferable or more honest than the surface society" (Tanner 43) employs Tristero's rival history and channels of communication to "keep in touch" (64). While the hopes invested by Oedipa in this secret form of communication seem excessive (she adopts the W.A.S.T.E. countersystem as her new religion, "that cry that might abolish the night"), she does make a persuasive case for its counterentropic role in pulling the disinherited back into a culturally significant structure.

But *The Crying of Lot 49* also illustrates the limitations of such subversion from the margins. However empowering, the "W.A.S.T.E." countersystem into which Oedipa and her disenfranchised friends are absorbed keeps them confined to their definition as "others." Their activism remains limited, illustrating the fate of—in Victor Turner's terminology—a "liminal community" of makeshift "edgemen" whose intervention is reaggregated by the larger society (*The Ritual Process* 128). Oedipa's newly found "motherly feelings" toward "all those nameless drifters" (181) are spent somewhat inefficiently, cradling a dying sailor in her arms or experiencing a nervous pregnancy that does not materialize. Even if we read that pregnancy metaphorically, as a symbolic renewal of the world, Oedipa seems mostly nauseated by it. The alternative meaning system she discovers may be no system at all, but a reverberating "Echo Court" that creates only the illusion of antiphonal meaning. Conversely, a countersystem can rigidify into a totalistic structure that allows the oppressed to become oppressors. *The Crying of Lot 49* cannot resolve this dilemma, remaining stuck in a binary matrix: "For it was like walking among matrices of a great digital computer, the zeroes and ones twinned above. [. . .] Behind the hieroglyphic streets there would either be a transcendent meaning, or only the earth" (181). But while Oedipa fails predictably to articulate some grand recuperative narrative, she manages to weave smaller narratives around her own interests and needs. These provisional stories challenge the knowledge bequeathed to her by patriarchy, outlining a better possible world for daughters.

3. BREAKING OUT OF THE "FUSSY BIEDERMEIER STRANGULATION" OF WESTERN THOUGHT: MARGINOCENTRIC CHARACTERS AND PROJECTS IN GRAVITY'S RAINBOW

Gravity's Rainbow cannot be understood without a sustained effort to accommodate *simultaneously* its conflicting propositions. As Leni Pökler, one of the novel's "sensitives," explains, the world's divergent signs have to be apprehended in "parallel, not series," as they "map on to different coordinate systems" (159). Human imagination must escape the "fussy Biedermeier strangulation" of Western polarizations. Events need not be perceived against each other, "but all together" in a precipitous convergence of "Metaphor, Signs and symptoms" (159), like the Rocket-firing at the end of the novel. Launched at high noon, the Rocket balances opposites, being bridal bed and tomb for Gottfried, a self-defining grail for Slothrop, a holy figure of reconciliation for the "Schwarzkommando," an economic and technological synthesis for the rocket scientists, a revealer of otherworldly orders for all.

As a figure of convergence "against centrifugal History" (737), Pynchon's Rocket challenges disconnected modes of thought, fusing technology and metaphysics, machinery and metaphor. And yet the Rocket is also a product of the polarized, dissociated rationality it challenges. The "ontogeny" (conception, production, launch) of the Rocket recapitulates the "philogeny" of Western culture, beginning with its premodern roots in life-denying Calvinism and ending with the bureaucratized rationality of modern Science. In Pynchon's gothic vision, modernity is at war with itself: "The War needs to divide this way, and to subdivide, though its propaganda will always stress unity, alliance, pulling together" (131). In addition, Western modernity is at war with everything it defines as its Other, "the problematic, 'marked' side, understandable only in terms of its distinction from the Western pattern of development, taken as normal" (Bauman, "The Fall of the Legislator" 129). *Gravity's Rainbow* deals with both aspects at length, offering extensive examples of self-polarization and repressive division.

Pynchon's characters are trained to expect "binary states," such as Slothrop's alternation of erections and flabbiness. These states are researched eagerly by Dr.

Edward Pointsman, who epitomizes the deficiencies of rigid binary thinking. Taking his cue from Pavlov's *Conditioned Reflexes and Psychiatry* (1941), Pointsman seeks a "physiological basis" (89) for the odd coincidence between Slothrop's erections and the V-2 rocket hits on London, reducing the mystery to a stimulus-and-response mechanism. He conceives his work as a confrontation between rational science and a figure of irrational entropy, "physiologically, historically, a monster" (144). The scientist's desire to master the "monster" has a sinister aspect that Pynchon's novel associates with the dark power of a "Führer [. . .] on the rise" (272). During World War II, Pointsman's research institute becomes a unit of psychological warfare against both freaks of nature like Slothrop and the "foreign" enemy.

Pointsman's repressive order of science is challenged by neurologist Kevin Spectro, who rejects the Pavlovian fascination with "ideas of the opposite" (55), seeing the cortex as an interface between Outside and Inside. Statistician Roger Mexico also questions the "virtues of [objective] analysis," proposing to replace antinomical models such as Pavlov's theory of "bi-stable points" (55) with intercrossings "*between* zero and one." His scientific disagreement with Pointsman has broader cultural implications: Roger's "yang-yin" sensibility makes him responsive to women, paranoids, or clairvoyants who challenge polarized thinking through "transmarginal leaps" (50). By contrast, Pointsman follows Pavlov strictly in trying to repress deviants, reducing their paradoxical imagination to "pathologically inert points on the cortex" (90). In the long run, neither position prevails. As he stares at his cubicle wall, "so pictureless, chartless, mapless" in the night vibrating with "bombers outward bound" (136), Pointsman has intimations of an irrational order in the universe that cannot be regularized by his science. His challengers are just as vulnerable to this irrational order: Dr. Spectro dies suspiciously in a V-2 rocket hit on the St. Veronica hospital (sole victim among the higher-ranking staff); and Roger is hounded by a sequence of events that separate him from those he loves.

The evil order that Pynchon's characters uncover is both political and metaphysical. The two manifestations are intermeshed, since they represent amplifications of the same conflictive ethos that Pynchon traces back to Calvinism and Enlightenment rationality. The evils of bureaucratized rationality are embodied in modern corporate science and militarized technology; the evils of metaphysical polarization are embodied in the more imprecise image of a supraterrestrial agency responsible for designing an antihuman order. Both promote division and death. Corporate science is represented by the sinister Dr. Laszlo Jamf whose activities connect not only the warring cultures of World War I and II but also terrestrial science with the supraterrestrial agency. After conditioning infant Tyrone Slothrop to have hard-ons in response to specific stimuli, Jamf returns to Germany where he is prodded by "the bureaucracies of the other side" to continue his research in plastics, looking for an alternative to the frail carbon connections that life depends on. His main contribution to the field of unnatural synthesis is Imipolex G., an aromatic polymer obtained by manipulating the structure of covalent organic bonds to mimic the stronger ionic bonds of metals wherein electrons are not shared but captured. With its hardness, transparency, and high resistance to temperature, Imipolex G. provides the Nazi State not only with a perfect material for the rocket program but also with a forceful metaphor of technological mastery over nature. Jamf's ideology is paralleled in the American corporate world by Lyle Bland, a Harvard man who achieves a successful "American synthesis" of new possibilities for control (581). The economic system he builds after World War I has tentacles everywhere, connecting German rocket scientists to American industrialists and ensuring the "grim rationalizing of the World" (588).

Entire cultures fall under the imperialistic sway of corporate science: The English bring nineteenth-century China into submission through the commercialization of illegal opium; a century later, Laszlo Jamf is "loaned" to the Americans to explore the analgesic potential of the morphine molecule. As Wimpe, Pynchon's cynical drug merchant, explains, the economy of the future will rest on the chemical cartel that offers to abolish pain at the expense of creating drug dependency (349). The other model of global techno-scientific domination is, of course, the Rocket Cartel. Like corporate chemistry, it knows "no real country, no side in any war, no specific face or heritage: tapping instead out of that global stratum, most deeply laid, from which all the appearances of corporate ownership really spring" (243). Composed not only of German companies involved in manufacturing Hitler's "vengeance weapons," but also of international corporations (Shell Oil, General Electric, etc.) associated with the Rocket program during and since World War II, the Rocket Cartel becomes "the very model of nations" foreseen by Wimpe, a "Structure cutting across every agency human and paper that ever touched it" (566). More than half of the novel is concerned with the emergence of this global "Structure" on the ruins of the Nazi program.

Already in its Nazi embodiment, the "Racketen-Stadt" brings together a technological and political elite with global ambitions. The German rocket program emerges at the conjunction of popular fantasies of flight and expansion: The same man (Herman Oberth) who built the model rocketship for Fritz Lang's movie *The Woman in the Moon* later designed the first experimental rocket with Werner von Braun and Willy Ley. The ensuing rocketmania spread rapidly among the German elites, populating their imagination (through Nietzsche, Rilke, Stefan George, and Franz Werfel) with "birds of passage" and nationalistic Avengers. This "Wandervögel idiocy" (162) brings together Franz Pökler, the chemical engineer responsible for developing the plastic fairing of Hitler's ultimate weapon, with his former college buddy Kurt Mondaugen and with Major Weissmann, the two characters involved in *V.*'s Southwest African events. If Weissmann's reappearance in *Gravity's Rainbow* was to be expected (the former instigator of Foppl's 1922 "siege-party" developing effortlessly into the sinister commander of a V-2 battery), Mondaugen's presence in the Nazi rocket program seems surprising. A reluctant witness to the horrors of genocide, Mondaugen becomes so disgusted with "everything European" in *V.* that he goes off to live with the bushmen. His scientific interests prevent him, however, from "going native." He leaves Africa with a theory of "electromysticism" in which the triode plays the role of magic regulator, integrating positive and negative values into an "informationless state of signal zero" (404). Believing that Europe can reach such a state of integrated serenity through the agency of the Rocket, he joins his former enemies in designing the ultimate instrument of transcendence/destruction.

Pökler himself invokes the Rocket's promise of transcendence ("We'll all use [the rocket] someday, to leave the earth. To transcend" [400]) as an excuse for his involvement in a program of mass destruction. What he fails to recognize is that the Rocket allows no real transcendence, moving along a controlled path that connects a "drive toward [. . .] freedom," to a downward movement into "enslavement gravity" (455). Pökler's own story as a scientist obsessed with the rocket mystique follows a predictable downward arch. He loses his wife, Leni, first to the "violent and shelterless street" (401), then to an SS "reeducation camp." His daughter, Ilse, becomes estranged to the point where he is no longer sure of her identity when he is allowed to take her once every year to the Zwölfkinderland. Ilse plays for him the role of Persephone, returning annually from her captivity in the Underworld (the reeducation camp Dora) to bear witness to the wicked efficiency of the Nazi State. A brief visit to the liberated Dora camp allows Pökler a glimpse of the "swimming orbits of pain" (428) he has been ignoring. He makes a gesture of belated atonement, slipping his gold ring on the finger of a dead "random woman" and freeing Ilse from the chore of visiting with him. Both Ilse and Pökler disappear in the interstices of the falling Rocket State, "out among the accidents" of a "drifting" humanity (610).

The apocalyptic 00000 Rocket also swallows Weissmann's lover boy Gottfried. In the symbolic-erotic game designed by Weissmann, Gottfried plays the roles of submissive Hänsel to Weissmann's bewitched Rocket-Oven, "plunging, burning, toward a terminal orgasm" (223). His relationship with Weissmann conforms to the Nazi code that requires the unconditional devotion of a disciple to his Leader. The disciple's task is to submit to the phallic Rocket as an antidote to entropy and irrationality: "Beyond simple steel erection, the Rocket was an entire system won, away from the feminine darkness, held against the entropies of lovable and scatterbrained Mother Nature" (324). But this phallic path to modernity entails a self-defeating paradox obvious at least since Max Weber's study *The Protestant Ethic and the Spirit of Capitalism* (1904), which suggested that "rationalized" capitalist society had its roots in irrational paternalism and that it needed the "iron cage" of bureaucratic control to maintain itself. The fiendish character of Weissmann-Blicero is a proof that the contradictions inherent in the concept of modern management are enhanced as Western societies pass from traditional power relations to corporate statehood. By substituting the informational power of his political organization for that of traditional economic bureaucracies, Major Weissmann heralds the strategies of the postindustrial age. Therefore, after the collapse of the Nazi Reich, Weissmann disappears presumably to join the supraterrestrial force that oversees the creation of a "new global order" (644). The "Weissmann's Tarot" section prompts us to look for him in the American nuclear program, "among the successful academics, the Presidential advisers, the token intellectuals who sit on boards of directors" (749).

The Rocket State's regenerative power depends on the perpetuation of class, race, and gender distinctions that translate into relations of domination. As a product of Western civilization, the Rocket is associated with names that suggest Northern whiteness and death: Nordhausen, Bleicheröde, Weissmann-Blicero (the bleacher), and so on. After Pökler escapes the British air raid on Peenemünde, he becomes convinced that the Rocket is used in a grand Western conspiracy against the weak, running "[a] strange gradient of death and wreckage, south to north, in which the poorest and most helpless got it worst—as, indeed, the gradient was to run east to west, in London a year later when the rockets began to fall" (423). Prentice also predicts that the Rockets of the future—such as the one that threatens to obliterate the novel's fictional world at the end—will continue to make victims among the "second sheep": drunks, old veterans, derelicts, "exhausted women with more children than it seems would belong to anyone" (4). Even more perplexing is Enzian's discovery of the well-preserved Jamf Synthetic Fuel Factory in the ruined German Zone. The only explanation that he can come up with is that the War is a conspiracy among the powerful to "take endlessly" (521), creating a permanent imbalance between the haves and the have-nots.

Through Oberst Enzian and his "Schwarzkommando" of Hereros brought from Southwest Africa to serve in the Nazi rocket program, *Gravity's Rainbow* links the rocket motif to Western imperialism. The novel provides us with a prehistory of the predatory raids of the North on the South, recounting the participation of Katje's Dutch ancestor, Frans van der Groov, in the extermination of dodo birds on the island of Mauritius. The dodoes are extinguished because they are deemed "so ugly as to embody argument against a Godly creation" (110). Even as he butchers them, Frans imagines—much like the mad priest in the catacombs of *V.*—a day when Dodoes endowed with speech will convert to Christianity. Three centuries later the Hereros are forced to play the role of converted "Dodo birds." The survivors of General von Trotha's genocide are submitted to rigorous indoctrination aimed at creating black juntas supportive of a German takeover in Africa. After contributing briefly to the German rocket program, the Schwarzkommando find themselves abandoned in the interstices of the war zone. As the threat of their extermination increases, the Hereros led by Enzian take up the work of reassembling the Rocket from pieces scattered across the Zone, seeing in it a model of self-configuration (318). Unfortunately, Enzian's effort to create an aggregate identity for the uprooted Hereros is compromised by the polarized thinking invested in the Rocket.

While ostensibly competing over the remnants of the Rocket State, the Allies reconstruct the "giant cartel" that puts all winners "in an amiable agreement to share what is there to be shared" (326). One thing they share is an imperialistic mentality that successfully thwarts plans such as Enzian's to create an alternative culture in 1945 Central Europe, outside traditional racial and social divisions. Part 3 ("In the Zone") focuses on the two emerging superpowers represented by cartoonish Major Marvy of U.S. Army Ordinance and a Russian intelligence operative, Vaslav Tchitcherine. The two establish a lucrative partnership, peddling furs, technology, and information across the ideological divide. They also bring Old Bloody Chiclitz into the spoils. This American "free-enterpriser" first introduced in *V.* comes to the Zone to scout German engineering for his pet projects: a toy factory that makes racist dolls and a huge aerospace conglomerate called Yoyodyne in *The Crying of Lot 49*. The "good Ruskie" Tchitcherine has his own racist stories to tell. The sections focalized through him recall the 1916 massacre of "darker refugees" by the Russian settlers in the Kirghiz steppes and Vaslav's own participation in the Stalinistic "alphabetization" of the Muslim cultures of Soviet Asia. The Turkic alphabet conceived by Russian linguists liquidates local traditions, replacing "speech, gesture, touch" with a rationalized script meant "to objectify, to confine, to imprison, to harden" the other (Fanon 44).

Several decades later, Vaslav Tchitcherine scoffs at the ethnic anxieties of his compatriots, but he himself fears his half brother Enzian as a dark force that can undo European history. Enzian is the product of a white man's temporary transgression of cultural boundaries. Tired of the "*familiar* unreality" of the North (351) and of the Russo-Japanese conflict in which he served as a gunner, Vaslav's father deserted his warship in a Southwest African port. He joined there a Herero girl who had lost her husband in the 1904 uprising and lived with her until the end of the Russo-Japanese war, enjoying her "honest blackness" as "a breath of life after a long confinement" (351). Then, as the two began to speak "a new language, a pidgin which they were perhaps the only two speakers of in the world" (351), the elder Tchitcherine returned to his country on the Baltic, leaving behind a conflicted cultural hybrid, Enzian. The younger Tchitcherine follows an equally ambivalent trajectory between a desire to cross boundaries and respect for the power structures. During his Asian mission, Tchitcherine has a mystical vision that he remembers as a coming together of opposites in a blaze of "Kirghiz Light" (358). Tchitcherine's twofold search for the magical experience of his youth, which he associates with that "unimagined creature of height and burning" (343), the Rocket, and for his half brother Enzian, takes him to the German Rocket State. The ambivalence of his position becomes conspicuous in the "German vacuum": As he rips off rocket technology for the Soviets, "mad scavenger Tchitcherine" behaves both as a master of anarchic improvisation, a "Red Doper" ready to bond with the "counterrevolutionary odds and ends of humanity" (346), and as a statist agent motivated by revenge against the competing cultures of the West. This ideological duplicity compromises Tchitcherine's vitality. As his body becomes "more metal than anything else," we are reminded that the European power mentality colonizes not only other cultures, but also its own people. Slothrop similarly discovers that secret technologies have invaded his body, turning it into "a colonial outpost" of "their white Metropolis far away" (285).

The Western mentality of power and division works both in the realm of ideology and that of metaphysics. The metaphysical aspect is introduced as a speculative proposition about a possible "bi-planar" structure in the world, with our daily and rational existence governed by an irrational otherworldly ontology. Pynchon's characters have recurring intimations of "another world's intrusion into this one" (*The Crying of Lot 49* 120), but never get conclusive proof for it. The "ominous logic" (*V.* 423) that Pynchon's questers beginning with Stencil discover in the coincidental structure of events can also be interpreted as a metaphor for the destructive "logic" of history itself. As other postmodern political novels—from Ishmael Reed's *Mumbo Jumbo* (1972) to Don DeLillo's *White Noise* (1985) and Umberto Eco's *Foucault's Pendulum* (1988)—have suggested, history has an uncanny way of realizing the most recondite metaphysical plots, turning ideological fictions into powerful tools of domination. Our metaphysical imaginary functions like the Oven-Rocket in *Gravity's Rainbow*: Its apocalyptic fires are fed by our most ordinary lusts, and its denouement can only be "corruption and ashes" (94). There is no escape from the hellish Hänsel-and-Gretel game devised by Weissmann for his two lovers, blue-eyed Gottfried and double agent Katje. Blicero himself is consumed by the "dark exterior Process" that culminates with the firing of the 00000 Rocket. Only double agent Katje Borgesius escapes "her game's prison" (97) by crossing over to the British lines. But her escape is also a capitulation, leaving the Oven intact (her departure thwarts British plans to bomb the rocket site) and co-opting Katje into the Allies' own "order of Analysis and Death" (722).

The strength of this "game's prison" is tested also by Tyrone Slothrop, the novel's undisputed *schlemiel.* Slothrop's belated intuition is that his entire life has been controlled by a transnational elite that includes sadistic scientists, baronial Foreign Service officers, half-crazed Russian agents, American "free enterprisers," and a ubiquitous "They" who haunt him "through the hole in the night" (217). The world is organized for him around "two orders of being, looking identical" and yet profoundly divided: a configuration of random appearances, and a supraworld "amply coded" (203). Since major decisions in history are ultimately left "for some angel stationed very high, watching us at our many perversities" (746), the development of a positive human agency—one that would include enlightened individuals who can resist the evil forces of polarization as well as metaphysically "sensitive" agents capable of reading the Hidden Design—remains a difficult task. And yet *Gravity's Rainbow* manages to find examples of both. Part 3 ("In the Zone") and part 4 ("The Counter-force") posit a fragile fifth column against imperialistic politics, composed of liminal characters such as Slothrop who acquire some subversive power at the interface of different planes, in a "free and experimental region of culture [. . .] where not only new elements but also new contradictory rules can be introduced" (Turner, "Liminal to Liminoid" 61). Most of these characters partake of the "sin" of double-dealing ("Everyone here seems to be at *least* a double agent" [543]). This duplicity allows them to defect from the system or fall through its cracks, upsetting momentarily the balance of power. The entire "Zone" section of the novel is populated with free-floaters, deserters, and migrants who straddle the polarized structure of wartime Europe. Gerhardt von Göll predicates his survival in the post-Nazi world on a shrewd interplay of his conflicting identities as filmmaker and knight of the black market. True to his nickname and emblem (a chess knight), "the Springer" overcomes divisions between "Elite and preterite, [. . .] darkness and light" (495) through oblique jumps across the reality field. Frau Gnahb, the "Pirate Queen of the Baltic" (496), imitates him on a more primitive level. Even Slothrop, after his "scattering" in the postwar "Zone," becomes "a living intersection" (625) through which everything flows.

In *Gravity's Rainbow* the subversion of polarities often starts from little more than a comic phrase, a Rossini or Puccini aria, or an extravagant metaphor. Other passages involve the reader in multiple breaches of the expected cultural order, as when a black Jamaican corporal sings a German fifteenth-century hymn to American pilots in an English church. The startling mix of cultures fills the "cold fieldmouse church" (129) with a life-asserting energy. Equally subversive is Slothrop's dream-trip into Roxbury's sewage system. His journey into the realm of Shinola ("Schein-Aula" or "Seeming-Hall" [687]) realigns the white scapegoat Slothrop with varieties of "blackness" that spring from a "single root, deeper than anyone has probed" (391). Already as a baby Tyrone was known to Jamf's associates as the "Schwarzknabe," the black boy. This code name fits even better the adult Slothrop whose hybridity defies the Western "mania of NAME-giving, dividing the Creation finer and finer, analyzing, setting namer more hopelessly apart from named" (391).

Slothrop's antics in a "Zone cleared, depolarized, [. . .] without elect, without preterite, without even nationality to fuck it up" (556), illustrate the subversive possibilities of identity switching on the margin. In order to elude controlling human and suprahuman agencies in time of war, this clerk in the Allied Clearing House flees to the German Zone, where he adopts a number of liminal roles: a trench-coated British war correspondent; a vaudeville entertainer with borrowed name; a black-market hero, "Raketemensch" (Rocketman); a Russian scout trekking across "the Imperial cauldron" teeming with uprooted nationalities (549); and mythic pig Plechazunga in a Northern village festival. The multicolored costume of the tenth-century pig-hero who routed a Viking invasion fits Slothrop's portly figure perfectly. This modern Falstaff epitomizes the metamorphic possibilities of a Zone that encourages surprising alliances and provisional arrangements for "warmth, love, food, simple movement" (291). But Slothrop's identity masquerade also suggests a political awakening. Once he enters the counter-culture of the Zone, Slothrop becomes a self-conscious outlaw and champion of the preterites: He warns the Schwarzkommando of the plot concocted against them; he defends the people of a small Baltic village from the attack of a German police reinforced with Russian agents; and he recovers a package of hashish from under the nose of the Potsdam Peace negotiators.

Other characters strive to transcend polarization through synthesis: Pirate Prentice grows "radiant yellow, humid green" bananas on the roof of an English maisonette that houses his commando group, turning his refectory table into "a southern island well across a tropic or two" (5–6). In the collapsed German Zone, Enzian's Hereros reassemble bits of rocketry in a symbolic effort to aggregate their own history: "The souls of the ancestors. All the same here. Birth, soul, fire, building. Male and female, together" (563). Gender and cultural oppositions are reconciled also through lovemaking that functions in this novel as "a long skin interface" (38), bonding individuals. Some characters use sexuality to bridge class distinctions: Prentice seeks access into the "glamorous silken-calved English real world" (36) through the wife of his aristocratic boss, Scorpia Mossmoon; Roger Mexico tries to win Jessica Swanlake away from her upper-class lover. These affairs create their own "sudden tropics in the hell breath of war and English December" (169). Other characters (e.g., Leni Pökler) use sexuality more subversively to defy the heterosexual "pornographies of love" (155); or they invoke their metaphysical healing powers (for example, Heli Tripping) to bring together irreconcilable enemies.

Pynchon's women are both facilitators and victims of an anarchic form of sexuality. Attracted by an "Outer Radiance" that proves hollow at the core, Aubade in "Entropy," Melanie l'Heuremandit in V., Greta Erdmann, Nora Dodson-Truck and Katje in *Gravity's Rainbow*, take with each erotic experience "a little more of the Zero into [themselves]" (150). But as "zeros" they remain ontologically versatile, playing simultaneously the role of agents and challengers of the terrible "Other Order of Being." In their enabling presence, male characters experience the play of "God-shadows" upon the universe (331) or have prophetic dreams that expose historical conspiracies and promote alternative choices, a "fork in the road [civilization] never took" (556).

In spite of Pynchon's obvious sympathy for resisters and transgressors, his "counterforce" is constantly overpowered by enemies it barely understands. Pynchon's characters cannot maintain indefinitely their "threshold period," being reaggragated back into corporate society. The modern capitalist system reabsorbs all manner of opposition, creating—in Roger Bartra's view—both its unified, homogeneous world and a controllable Other ("marginal minorities, abnormal beings and deviations") to define itself against. As a result, social identity is locked into an antinomical system whose sides annihilate each other: "at one end the man of the streets, normal and anodyne, silent and integrated, one-dimensional; and at the other end the marginal man, sick or perverted, guerilla or terrorist" (Bartra 8, 13) Not only the Rocket State but also its interstitial Zones are poisoned by the frigid divisions invented by "the cutters and the sleek hermaphrodites of the law" (37). As a consequence, love gives way to sadomasochism and vampirism: Brigadier Pudding submits to a punitive co-prophilic relationship with Katje, Weissmann-Blicero experiments with "terminal orgasm," and Pynchon's arch-lover, Slothrop, behaves like "a vampire whose sex life actually *fed* on the terror of that Rocket Blitz" (738). Women have their own masochistic roles to play: those of the fairy-tale Gretel, Euridice, or Fay Wray (King Kong's famous quarry). The latter role is enacted in countless versions by Margherita Erdmann, the "Anti-Dietrich" star (394) of Von Göll's S&M movies. Film, whose technology relies on another of Jamf's diabolic inventions, "Emulsion J," becomes Pynchon's choice metaphor for an insidious form of control that accentuates shadows and divisions (see Clerc, 103–51; Moore 30–62; Berressem 151–92).

The System is difficult to combat also because it has been "put inside" (30), its repression internalized. Just as the Rocket creates "its own great wind" of destruction, the subversive economy of the postwar Zone—peddling dope, sexuality, and multiple identities—becomes entirely self-regulated. The "counterforce" has difficulty in reinserting itself in the network of (dis)information emerging on the "ruins [. . .] of an ancient European order" (436). In response to this new informational economy that—as *Vineland* suggests—is a subtler version of the totalitarian "Rocket State" aiming to "restore fascism at home and in the world" (12), Pynchon's resisters create their own paranoid "We-system" as opposed a "They-system" (638); or they take refuge in "the swamp between the worlds" (217).

And yet the Industrial-Military complex arising at the end of World War II cannot entirely erase the Zone. The rocket that threatens to fall on the Orpheus Movie Theater that holds all of us, readers, is left suspended at the end of the novel in "the last unmeasurable gap above the roof of this old theater, the last delta-t" (760). There is a slim chance that the Rocket may "strike *beside* this old theater,'" missing its intended target like earlier rockets (Fludernik, "Hänsel und Gretel, and Dante" 50). Even if it does not, the suspended time or "delta-t" effect we experience in the end functions like an infinitesimal Zone that technology has not yet managed to cancel. This "crippl'd Zone" still allows us to "touch the person next to [us]" (760) and move away from the destructive Center into our own "alternative Zones." A postmodern social ethos is implicated in the novel's final segments, especially those that emphasize the subversive potential of Slothrop's pluralization ("Some believe that fragments of Slothrop have grown into consistent personae of their own. If so, there's no telling which of the Zone's present-day population are offshoots of his original scattering" [742]) or envision a constellation of fragile, whimsical states like the one imagined by Germany's village idiots. Even von Göll contributes an unfinished movie, "New Dope," to this postmodern utopia, depicting a world-in-reverse where guns like vacuum cleaners suck their bullets back, "operating in the direction of life" (745). At first glance, these fanciful notions of reversed ontology and microsocial pluralization strike us as weak answers to the grim questions raised by the novel. But as the short section "Orpheus Puts Down Harp" indicates, there is power in aesthetic and social dissonance. In this political flash-forward, Richard M. Zhlubb ("night manager" of the Orpheus Theatre and a parody of Richard Nixon) is confronted on the Santa Monica freeway by "freaks" in an "unauthorized state of mind" (755) who play dissonant music on their harmonicas. The same Nixon who in the epigraph to part 4 rejected incredulously the notion of a "counterforce" now rails against the "irresponsible use of harmonicas" that he blames for the country's defiant mood. Aesthetic and social liminality can threaten the most complacent power system.

4. NARRATIVE TRANSGRESSION AND REARTICULATION IN THE AGE OF GLOBAL NETWORKS: VINELAND

If *Gravity's Rainbow* attempted to recover the interstitial zones of a world dominated by an imperial, systematizing imagination, *Vineland* returns us to the dissonant female hermeneutic and personal narration of *The Crying of Lot 49* while also interrogating critically the political claims of liminality. *Vineland* opens on a summer morning in 1984, but this Orwellian *annus terribilis* has historical antecedents not only in the turbulent 1960s, but also in the Cold War 1950s and the fascistic 1930s and 1940s that Pynchon explored in detail in his previous fiction. History intrudes in the fictional Vineland like the annoying noise of blue jays that wakes up Zoyd ("Dozy"?) Wheeler and distresses his old dog Desmond. Upon awaking from his dream of "carrier pigeons from someplace far across the ocean, landing and taking off one by one" (3), Zoyd remembers the letter he received the previous day with his mental disability check, announcing him that "unless he did something publicly crazy before a date now less than a week away, he would no longer qualify for benefits" (3). This rude intrusion of official reality triggers a process of narrative retrospection at the end of which the novel's protagonists feel like Desmond, "roughened by the miles, face full of blue-jay feathers" (385).

Vineland pits the recent past (the 1960s) against the diegetic present (the 1980s) and asks "Who was saved?" (28). This question receives no simple answer, for if the 1980s are "comfortably numb," the 1960s are partly to blame for failing to carry out their revolution. The survivors of the 1960s counterculture have allowed themselves to be turned into a liminal community, but one that has lost much of the subversive potential attributed by Victor Turner to such "place[s] of fruitful chaos and possibility" ("Liminal to Liminoid" 60). Zoyd's marginality as gypsy roofer and official "mental case" in alleged "refuge from government" (306) illustrates the parochial tendencies of postmodernism to the point of caricature. There is no gain in this liminal condition: The enclave Zoyd shares with his marijuana-growing friends cannot escape or transform history, becoming entangled in its insidious vines.

The history reconstructed in Vineland is both oral and written, "analog" and "digital." Pynchon's characters follow diverging paths in their explorations. Prairie Wheeler struggles to understand her matrilineal history, seeking out her lost mother, Frenesi. Federal agents Brock Vond and Hector Zuñiga are also after Frenesi, but in order to exploit her biography as "a legendary observer-participant from those times" (51) in their "Political Re-Education Program." Zoyd Wheeler recalls moments in his relationship with ex-wife Frenesi, groping for an understanding of their familial and generational history. D(arryl) L(ouise) Chastain, former security head of the 24fps film collective, tracks down Brock Vond in order to avenge her friend Frenesi and her own wasted idealism. These pursuits take place in microworlds that intersect periodically, creating new historical complications. The insertion of a mythic-symbolic plane, represented primarily by the legends of the Yurok Indians, creates a dialogic "hesitation" between contemporary history and myth, official representations and alternative ontologies. The novel makes frequent references to rivers (especially a "river of ghosts") that separate out distinct spheres of existence, such as those of the living and the dead. Some characters have the ability to shuttle between the two worlds. (Vietnam veterans Eusebio "Vato" Gomez and Cleveland "Blood" Bonnifoy act as Vond's "ferrymen," taking him into the land of death.) But only a very few can transcend oppositions, like the Woge spirits imagined by the Yuroks who dwell in features of landscape, reconciling their spiritual and earthly condition.

By contrast to this integrative Yurok myth, popular culture enhances the ontological and political division of Vineland. Bearing out Ihab Hassan's adage that we all "love funk and horror, so, like Mom and Dad, Television and the Mafia give us what we want" (The Postmodern Turn 161), the Vinelanders gladly embrace both. The motherly "funk" is embodied for them in the universal babysitter of Prairie's generation, the TV, but also in the passion for movies of her mother's generation. The fatherly "horror" is represented by corporate conspiracies, government surveillance, gang warfare, and the Cold War. Takeshi sums up the 1980s' world as a "planetwide struggle [that] had been going on for years, power accumulating, lives worthless, personnel changing, still governed by the rules of gang war and blood feud" (146). *Vineland* contrasts two successive generations, trying to understand why the historical activism of the 1960s was replaced by the self-indulgent apathy of the 1980s.

Christopher Lasch's diagnosis of postindustrial America as a culture that "gives increasing prominence and encouragement to narcissistic traits" such as self-indulgent craving, disregard for past or future, and misgivings about growing up (xvi-xvii), fits the collective malaise of the Vinelanders. A form of "panic narcissism" is experienced by all major characters. By playing the role of the local loony, Zoyd Wheeler gains a comfortable invisibility at the state's expense. But his need to earn back his meager sinecure benefits through an annual display of insanity (his leap-through-the-window act) throws him back into the public arena, in front of TV cameras. Zoyd's narcissism accrues ritualistic, quasi-redemptive implications when he cross-dresses imitating the berdachist practices of the Yuroks, or dons a polychromatic Day-Glow outfit reminiscent of Slothrop's Hawaiian shirts. A quasi-mythic form of self-absorption is experienced also by the Thanatoids, an assortment of "dead heads" and Vietnam vets who "spent at least part of every waking hour with an eye on the Tube" (171) but who finally achieve some "Karmic Adjustment" by marshalling Brock to his death. If Pynchon's fringe groups cultivate a narcissism that is not entirely destructive, representatives of the power structure suffer from more vicious forms of the ills they try to control. Drug-enforcing agent Hector Zuñiga, who keeps calling everybody "li'l buddy" and "Gilligan," is a walking proof that television is addictive. The overdose of TV programming he receives at the Tubaldetox center ("the aim being Transcendence through Saturation" [335]) ruins his ability to differentiate between "tubal" stories and reality. Since Pynchon typically associates loss of differentiation with entropy in closed systems, the larger suggestion here is that narcissistic mass culture is a paragon of closeness. We need only recall Marshall McLuhan's description of Narcissus as a "servomechanic of his own extended or repeated image" who "had adapted to his extension of himself and had become a closed system" (51).

Popular culture is only one of the closed circuits that reinforce the power grid. The system also relies on militaristic government agents and corporate plotters to reproduce its power. After invading Vineland "like some helpless land far away" (357), federal agents Brock Vond and Hector Zuñiga keep leftover hippies like Zoyd under continued surveillance in the name of their "War on Drugs." But the agents themselves are "no more than [. . .] hired thugs" (12) for a power system that runs Vineland like a "garrison state" (314), with reeducation camps, house arrests, and emergency exercises. This concentrational remapping of Reagan's America is made possible through a shrewd "endo-colonization" of one's own people "by imposing the terms in which people understand themselves and their social reality" (Maltby, *Dissident Postmodernists* 176). By Brock's own admission, television plays a crucial role in this process, turning the youth revolution of the 1960s into an infantilizing spectacle: "While the Tube was proclaiming youth revolution against parents of all kinds and most viewers were accepting this story, Brock saw the deep [. . .] need to stay children forever, safe inside some extended national Family" (269). Brock's vision of a firm paternalistic state, which he shares with Captain Blicero of *Gravity's Rainbow,* is a version of the crypto-fascistic ideology that three generations in Frenesi's family have vainly fought. Pynchon's postrevolutionary "Vineland" moves toward a "prefascist twilight" (371) in which mass consumerism and reactionary politics collaborate to suppress individual will.

Arguably, the flower generation "didn't understand much about the Tube. Minute the Tube got hold of you folks that was it, that whole alternative America, [. . .] just like th' Indians, sold it all to your real enemies" (373). But Prairie's postrevolutionary generation is just as culpable for allowing the "cold" medium of television to co-opt the "hot" universe of cultural activism. If in 1964 Oedipa took seriously Thoth's warning against television ("It comes into your dreams, you know. Filthy machine" [*The Crying of Lot 49*, 63]), by 1984 Prairie can hear the warning only in the form of a commercial jingle (TV "sees you in your bedroom / And—on th'-toilet too! [. . .] It knows your ev'ry thought / [. . .] It's plugged right in, to you!" [336–37]). *Vineland*'s saturation with references to sitcoms, game shows, TV commercials, and New Age fads suggests that the entire culture has been kidnapped by pop entertainment and immersed in "rays of radiation that cause us to forget" (218). No facet of public or private life escapes the narcissistic "invasion of spectacle" (Lasch 122). Devoid of historical memory beyond the rudimentary "remembering" that a TV series allows, the Vinelanders lead disconnected lives in an episodic present, borrowing plots from popular culture. As Brian McHale notes, specific TV genres are "keyed on" particular characters and subworlds: "Thus, the region of the fictional world around Zoyd Wheeler seems to conform to the genre norms of TV sitcoms, the region around Brock Vond to the norms of TV cop-shows, the region around Frenesi and her husband Flash to those of soap-operas, and so on" (*Constructing Postmodernism* 135). In other episodes it is difficult to identify a particular show, Pynchon's narrative suggesting an insidious merging of plots and levels of existence into a "seamless [television] world in which all things are related to all other things and in which nothing stands out" (Hart 99). This entropic effect is enhanced by a number of "Tubefreak miracle[s]" (84) that break down the boundary between empirical reality and fantasy: The police officers Frenesi watches on the TV materialize outside her door; Zuñiga's martial law fantasy becomes "real" as an "emergency exercise" is broadcast to the entire country by mistake. In a zanier strain, Zoyd experiences a movie-inspired UFO invasion on his Hawaiian flight, and DL is whisked by the Mob to hyperreal Japan.

Pynchon's Japan and United States are glitzy showrooms of postmodern information technologies, competing for control over the world's communication channels. By involving both in his extravagant plot, Pynchon shows an uncanny understanding of the political arrangements of the (post -)Cold War world. Japan's self-advertised postmodernity, mixing "racism, emperorism, consumerism, stupefying anti-intellectualism, governmental depoliticization of culture, institutional defanging of criticism, diffusing of environmental programs, tattering of information and analysis" (Miyoshi and Harootunian 390), reflects the contradictions of the First World with a vengeance. The cross between Western and Eastern postmodernities globalizes consumerism but removes the possibilities for "meaningful and purposive action" (Miyoshi and Harootunian 392, 394). Pynchon's novel bears this out, focusing on the quietist New Age "zone" created at the intersection of American and Japanese consumer cultures. Mixing predigested oriental myth with ultramodern consumerism (Prairie works in the Bodhi Dharma Pizza Temple, befriends

Ninjette warriors, and witnesses Takeshi's "karmic adjustment" sessions), the "New Age mindbarf" (354) trivializes the experimental ethos of the 1960s. In the America of the 1980s, countercultural resistance has given way to gratuitous violence: The heavy metal band leader Isaiah Two Four, whose name alludes to the Biblical exhortation to turn swords into plowshares, plans to open a chain of "violence centers," and his band plays happily at a Mafia wedding.

Individual initiative gets lost in the postmodern spaces of "America" and "Japan" that turn action into absurd peripety. Still, while Pynchon's Japan absorbs opposition almost entirely, treating especially women as commodities, Sister Rochelle's Californian retreat allows Ninjette-nuns of various races to experiment with a liminal form of freedom. As a member of the Kunoichi Retreat and "lady asskicker" (107) in a world whose "first man was the serpent" (166), DL Chastain comes closest to attaining heroic status. Her role in the novel is to offer a somewhat naive though not entirely ineffective model of marginocentric resistance, "reclaiming her body" (128) and challenging the patriarchy with her "what-is-reality exercises" (141). She also acts as Prairie's mentor, bringing her to the Kunoichi Retreat and teaching her to pay attention to details "out at the margins" (114).

There are similar pockets of resistance within techno-pop culture itself. While Pynchon's novel bears out Guy Debord's argument that the electronic media turn everything into a "spectacle, whose function it is to *make history forgotten within culture*" (186), it also concedes that these technologies can inspire shrewder uses than the intoxicated consumption illustrated by Hector Zuñiga. The fact that the ghost of a 1960s radical, Weed Atman, walks among the Thanatoids suggests that the "benumbed" 1980s have not been definitively severed from the heretic temper of the earlier decades. Pynchon's characters submit television to "unauthorized" uses, smoking dope in front of the Tube, watching it with the volume turned off, or making love with the TV on. As McHale argues, these "mindless pleasure[s]" are not very different from "those which throughout *Gravity's Rainbow* [. . .] are associated with the evasion of Puritanism, bureaucratic rationalization, and technologization" (*Constructing Postmodernism* 124). Prairie herself uses her familiarity with images to actively reread the stories bequeathed to her by her parents' generation. Her quest for her absent mother takes her out of the "tubelight" into the field of history, immersing her in the written, spoken, and photographed past.

At first glance, Prairie is as an unlikely candidate for the important counterentropic function that Pynchon assigns to her. Still, her name suggests a pioneering spirit that opposes the defeatism of her parents' generation. She also seems fairly well suited for the task of interpretation in a multimedia age. As the child of the "Tubal" 1980s, she is a skillful reader of message screens; in time, she becomes also a good "bricoleur," assembling and rearranging bits of narrative information. Unlike other women in the novel, Prairie is not defined socially beyond the elementary categories of daughter and granddaughter. Abandoned by her mother at a young age, Prairie grows up in Zoyd's structureless trailer home, escaping a more rigorous socialization. This lack of definition allows her to "empty" her mind and let herself be absorbed in other people's stories, "conditionally becom[ing] Frenesi, shar[ing] her eyes, feel, when the frame shook with fear or fatigue" (199).

By contrast, Prairie's mother is defined by others and by history. As her name suggests, Frenesi Gates is the epitome of her frenzied times, combining the roles of Revolutionary, Betrayer, Pretty Face, and Fool who is "always surprised when she finds out who she's been working for" (266). As a "third generation Red" (grand-daughter of labor organizers in the 1930s, and daughter of Hollywood union ac-tivists in the 1950s), Frenesi "absorbed politics all through her childhood" (81). Inspired by the principled activism of her parents, Frenesi came to believe in the power of film "to reveal and devastate":

> When power corrupts, it keeps a log of its progress, written into that most sensitive memory device, the human face. Who could withstand the light? What viewer could believe in the war, the system, the countless lies about American freedom, looking into these mug shots of the bought and sold. (195)

In joining the radical docudrama group 24fps, Frenesi hoped to define herself as an agent of history, "liv[ing] out the metaphor of movie camera as weapon" (197).

Frenesi's aesthetic beliefs seem to contradict Pynchon's previous use of film as a metaphor for manipulation, ascribing cinema a subversive function. Unlike feminist "countercinema," however, which emphasizes the revisionistic capabilities of film, Frenesi takes a more naive approach that undermines the political efficacy of her work. The "hit-and-run filmmaking" of the 24fps collective reduces the creative process to an impersonal "record," "removed from everyday actualities" (Olster 121–22). Frenesi's obsession with the voyeuristic power of her camera objectifies peo-ple and compromises her own historical agency. As soon as her camera tracks Brock Vond's handsome face at a political rally, Frenesi herself is "pleasantly framed" (258) into a relationship with a government agent and the "dark joys of social control" (83). Her camera becomes Brock's tool, double-crossing fellow activist Weed Atman and facilitating his murder. After she is rescued by DL from a "National Security Reser-vation" marked with plaques memorializing "American Martyr[s] in the Crusade against Communism," Frenesi solves the conflict between her desire to become a se-rious artist and her commitment to Zoyd and newborn daughter Prairie by betraying both and returning to Brock's "Witness Protection" program. Her subsequent career strays even further from the 1960s' idealism, joining an older line of capitulators to the "Cosmic Fascist" (83) that includes Katje Borgesius and Greta Erdmann.

Pynchon allows an older Frenesi (now remarried to a "reeducated" informer turned agent, Flash) to reflect on her own "entanglement in the world" (108). Though as a protected witness, Frenesi enjoys the illusive freedom "granted to a few [...] to ignore history and the dead, to imagine no future," she continues to be haunted by the past, "the zombie at her back, the enemy no one wanted to see" (71). She does not have, however, the will or tools to explore it in order to learn from its mistakes. Just as an earlier Frenesi was reduced to passiveness by her superstitious faith in the capacity of images to speak for themselves, the Frenesi of the 1980s sim-ply expects "the rays coming out of the TV screen [to] act as a broom to sweep the room clear of all spirits" (83). Her apathy is disrupted when the Witness Protection program collapses under the "ax blades [of Reaganomics]" (90). In a "moment of undeniable clairvoyance" (90), she suspects that she and Flash have become a "string of ones and zeroes" in the government's computers, in imminent danger of being erased by "the hacker we call God" (91).

Prairie does not share Frenesi's fatalism, using her computer skills to rescue fam-ily records from entropic archives. On the surface, she behaves like a "systems per-sona" in Tom LeClair's description: "a collector rather than a creator, and editor rather than an artist, an 'orchestrator' [. . .] rather than an inventor" (*The Art of Excess* 23). As a product of the age of digitalized technology, Prairie is perhaps better positioned to perform the shift advocated by Paul Virilio from a traditional hermeneutics to information processing that promotes nonlinear "relationships between phenomena" (100, 138, 140). However imperfectly, Prairie plays the role of an information age historian who navigates a hypertextual field assembled from unlike sources: "letterhead memoranda from the FBI [. . .] clippings from 'underground' newspapers [. . .] transcripts of Frenesi's radio interviews [. . .] photographs" (113–14), film footage from the 24fps archives, and microchip databases. But she also stretches the boundaries of this information system, supplementing it with the oral stories told by Frenesi's family and friends and applying to it a dialogic reading that allows the digitalized images to come to life in a "definable space" across data banks and "low-lit campuses" (94).

There is some uncertainty in the novel as to the extent of Prairie's historical reinterpretation. While Prairie's conversion into a careful rereader is not in question, it is more difficult to argue that she is a genuine "rewriter" of her culture. For one thing, she seems to have inherited Frenesi's obsession with male authority that brings her in dangerous proximity to the novel's villain, Brock. Since Brock introduces himself as her real father, Prairie's interest in him can be interpreted as a temporary defeat of the mother quest by a "longing for the lost father" and the "father's [repressive] order" (Ellen G. Friedman 241). While there is a point to her delivery in the end from Brock's phallic order (Prairie survives because of her willingness to reexamine/rearrange the cultural narratives she has inherited), her pursuit of a truth beyond the self-reflexive mirrors of her narcissistic culture is more naïve and hesitant than Oedipa's. The very act of symbolic "transfenestration" is rendered problematic by Zoyd's dive-through-the-window routine at the beginning of the novel. Unbeknownst to him, Zoyd leaps through a pane of candy glass, back into (rather than out of) the specular TV mirror. The production staffers package this deception as a "real escape" ("On the Tube, Zoyd came blasting out of the window, along with dubbed-in sounds now of real glass breaking. Police cruisers and fire equipments contributed cheery chrome elements" [15]), rewriting events into a telegenic plot. The act of "rewriting" plays thus an ambiguous role in *Vineland*. As it is passed down from one generation to the next, Vineland's history becomes a "multitude-of-hands rewrite [. . .] no more worthy of respect than the average movie script" in which "characters and deeds get shifted around, and heartfelt language gets pounded flat" (81). Every narrative form is prone to manipulation. As she pursues her "mother's ghost" (114) through computer files, film footage, and newspaper clips, Prairie has a hard time extricating the human story from the frames that "conceal" rather than "illuminate [. . .] the deed" (261).

And yet something does emerge from her mother's "tale of dispossession and betrayal" (172): a new sense of female history. Not surprisingly, Prairie's best informants are women: her grandmother Sasha, Sister Rochelle, Frenesi's associates in the 24fps film collective, and Frenesi herself through her film preserve. Under their influence, Prairie engages not only her imagination but also her emotions in a reconstruction of recent history as a "mother story." Despite the strangeness of its plot, pitting the "kinship system" against the "snitch system" and "networks of government agents that seek to gain information, incarcerate dissidents, and control the population" (Hayles 15), this story manages to recover the ignored, interpersonal aspects of (post -)Cold

War culture. It also makes the reunion of three generations of women (Sasha, Frenesi, and Prairie) finally possible.

Prairie thus performs an important counterentropic function in Pynchon's novel, recuperating a sense of the past and enacting—however obliquely—transformations on the present. One such transformation is the expulsion of federal prosecutor Brock Vond from a reaffirmed Vineland. Brock, who has just been slashed from Reagan's Department of Justice budget, is kidnapped by the Thanatoids and trucked to the Yurok "river of ghosts," where a tribe of "third-worlders" (380) prepare to remove his bones before allowing him into "Tsorrek, the world of the dead" (186). This entire episode falls outside the realistic narrative of postindustrial United States, returning us to a preindustrial (pre-Western) mythos that connects America to the pristine "Vinland" discovered by the eleventh-century Vikings (Magnusson and Pálsson 93), and to various liminal populations. As Sister Rochelle's retelling of the myth of the Fall suggests, the earth suffers a Third World syndrome, feeling abandoned by the empires of Heaven and Hell. Human civilization can compensate for this feeling of rejection by embracing its earthly condition, "the true, long-forgotten metropolis of Earth Unredeemed" (383).

If politically, this ending appears overly optimistic in its prediction of a postideological age that submits even the most reactionary forces to economic cuts and returns us to ground structures, narratologically it makes more sense. Pynchon's "experiment in literary hybridization" (Cowart, "Attenuated Postmodernism" 3), blending traditional and postmodern, realistic and magic forms of narrative, accommodates the dispossessed Vinelanders in an "alternate universe" (334) whose definition has radically changed, suggesting now a real marginocentric space of "common sense and hard work" (112) rather than a metaphysical "borderline [. . .] between worlds" (105).

Vineland thus participates in an important polysystemic rethinking of the categories of center and marginality. Closer to the philosophies of postcolonial criticism, it redefines the Vineland community as a marginocentric "third world" (49), that is, as a space of sociocultural limitations and of imaginative bifurcations simultaneously. As long as they remain separated from one another and from the larger world, the Vinelanders are condemned to repeat the stereotypic plots borrowed from popular culture. However, as soon as they begin to "criss-cross" the "vineland" of collective reminiscences, the Vinelanders gain some control of their stories. As the novel's pioneer in hypertextual spaces, Prairie gets at least an intuitive understanding of what it means to navigate the bifurcated paths of history. Through her, the disjointed narratives of Vineland begin to talk to each other, reconnecting themselves to a future stretching ahead like "*a field of possibilities*" in which the "might-have-beens and the might-bes" are made "simultaneously visible" (Morson 117, 121).

5. SURVEYING MODERNITY'S "CRIMES OF DEMARCATION": REVISIONISTIC CARTOGRAPHIES IN PYNCHON'S (POST-)COLONIAL EPIC MASON & DIXON

Conceived around the same time *Vineland* was written, *Mason & Dixon* (1997) shares with it not only some of its key marginocentric metaphors (the mythic "Vinland," the "long-forgotten metropolis[es]" of the Earth, First World and Third World interfaces) but also a more general concern with narrative dialogics. History as a unified narrative is replaced with a multivoice account coauthored by the book's raconteur, Reverend Wicks Cherrycoke, with other storytellers and narratees. The narratees play a significant role in eliciting, supporting, or challenging Cherrycoke's narration. Gathered in the parlor of Cherrycoke's sister Elizabeth and her husband, Philadelphia merchant J. Wade LeSpark, to listen to another of their "far-traveled Uncle's" tales (7), the narratees include twins Pitt and Pliny, their older sister Tenebrae, cousins Ethelmer and De Pugh, and Aunt Euphrenia, Cherrycoke's musical sister. J. Wade's brothers—"Uncle Lomax," maker of low-quality "Philadephia Soap," and "Uncle Ives," father to Ethelmer and De Pugh—enter the room later to question "a bit smug[ly]" (104) the veracity of their brother-in-law's narration. Cherrycoke protests that he is no Baron "Munchausen" (720), but his role often goes beyond that of a faithful recorder who "scribbled everything down just like Boswell" did for Dr. Johnson (747), imagining situations and postulating alternatives.

While the three "Uncles" censor Cherrycoke's flights of imagination, his younger narratees whose names suggest an artistic investment in history (the young twins are named for famous statesmen and writers; Ethelmer's name recalls the Anglo-Saxon tradition of chivalry; and his embroidering cousin Tenebrae evokes the sixteenth-century Tenebrist style that made dramatic use of chiaroscuro in painting) encourage him to take liberties with his material. Expanding on Cherrycoke's observation that "Facts are but the Play-things of Lawyers" (349), Ethelmer argues that the "Historian's duty [is] to seek the Truth, yet [. . .] not to tell it" since "Who claims Truth, truth abandons" (349). Ives LeSpark counters his son's proto-Foucauldian understanding of History as "hir'd, or coerc'd" by those in power (350), with a dogmatic reassertion that "Facts are facts [. . .]. No one has time, for more than one Version of the Truth" (350). He further denounces "storybooks,—in particular those known as 'Novel'" for irresponsibly mixing fact and fiction (351). Cherrycoke's own sympathies side clearly with the artist rather than the factologist. Encouraged by his narratees who commit their own acts of transgression (as when Ethelmer and Tenebrae abandon their listening level, to read for us Captive Eliza's Tale [529]), Cherrycoke infringes the boundary between reality and fiction, exploring "borderland" events and "enigmata" (246) that prove "that Miracles might yet occur" in America (353).

The younger narratees also influence the content of Cherrycoke's narration, challenging their uncle to tell them a story about America. Cherrycoke obliges, recalling his association with Mason and Dixon twenty years before: "[W]hat we were doing out in that Country together was brave, scientifick beyond my understanding, and ultimately meaningless,—we were putting a line straight through the heart of the Wilderness, eight yards wide and due west, in order to separate two Proprietorships, granted when the World was yet feudal" (8). Cherrycoke is known to his nephews as a weaver of "Herodotic web[s] of Adventures and Curiosities selected [. . .] for their moral usefulness" (7), and his story of Mason and Dixon is no exception. Begun on Christmastide of 1786, "with the War settled and the Nation bickering itself into Fragments" (6), Cherrycoke's 773-page-long tale reveals the fact that the "wounds bodily and ghostly, great and small [that] go aching on" (6) after the Revolution have a lot to do with the infamous Mason and Dixon line that divided the freed American states. The first twenty-five chapters that recount Mason and Dixon pre-American assignments in London, Cape Town, and St. Helena further suggest that the story of America cannot be told without a detour through the history of the "Age of Reason" as experienced both in the metropolitan centers and on the fringes of the colonial empire. The narrator himself is a product/outcast of the Enlightenment: Banned from his country because of his avocations on behalf of the unprivileged, he becomes a spiritual nomad in search of "something of the Peace and Godhead, which British Civilization, in venturing Westward, had left behind" (10). His quest "cross[es] Tracks" (352) with Mason and Dixon several times in the novel.

Conflict, division, and "Reason run amock" (170) form the background of both Mason and Dixon's story and Cherrycoke's postrevolutionary narration. Wherever their assignments take them, Mason and Dixon experience the reverberations of war—from the Seven Years' War (1756–63) that involved the French and the British in a struggle for colonies, but also the expansionist ambitions of Austria, Sweden, Russia, Spain, and Prussia, to the war of the whites on people of different color, and to class and cultural warfare. (The massacre of the British Weavers is mentioned several times.) In America, as they witness new racial and religious antagonisms, Mason and Dixon are reminded of two older conflicts—England's 1739 War with Spain and the Jacobite Rebellion of 1745 in which the Scottish Highlanders rose against the English in a last attempt to restore the Stuarts to the throne. The ubiquity of war in a novel about two surveyors is not surprising: As Henri Lefebvre put it, geography is "the art of war," of containment of "space, as knowledge and action" (11).

With few exceptions, commerce and science confirm in Pynchon's novel the cartographies of polarization. *Mason & Dixon* deplores what Pynchon in a 1984 article called the conversion of eighteenth-century "magic" into "mere machinery" by the "emerging technopolitical order" ("Is It O.K. to Be a Luddite?" 40–41). Everywhere they go, Mason and Dixon get proof that the "new world order" built by the "Charter's Companies" (252) relies on racial violence and slavery, "for Commerce without Slavery is unthinkable, whilst Slavery must ever include, as an essential Term, the Gallows,—Slavery without Gallows being as Hollow and Waste a Proceeding, as a Crusade without a Cross" (108). On their first assignment to Cape Town to observe the "transit" of Venus between the sun and the earth, Mason and Dixon find this remote place haunted by "the Wrongs committed Daily against the Slaves, petty and grave ones alike, going unrecorded" (68). St. Helena, the island in the south Atlantic Ocean where Mason carries out his second assignment, thrives on various forms of enslavement, from the sexual subjugation of women of all color ("African . . . Malay . . . the odd Irish Rose . . ." [129]) to the more insidious mental slavery that company soldiers are submitted to. By the time they reach America, Mason and Dixon are no longer surprised to hear that the sugar they enjoy in coffeehouses is "bought [. . .] with the lives of African slaves, untallied black lives broken upon the greedy engines of the Barbados" (329). In settling the dispute between the slave-holding colony of Maryland and the free colony of Pennsylvania, the surveyors ratify the "line between [. . .] Slave-Keepers" and "Wage-Payers" (629). Their work lays the foundation for a future America of "two nations" (Wills 35).

Unable to correct "this shameful Core" of slavery, Pynchon's surveyors are "doom'd to re-encounter [it] thro' the World" (692). And yet Mason and Dixon play in the novel a more complex role in relation to a geographic and cultural "Other." Despite their contribution to the work of colonial partitioning, they surprise us periodically with their capacity to "to cross that grimly patroll'd Line, that very essence of Division" in order "to reach some happy Medium" (703), "poise[d] upon a Cusp" (705). Mason and Dixon's journeys, beginning with the time they first crossed the Equator on their way to Cape Town, can be summed up as a "passage" into the "haunted and *other* half of ev'rything known" (58). The places they visit during their varied assignments are cross-cultural and even "otherworldly." Cape Town is a city with a "precarious Hold upon the Continent, planted as upon another World by the sepia-shadow'd Herren XVII back in Holland" (58). James's Town on St. Helena is likewise "a very small town [that] clings to the edge of an interior that must be reckoned part of the Other World" (107). Cape Town is the novel's first example of a liminal city but not in the narrow postmodern sense that, as Kevin Robins has pointed out, exults the pleasures of transgression from the perspective of a privileged middle-class "flâneur" who carefully distinguishes himself from the "'have-nots' in the abandoned zones of the city" (323). Cape Town illustrates the category of what I would call a "marginocentric" city, functioning both as an interchange in the imperialist order that segregates the "haves" from the "have-nots" and as a ground for the manifestation of unruly energies and "multi-spiced," oriental influences. Precisely because of its marginocentric position at the "End of the World" (78), Cape Town encourages the overstepping of boundaries both by the colonizers and the colonized. Zeemanns' kitchen slaves defect to the northern mountains, leaving Mason and Dixon without meals. The two astronomers welcome their delivery from the unimaginative Dutch food and dabble in exotic Malay staples. As the narrator Cherrycoke observes, "the cuisine of a people whose recreations include running *Amok* is necessarily magickal in its purpose and effect, and no one is altogether exempt" (86). Even without partaking of this energizing food, Vroom's nubile daughters devote themselves "most unreflectively to the Possibilities of Love [. . .] for they are the Girls of the end of the world" (80). Later on, as Dixon accompanies Cornelius Vroom to his Company Lodge, he finds that the somber Dutch burghers have their own "Garden of Amusement" (150) inside which history is being replayed as subversive farce: Lodge-nymphs with scimitars command their naked "captives" to squeeze together into a tight replica of the cell in Calcutta where 146 Europeans were condemned by the Nawab of Bengal to spend the night of June 20–21, 1756. As the narrator suggests, this scene of horror eroticized allows the Europeans to experience on their own skin the viciousness of their imperial power that "encourage[s] the teeming populations they rule to teem as much as they like, whilst taking their land for themselves, and then restricting the parts of it the People will be permitted to team upon" (153).

James's Town on St. Helena, where Mason carries out his second assignment while Dixon stays back in Cape Town, is another marginocentric place on an island that functions as a transit station for the empire's "Birds of passage"—convicts on their way to the South Seas, young wives going to India to join husbands in the army and navy, "Company Perpetuals, headed out, headed home" (109). At night this town, "which proves to be as Mazy as an European City" (126), becomes an "unlit riot of spices, pastry, fish and shellfish. [. . .] Smells of Eastern cooking pour out of the kitchen vents of the boarding-houses, and mix with that of the Ocean" (114). Believing that James's Town is still too infused with the power of the Company, Mason's employer—the Assistant Astronomer Royal Maskelyne—takes the puzzled Mason to the "Windward Side" of the island (163) where the Wind, "that first Voice, not yet inflected [. . .] of the very Planet" (159), makes music that is not British, but "Viennese, perhaps, Hungarian, even Moorish" (173). Maskelyne welcomes it because "its properties of transformation" (163) help him "imagine things, that may not be so" (161). Mason himself experiences there the visitations of his dead wife, Rebekah. For both Mason and Maskelyne St. Helena is a place of haunting, underwritten by the "infernal" (132) authorship of the unknown.

European imperialism reabsorbs periodically these liminal places into its global cartographies. Cities are more susceptible to recuperation, acting "like capacitors [...] plugged into the globe of history [...]: they condense and conduct the currents of social time" (Holston 65). After Venus's transit on June 5–6, 1761, Cape Town returns to "colorless rectitude," with "Impulse, chasten'd, increasingly defer[ring] to Stolidity" (*Mason & Dixon* 99). The astronomers view this retrenchment ruefully, realizing that what drew them to Cape Town was not the celestial event itself, "but rather that unshining Assembly old Human Needs, of which Venus, at the instance of going dark, is the Prime Object" (102). What they witnessed was an encounter with "a Goddess descended from light to Matter" (92); in other words an encounter with a mythic Other (Lacan's "grand-autre" [68]).

Mason and Dixon welcome such encounters with a mythic or real Other because they promise "another great Turning," a moment of "Grace" (*Mason & Dixon* 101). What they seek are "stories of contact," in Susan Stanford Friedman's sense of the word, animated by a "desire for the other, the different, the alien; desire to connect across or bridge difference; [...] desire to fabulate, fantasize, dream or create" (*Mappings* 134). Dixon alienates himself from the Dutch colonists by pursuing the food, music, and habits of the Malays. Mason cautions Dixon against the sort of magic that the Malays practice, but he himself sails to the end of the world to "have a chance to watch Venus, Love Herself" from these "parts exotic even in [their] workday earth tones" (61). The exposure to other cultures enriches Mason and Dixon's outlook, challenging their at times complacent first-worldist viewpoint.

The two astronomers also come into contact with "preternatural" creatures and events that disrupt the flat ontology of the Age of Reason. Their very first meeting in Portsmouth, before they embark for the South Seas, is attended by a Norfolk terrier who gives advice and sings the song of the "Learned English Dog." Later on, in America, Mason and Dixon watch in amazement Vaucason's animated mechanical duck turn their West Line into its favorite haunt. The Duck becomes a magical messenger for those who believe in "the shifts of Breeze between the Worlds" (448). Mason feels encouraged by these encounters, hoping he can gain "safe-conduct [...] to the realm of Death," to "visit [his dead Rebekah], and come back, his Faith resurrected" (25). Though he takes pride in his rationality, Mason defends his right to imagine other "manifestations" (165) that defy his earthly senses. Dixon chides Mason for his superstitions, but he also dabbles in otherworldly magic. While the astronomer Mason finds models of the otherworldly primarily in the celestial sphere, the surveyor Dixon seeks them in the bowels of the earth, as taught by the inspired idiot Lud Oafery. In response to the gradual diminishment of the surface world due "to Enclosure, Sub-Division, and the simple Exhaustion of Space," Lud perfected his art of "Tunneling," seeking "Down Below, where no Property Lines existed, [...] a World as yet untravers'd" (233). Even though most of his own efforts as a surveyor are spent mapping the surface of the earth, Dixon wields his compass as if it were a "*Cryptoscope,* into Powers hidden and waiting the Needles of Intruders" (301).

Mason and Dixon's next encounter with otherness takes place in America. They arrive there in November 1763 to settle the boundary dispute between Maryland and Pennsylvania by "mark[ing] the Earth with geometrick scars" (257). Even though neither has illusions about America (Dixon expects to encounter just another "slave-colony" [248]), they find the New World more open-ended topographically and politically than other places they have visited. Philadelphia illustrates on a larger scale the role played by Cape Town and James's Town in the first third of the novel, that of a resourceful marginocentric node. The first time Philadelphia is introduced, in a quotation from Timothy Tox's *Pennsylvaniad* (both poet and poem are fictitious), it is described as one of the "pelfiest" towns, helping "A young man seeking to advance himself [. . .] get [. . .] to the nearest Source of Pelf" (217). As the narrator confirms, at that time, "Philadelphia was second only to London, as the greatest of English-speaking cities" (258). In the middle of November, Philadelphia is bustling with sailors, stevedores, traders, charlatans, evangelists, and coquettes that "put all the stoep-sitters of Cape Town quite in Eclipse" (259). Its coffeehouses are alive with debates on the most varied subjects, from the techniques of land surveying, to paranormal phenomena, religious divisions, and the British politics in the colonies. In an apothecary, Mason and Dixon come across a resourceful Benjamin Franklin who offers to sell them a solution that imitates laudanum. But the philosopher also gives Mason and Dixon proof that the rising colonial capital may not be too different from the imperial metropolis, competing with the latter in its perverse, "Italian in intricacies" politics (269).

As the narrator suggests at the beginning of chapter 29, the function of colonial cities like Philadelphia is to "screen away Blood and Blood-letting, Animals' cries, Smells and Soil, from residents already grown fragile before Country Realities" (289). But these civilizing buffers are also implicated in the commerce of death, selling rifles to anyone interested, including the Indians. This policy backfires periodically, as the backwoods people take their war back to the eastern cities. The latest installment in this war between "Cits" and Country people (344) involves a group of Presbyterian frontiersmen known as the "Paxton Boys," who massacre twenty-six peaceful Canestoga Indians in Lancaster and threaten Philadelphia but are persuaded by Franklin to return home. Mason finds it odd that "the first mortal acts of Savagery in America after their arrival should be committed by Whites against Indians," but Dixon reminds him of the colonial violence witnessed at the Cape of Good Hope.

As Mason and Dixon take their West Line into the back country, they realize that "[n]ot all Roads lead to Philadelphia," nor do the country roads respect the West Line, running "rather athwart it" (484). The territory beyond the metropolis is inhabited by mystical sects such as the Chesapeake Kabalistic community that believes in the existence of an Indian Golem created presumably by one of the ten lost tribes of Israel. The New World beyond Philadelphia unfolds as a secret text at the intersection of Jewish, Indian, Chinese, and European myths. As they take in the mysterious otherness of the Western world, the surveyors cannot decide whether America is a real alternative to the metropolitan-colonialist world left behind or an outpost for the manifestation of its repressed impulses:

Is America [Britannia's] dream?—in which all that cannot pass in the metropolitan Wakefulness is allow'd Expression away in the restless Slumber of these Provinces, and on West-ward, wherever 'tis not yet mapp'd, nor written down, nor ever, by the majority of Mankind, seen,—serving as a very Rubbish-Tip for subjunctive Hopes, for all that *yet may be true,*—Earthly Paradise, Fountain of Youth, Realms of Prester John, Christ's Kingdom, [. . .] safe till the next Territory to the West be seen and recorded, measur'd and tied in, back into the Net-Work of Points already known, that slowly triangulates its Way into the Continent, changing all from subjunctive to declarative, reducing Possibilities to Simplicities that serve the ends of Governments,—winning away from the realm of the Sacred its Borderlands one by one, and assuming them unto the bare mortal World that is our Home, and our Despair. (345)

Going West, away from the repressive white civilization, the surveyors "believe themselves pass'd permanently into Dream" (477)—a dream of pristine otherness (499); until they cross the Conococheague River and discover new atrocities committed against the Indians. Instead of leaving behind the tensions of colonial culture, the West Line runs parallel to the "path scarlet with hundreds of small innocent lives wild and domestic" (501) left by Edward Braddock's army. As Pynchon's novel makes clear, the white surveyors' "story of contact" with otherness translates into a *story of violent delineation,* approaching the other with the instruments for measuring and partitioning. Utopia and devastation are the intertwined facets of the colonial project.

Mason and Dixon attribute the destruction of the pastoral borderlands of America to the colonialist mentality that reduces "Possibilities to Simplicities that serve the ends of Governments" (345). "Subjunctive Hopes" are tied back into a "Net-Work" of binary distinctions that reassert the superiority of the white race. When Colonel Washington is ridiculed for speaking like an African, he defends himself by claming that Virginian gentlemen may be "insensibly sliding into their [slaves'] speech" but not also into "their Ways" (276). He reassures the surveyors of his commitment to the colonization of the Western territory, praising the stalwart Ohio Company for showing, alone "in the wild Anarchy of the Forest, [. . .] the coherence and discipline to see this land develp'd as it should be" (281). Prejudice against the "other" is pervasive also in the free state of Pennsylvania. Even before they start work on the West Line, Mason and Dixon become aware of the racial and religious divisions that undermine the colonies, with Quakers, Presbyterians, Anglicans, Catholics, and "Reborns a-dazzle with the New Light" (293) fighting each other or taking it out on the Indians.

While they are quickly apprised of the rifts around them, the surveyors overlook initially the divisive implications of their own boundary-making that begins in the "Delaware triangle" (323) disputed by "the most litigious people on Earth,—Pennsylvanians of all faiths" (324). On May 29, 1765, Mason and Dixon finish calculating the northeast corner of Maryland, establishing the "purest of intersections mark'd so far upon America." But their chaste geometry is marred by the "failure of the Tangent Point to be exactly at this corner of Maryland, but rather some five miles south, creating a semi-cusp of Thorn of that Length, and doubtful ownership [. . .] occupied by all whose Wish, hardly uncommon in this Era of fluid Identity, is not to reside anywhere" (469). This unmappable corner is just one of the hurdles encountered by Mason and Dixon in their path. As they draw their line westward, the surveyors come across anomalous magnetic mountains, mysterious burial mounds, and the forbidding Warrior's Path that turn the West Line it into a "Message of uncertain length, apt to be interrupted at any Moment" (487). Mason and Dixon's concentration is also broken by signs of cultural violence that contaminate the space they map with "Remnants of Evil" (346). As they move farther West, they begin to understand the extent to which their own boundary adds to this geometry of divisions. Their party is joined by victims of partitioning, from Frau Luise

Redzinger, who flees the double taxation levied on her farm cut in two by Mason and Dixon's line, to Eliza Fields and Captain Zhang, who escape one of the strongholds of the ideology of division, the Jesuit College in Quebec.

Mason and Dixon's West Line becomes a battlefield for adepts and adversaries of partitioning. The sinister forces that conspire to divide the virgin heart of the New Continent are suggested in the inset tale of Eliza Fields of Conestoga (chapter 53). Eliza is kidnapped by a band of Shawanese Indians, who take her to the Jesuit "headquarters for all operations in North America," the Monastery and College of Quebec (515). There she is recruited by the sinister teacher-inquisitor Zarpazo ("blow" in Spanish) for his special contingent of prostitute nuns whose job is to beguile those "Jews, [. . .] Savages, English Wives, [and] Chinese" (524) who "seek God without passing through the toll-gate of Jesus" (543). Against the infectious influence of these "heretics," Father Zarpazo a.k.a. "Wolf of Jesus" proposes a range of strategies, from sexual enticement to brainwashing, that "remove[s] certain memories, and substitute[s] others,—thus controlling the very Stuff of History" (530). When everything else fails, Zarpazo advocates inquisitorial immurement:

> If we may not have love, we will accept Consent,—if we may not obtain Consent, we will build Walls. As a Wall, projected upon the Earth's Surface, becomes a right Line, so shall we find that we may shape, with arrangements of such Lines, all we may need, be it in the Crofter's hut or a great Mother-City [. . .]. (522)

Eliza's narrative confronts a dogmatic version of the ideology of division (the Jesuits play in this novel the role assigned to Nazis in *Gravity's Rainbow*), but also suggests ways in which that ideology can be resisted. This story begins with a cultural cross-over (while in the Indians' care, Eliza experiences a liberation of body and soul that no amount of Jesuit training can erase) and is later followed by a gender cross-over (she escapes the Jesuit College dressed as a boy and is tempted by both heterosexual and homosexual relations). Eliza is joined in her escape by Zhang, who flees Father Zarpazo's brainwashing. Guided by Zhang's miraculous "Luo-Pan" (531)—a compass that measures the flow of natural energy ("Ch'i")—the two refugees cross the Canadian border into America and wander south to where Mason and Dixon cut their line. There they join the surveyors' motley crew of "Tailors, Oracles, Pastrymen, Musicians, Gaming-Pitches, Opera-Girls, Exhibitors of Panoramic Models" (384), who seek a redefinition of their identity in the wilderness.

Inspired by the ancient Chinese lore of *Feng Shui*, Zhang redefines himself as a challenger of the Western ideology of partitioning. For "Captain" Zhang, the world comes down to a confrontation between a spiritual geomancy like Feng Shui that shows a "postcolonial" respect for the integrity of a place and the arrogant European science that inflicts "sword-slashes" upon the sacred Dragon of the Earth (542). The latter approach is illustrated by Stig, the Swedish axman who chops down trees with the staggering skill of a Prussian soldier, conjoining "Science and Slaughter" (614). By his own admission, he joins the West Line on behalf of "certain Principals in Sweden" (611) interested in reclaiming the mythic colony of "Vinland," established by the Northmen through an act of violence against Indians, but he may well be the agent of a more insidious power, a "Corps of Intermediaries for Hire" (613) who descend into the "Sin-laden World, posing as Finns, Swedes, the odd Hungarians" (613).

But Stig's Prussian commitment to efficiency remains an exception in Mason and Dixon's party. While eighteenth-century Europe moves toward the terrifying "German precision" of science and war, America makes "tidy" divisions break down "among the Leagues of Trees unending" (551). Several members of the surveyors' party find magic in the wilderness. Though he laments his exile among "foreign Peasants and skin-wearing Primitives" (369), the French chef Armand Allègre enjoys a mysterious relationship with the run-away mechanical duck created by Jacques de Vaucanson. The duck follows him into the wilderness, continuing "her strange Orbit of Escape from the known World, whilst growing more powerful within it" (380). Another oddball who understands the magic of place is Professor Voam, the "camp naturalist" (321) who employs an electric ray called Felipe in his studies of earth's magnetism. Mason welcomes Felipe's presence as a proof that their expedition westward is a journey back to "Innocence,—approaching, as a Limit, the innocence of the Animals" (427). Mason and Dixon themselves experience an alternative sense of time and space on the American frontier and in the "roguish" spaces of Philadelphia and New York. The surveyors watch in amazement as "something styling itself 'America,' [is indeed] coming into being, [. . .] something no one in London, however plac'd in the Web of Privilege, [. . .] seems to know much about" (405). This embryonic postcolonial culture defies older maps, relocating "Sons of Liberty" with Massachusetts accents out in the Allegheny, "New-Yorkers in Georgia, Pennsylvanians in the Carolinas, Virginians ev'rywhere," creating a mixed music that is "already, unmistakably, American" (570–71).

The frontier has been traditionally understood as a free zone beyond the western edge of settlement, a liminal space that promotes both "social atomization" and "cultural regeneration" (Thomas 118, 127). From this point of view, it functions like Pynchon's other liminal zones. Seduced by the "great current of Westering" (671), Mason and Dixon's expedition experiences a double transgression atop the Allegheny Divide: geographic (by crossing this line they violate general Bouquet's Proclamation that forbade white settlements beyond the Allegheny Crest) but also ontological (the axman Zepho Beck turns into a giant beaver under the full moon, undoing Stig's record in tree-cutting). As the line draws to its end in the midst of this magic land, the surveyors enjoy a "Holiday from Reason" (675).

The frontier advertises itself as "a damn U-topia" that nobody owns—"No King, No Governor, naught but the Sheriff, whose delight is to leave you alone, for as long as you do not actively seek his attention" (638). But its magic is already co-opted by the "Projectors, Brokers of Capital, Insurancers, Peddlars upon the global Scale" (487–88). As Brook Thomas also reminds us, the frontier was not a *tabula rasa* into which American settlers could advance unimpeded but a "space of displacement in which something or someone is reconstructed as something or someone else" (118). Against the classic Frederick Jackson Turner argument, according to which "the frontier creates a space in which the United States can avoid conflicts that have plagued European history" (Thomas 128), Pynchon's novel emphasizes the violence of the frontier as a result of cultural displacements (of Native Americans by European settlers, of the margin by the metropolitan center, of a natural economy by global trade, and so on). Thomas Cresap's story is a case in point: Settling too close to the Maryland border and refusing to pay double taxes, his farm is attacked by an army unit. After he moves to the frontier town of Antiem, he continues to be harassed by Eastern authorities who intercept and destroy his pelts. The agents of the metropolitan world hound the remotest places. Mason and Dixon's party meet three land speculators from Philadelphia beyond the Allegheny line, trying to purchase Indian land for the "Metropolitan cabal" (658). The Indians themselves play an ambiguous role on the frontier. While they tell stories of the "great unbroken meadow of the West" 674), they uphold their own divisions. Mason and Dixon's own line cannot proceed further because of the "interdiction" that the "Great Warrior Path" places across their track. This important Indian highway running north to south becomes an uncrossable boundary not only for the whites but also for the Indian tribes.

As Captain Zhang muses, nothing produces "Bad History more directly [or] brutally, than drawing a Line, in particular a Right Line, the very Shape of Contempt, through the midst of a People,—to create thus a Distinction betwixt 'em [...]—All else will follow as if predestin'd, unto War and devastation" (615). Mason and Dixon's boundary confirms the connection between topological and cultural violence. From atop the Savage Mountain, the pastoral green appears irredeemably divided both by Mason and Dixon's line and by the "Theater of the late War, where Indians are still being shot by white men, and white scalp'd by Indians" (614). As the surveyors wrap up their work in November 1767, Mason wonders how "[s]hall wise Doctors one day write History's assessment of the Good resulting from this Line, *vis à vis* the not-so-good" (666). The immediate signs are not encouraging: several axmen die while felling trees as a reminder of the subtle connection "Between the line in all its Purity" and mortality (673). Even after the party turns around to head back east, misfortunes continue to attend it. New suspicions arise between the surveyors. As the narrator reflects, the "interdiction" visited upon Mason and Dixon at the end of their eight years together comes from inside rather than outside, having to do with their inability to cross the "perilous Boundaries between themselves" (689).

When Mason and Dixon meet again, years later, America is their favorite topic of nostalgic conversation. After Dixon's death in 1779, Mason returns to America where he gradually sinks into a "melancholy aberration of the mind" (762). In his most paranoid moments, he suspects that the world has fallen under the control of "a great single Engine, the size of a Continent" (772) that disseminates messages everywhere. Mason misses, however, the real significance of his insight. The "great single Engine, the size of a Continent" is none other than the emerging America, whose rough outline he helped map but whose connections are not yet all made, "that's why some of it is invisible" (772). Mason dies leaving this country on the cusp of history, poised on becoming either a place of "abundance" and "complexity" or the unified Engine dictating messages to the world. Mason's older sons, Doctor Isaac and Williams, who choose to live in "the realm of [America's] Sacred [...] Borderlands," increase slightly the new country's chances to remain a place of magic "Possibilities" (345).

If "*Gravity's Rainbow* ends with disturbing, apocalyptic power, *Mason & Dixon* ends touchingly, with nostalgia for what's been lost and hope for what might, even now, be found" (McLaughlin 217). Pynchon's novel suggests some ways in which America can reclaim a more complex mapping. Mason and Dixon's West Line is challenged not only by geomancers like Zhang, but also by fellow surveyors like Captain Evan Shelby, who want America's space to follow extravagant "polygon[ies]" rather than "simple Quadrilaterals" (586). His concept of flexible mapping defined by as many "exhilarating instrumental Sweeps, as possible" (587) is realized by the more daring people in his county who move at night some of the boundary stones, making the line lose its "pretence to Orthogony" (711). Pynchon's own prose aspires to an extravagant polygony, "zigging and zagging" (586) through interminable em dashes and semicolons, refusing to settle down into a simple, noncontradictory line.

The power of narration to imagine an alternative mapping of the world is well illustrated in Dixon's dream trip to the North Pole. Guided by a mysterious apparition, Dixon enters the "great circumpolar Emptiness" (739) to discover an upside-down world, hundreds of miles below the outer surface. The population of this "Terra Concava" (740) makes shrewd use of telluric forces unknown to man. Though they fear that one day their hidden world will be discovered and standardized by human science, they refuse to join Dixon's world because living on the convex surface of the earth points people away from one another. By contrast, in the "Earth Concave, everyone is pointed *at* everyone else,—everybody's axes converge,— forc'd at least thus to acknowledge one another" (741).

Mason and Dixon miss the possibility of imagining a similar world of convergence in America, but chapter 73 offers a taste of this ignored alternative: "Suppose that Mason and Dixon [. . .] cross Ohio after all, and continue West by the customary ten-minute increments" (706). What they would have discovered is a "perfect Latitude," a new frontier of multicultural borderlands "coming their way, with entirely different Histories,—Cathedrals, Spanish Musick in the Streets, Chinese Acrobats and Russian Mysticks" (708). Like the geopolitical borderland between Mexico and the United States explored by Gloria Anzaldúa, this imaginary borderland as yet unabsorbed by America's exceptionalist project would have played the role of a genuine "contact zone" between civilizations, challenging "dualistic thinking in individual and collective consciousness" and "healing the split [. . .] between the white race and the colored, between males and females" (*Borderlands* 79).

The project of a multicultural, "polygonic" mapping of America that failed in Mason and Dixon's time is brought back by Cherrycoke's postrevolutionary narration. What the narrator foregrounds is an alternative "America of the Soul" pieced together by "one heresy after another, [. . .] ever away from the Sea, from the Harbor, from all that was serene and certain, into an Interior unmapp'd, a Realm of Doubt" (511). On a more positive note, Reverend Cherrycoke tries to envision a "planet-wide Syncretism, among the Deistick, the Oriental, the Kabbalist, and the Savage, that is to be" (356). This utopia is partly realized narratologically, as the space of storytelling is invaded, after the narrator and his audience go to bed, by a veritable "Bedlam of America": "[S]lowly into the Room begin to walk the Black servants, the Indian poor, the Irish runaways, the Chinese sailors, the overflow'd from the mad Hospital, all unchosen Philadelphia" (759). This invasion infringes the narrative line, re-placing it with a bedlamic web. Cherrycoke's own narrative has illustrated in part that model, suggesting that "[n]o one thread [. . .] can be followed to a central point where it provides a means of overseeing, controlling and understanding the whole. Instead it reaches sooner or later, a crossroad, a blunt fork" (Miller, *Ariadne's Thread* 21). Historical narrative becomes by necessity a "complex knot of many crossings" (22), a "web of storytelling" (23) within which the reader is invited to explore alternative paths.

CHAPTER 4

Interventive Writing in the "Post-Human" Age

EXPERIENTIAL AND CULTURAL REARTICULATION IN RONALD SUKENICK'S FICTION

[Noncausal narrative] is one of the mind's most formidable methods of organizing the disparities of experience. It has the virtue of generating unforeseen connections, and is particularly useful in a time when traditional causes no longer seem adequate to account for observed effects. [. . .] Since it starts beyond system it is capable of including kinds of experience that given systems might exclude. Or if such a method of organization is considered as a system, then it is an open system. [. . .] Noncausal narrative implies discontinuity and fragmentation reaching toward continuity and wholeness, which seems more appropriate to a time when mystiques and their processes are laid bare.

—Ronald Sukenick, *In Form* (14–15)

For the modernist predicament, often epitomized in Yeats's words—"Things fall apart: the center cannot hold"—we have the dialectal answer: "Things fall together, and there is no center, but connections."

—Charles Jencks, *Postmodernism* (350)

I. REALIGNING THE TRUTH OF EXPERIENCE WITH THE TRUTH OF THE PAGE: SUKENICK'S GENERATIVE CONCEPT OF FICTION

Though not yet fully recognized, Ronald Sukenick's work has contributed to every aspect of the postmodern scene, from its aesthetic and sociocultural debates, to the development of new discursive practices. His theoretical essays, gathered in two collections, *In Form: Digressions on the Act of Fiction* (1985) and *Narralogues* (2000), suggest a cumulative, polemical effort to redefine fiction "as thought and articulation" rather than as "mind-numbing make-believe" (*Narralogues* 6). Challenging both the mass-market taboo against "thought" in fiction and the prevailing conventions of literary articulation, Sukenick's essays engage the issues of narration in the oblique, self-questioning way of postmodern theorizing, positing a flexible concept of fiction as a "concrete structure, rather than an allegory, existing in the realm of experience" (*In Form* 206). Sukenick dissociates this form of revisionist thinking from the "hierophantic complications" of "formal thinking" (4). His emphasis remains on "the way [fiction] is composed rather than on the way it is interpreted" (xix), a fact that further distinguishes this "in-formal" theorizing from "academic theories of reading and interpretation" (xvi). But this separation is itself submitted to revisions that negotiate the poles of cultural theory and narrative practice under a common agenda.

Much of Sukenick's "new thinking" about fiction is incorporated in his novels in the form of contending attitudes that, while "not very well thought out on a theoretical basis" (*Long Talking Bad Conditions Blues* 2), allow the reader to participate in a reconsideration of the genre. Sukenick's narratives unfold in a "questioning" rather than "answering" mode, addressing simultaneously the theory and practice of writing. Conversely, his essays are "narrative arguments" or "experimental enactments" of "contingent truth[s]" (*Narralogues* 3, 11). These "agonistic, sophistic, sophisticated, fluid, unpredictable, rhizomatic, affective, inconsistent, and even contradictory, improvisational and provisional" "narralogues" redefine fiction as a form of thinking/talking that "is quick enough, mutable enough, and flexible enough to catch the stream of experience, including our experience of the arts" ("Introduction," *Narralogues* 1). They do not try to settle the issues, leaving them suspended between contending points of view. The interplay of perspectives establishes a transcultural dialogue, confronting traditional positions in Western culture in a comic-revisionist spirit that "Chat" defines through the notion of "mutiny" (*Narralogues* 22). Sukenick's "mutinous narralogues" engage our culture critically, testing alternative styles of narration focused on "the little things" of daily life.

Common to both Sukenick's novels and his critical "narralogues" is a concept of "generation" according to which the literary text "is generated by the activity of composition in an ongoing interchange between the mind and the page" (*In Form* 8). First sketched in "The New Tradition" essay (1972), this generative theory of fiction allows the novel to define itself in the process of composition, as "a form of invention, a way of bringing into being that which it did not previously exist" (*In Form* 8). Instead of working from a "presiding idea or a preexisting scenario," the innovative writer relies on improvisational/combinatorial techniques whose precedents can be found, according to Bruce Morrissette, in an alternative literary tradition running "from antiquity through the Grand Rhétoriqueurs, the Gongorists and baroque poets" down to the French avant-garde—Roussel, Perec, Ricardou (1–2). Though overshadowed by the Aristotelian poetics of character and action, this "generative" tradition has managed to counterbalance the causal and motivational model of narration with an emphasis on the process of narrative production. While some of Sukenick's procedures (word and event permutations, divergent endings, textual cross-referencing) recall the linguistic and "situational" generators of Raymond Roussel, Georges Perec, and Alain Robbe-Grillet, his approach is for all intents and purposes different from the "poetics of formal constraint" (Motte Jr. 19) practiced by the French. Sukenick's concept of generation has more in common with action painting and "the Williams-Olson idea of composition as an open field," a "nexus of various kinds of energy, image and experience" (*In Form* 8, 11). Sukenick has followed the spirit and only rarely the techniques of "action" painting or poetry. A reviewer called Sukenick's first novel, *Up*, "action painting in words." His more recent work from *The Endless Short Story* (1986) to *Mosaic Man* (1999) celebrates "open narrative form," calling attention to the dynamic of words and spaces.

In addition to this "open-field" approach, Sukenick's fiction has valorized an older line of narrative improvisation from Rabelais and Sterne to Henry Miller and Jack Kerouac. This tradition, as Sukenick understands it, has emphasized narrative "immanence" or the novelist's immersion in the historicity of the experiential-compositional process. As Victor Shklovsky put it in 1916, "in art, it is our experience of the process of composition that counts, not the finished product" ("Art as Device," qtd. in Scholes 84). Fiction, Sukenick has likewise argued, should concern itself not only with the truth of experience, but also with the "truth of the page," with the performance of "someone sitting there writing the page" (*In Form* 25). The reader "is forced to recognize the reality of the reading situation [. . .] and the work, instead of allowing him to escape the truth of his own life, keeps returning him to it but, one hopes, with his own imagination activated and revitalized" (25).

This concept of narrative "immanence" also foregrounds the temporality of the creative act, returning us to a "rhythmic" compositional time that redefines the novel as an "experience of completion and departure taking place over and over again" (Ermarth 200). The novel's retemporalized field modifies whatever enters it, producing new relationships between its signs, agents, and codes. A rhythm develops as the "stream of pure nothing" (the novel's "movement in time") is interrupted by "intermittent beeps of pure something" (moments of significance) (*The Endless Short Story* 103). Then, as the field gets "organized, the shaping influence of personality, and of any other single element, becomes less and less until finally it is the structured field itself that becomes the organizing power" (*In Form* 14). The novel retains the character of "event" for each reading, "both demostrat[ing] and partak[ing] in the universal process by which things unfold" (*In Form* 9). Even the novel's "spatial structures" (mathematical ratios of lines, variable blocks of text, "quirky punctuation" used as "a kind of scoring" ["Autogyro" 284]) participate in this temporal dynamic, reversing the modernist spatialization of time.

The rationale for this approach was worked out by Sukenick already in his dissertation on Wallace Stevens (1967): Since the artist's response to fluid experience is "by nature fugitive, [. . .] there can be no formulation of it that one can repeat to summon it up; nothing avails but improvisation" (*Wallace Stevens* 32). Improvisation is not "unchecked playfulness" (Alter 182) but a form of experiential thinking materialized in a purposeful act of composition. The presence of an author as performer-rewriter remains important. In contrast to Derridean poststructuralism, Sukenick predicates his notion of narrative generation on the experiences of recognizable discursive agents (authorial surrogates, narrators, critical respondents) rather than on the self-begetting structures of textuality. The generative potential of language is not overlooked, but its aim is to involve us in a dynamics of articulation that transcends the textual. As "surfictions," Sukenick's books

maintain the obdurate presence of the printed text, where life and fiction intersect, but they also engage us in what becomes the dynamic of the novel's surface—the surf of their fiction. [. . .] Surfiction [. . .] evokes a world that is [. . .] all surf—encompassing facts of contemporary reality and fictions that are not so discernible from facts, events that are possible and events made possible by the power of language, the author's actualities, actual inventions, and actual acts of composition. (Pearce, *The Novel in Motion* 119, 123)

The compositional dynamic of Sukenick's novels supports a broader sociocultural dynamic that negotiates between structuring agents and structured field, reflecting consciousness and world.

Sukenick's concept of "generative fiction" departs polemically from both mimetic realism and self-reflexive formalism. For William Gass, the novel is a metaphoric art in paradoxical relation with its world, both richly figural and referentially empty, a "circle" and a "blank" (*Habitations of the Word* 73–112). By contrast, Sukenick regards fiction as an experiential medium: "Its truth is poetic: a statement of a particular rapport with reality sufficiently persuasive that we may for a time share it" (*In Form* 31). This rapport, Sukenick's essays suggest, has to be conceived outside the traditional opposition mimesis vs. poesis that continues to haunt the metafictional project. The solution proposed by Sukenick in his early "New Tradition" essay was a radical version of non-Aristotelian fiction with "no plot, no story, no character, no chronological sequence, no verisimilitude, no imitation, no allegory, no symbolism, no subject matter, no 'meaning.' [. . .] The Bossa Nova is non-representational—it represents itself" (*In Form* 211). This essay usually has been read as a repudiation of referentiality: "The purpose of all such anti-illusionistic strategies would be to prevent the reader from suspending disbelief by reminding him constantly [. . .] that there is no reality to plunge into beyond the surface of the work" (Kutnik 65–66). But Sukenick's other "digressions on the act of fiction" carefully qualify the antirepresentationalism of this early piece intended, by the author's own admission, to shock the New York "intellocrats" ("Autogyro" 291). The anti-Aristotelian temper of Sukenick's later essays is subtler, proposing to restructure and rehistoricize narrative reference. The referential relation is not discarded, but rather submitted to a critical revision that renders its opposing terms (model/image, mimesis/poesis, fiction/reality) unstable and problematic. Representation" is reconceived as an "interventive, i.e., aggressive, interactive" performance ("The Rival Tradition" 4) whose purpose is to immerse fiction in life, reclaiming segments of experience obfuscated by conventional realism:

> [Surfiction is] an attempt to get at the truth of experience beyond our fossilized formulas of discourse, to get at a new and more inclusive "reality," if you will. This is a reality that includes what the conventional novel tends to exclude and that encompasses the vagaries of unofficial experience, the cryptic trivia of the quotidian that help shape our fate, and the tabooed details of life—class, ethnic, sexual—beyond sanctioned descriptions of life. ("Autogyro" 294)

Sukenick's revisionist emphasis has to be understood in relation to this effort to readjust the "truth of the page" to the truth of experience (*In Form* 212). As the writer found early in his career, "An adequate adjustment to the present can only be achieved through ever fresh perceptions of it" (*Wallace Stevens* 3). As "a way of looking at the world" (*The Death of the Novel* 41), fiction needs to reformulate continually its modes of articulation in order to allow more "reality" into the text. From this point of view, Sukenick's pursuit is closer to the reconstructive notion of postmodern representation proposed by Wolfgang Iser and Umberto Eco than to any deconstructive self-reflection. For Wolfgang Iser, "Representation is first and foremost an act of performance, bringing forth in the mode of staging something which itself is not given" (232), "something that hitherto did not exist as a given object" (217). For

Sukenick, as well, "the job of narrative fiction is not to record some preexisting reality but to contribute to the ongoing process of cultural building in and through the process of writing itself" (*In Form* 79–80). Sukenick reconceives representation as an experimental relating—rather than a rigid matching—of two signifiers, "text" and "world." Sukenick can thus agree with Umberto Eco that the "proper effect" of postmodern representation is "a sense of logical uneasiness and of narrative discomfort. [...] [These narrative constructions] arouse a sense of suspicion in respect to our common beliefs and affect our disposition to trust the most credited laws of the world of our encyclopedias. They *undermine* the world of our encyclopedia rather than build up another self-sustaining world" (*The Role of the Reader* 234).

Sukenick's own fiction hopes to produce not just a crisis but also a restructuring in our axioms of reality, through a self-conscious form of narration that continues to articulate its provisional "world reference." The only other alternative is total deconstruction, the recourse to a visceral (non)language of "growls squeaks farts gargles clicks and chuckles" like "Bjorsq" (*98.6* 161). Sukenick's narrators dream periodically of "destroy[ing] the English language" (*98.6* 75) as a tool of conventional communication, but their better efforts go to "stitching together" the novel's "bungled fragments" (187–88). Even when they confessedly fail to produce coherent narrative orders, these narrators convert a narrative crisis into an experiential proposition, negotiating some space for those "leftovers of reality" that "make another kind of sense, a sense that you don't, can't, or don't want to see" (*Doggy Bag* 65, 66). The lack of a traditional "beginning-middle-end kind of closure" in Sukenick's fiction becomes an asset because it validates the "ongoing process of cultural creation" (*In Form* 78, 79) and teaches us to make "another sense" of reality.

In *Narralogues,* Sukenick's capacity to think and write "otherwise" is trained on the most diverse cultural topics: the competing traditions underlying Western culture ("native" and "exiled," "Judaic" and "Greek-Christian") in "Gorgeous" (10); the "postcolonial" position of American culture in relation to European cultures, and of European cultures in relation to U.S. postmodernity in "Chat"; the absurdities of contemporary academic life, divided between "anarcho-socialist" and "neocon antideconstructionist" dogmas in "A la Bastille" (29); the dissociation of literary sensibilities (oral vs. visual, printed vs. electronic) in "Art Brute"; or the entropic tendencies of a postindustrial culture caught between West European decadence and American pop consumerism in "Death on the Supply Side" and "Name of the Dog," reprinted from *Doggy Bag*. Against these aesthetic and cultural dissociations, Waldo (a cross between the writer's Emersonian persona and a younger-generation innovator who "steals" his mentor's ideas) proposes a nonlinear form of narrative that "ditches static abstraction for experiential dynamism" ("Narralogue on Everything," *Narralogues* 62). Expanding the "concept of writing to include kinds of expression beyond the reach of language alone," Waldo's improvisational "Art Brute" breaks down conventional orders, maximizing sensuous participation and minimizing "rational info academedia" ("Art Brute," *Narralogues* 45, 48). As two other previously published pieces—"What's Watts" and "Divide" (*The Endless Short Story* 1–3, 25–33)—suggest, the efficacy of this art is greatly enhanced by the "quirkiness" of its structures (like Simon Rodia's disorderly Watts Towers, innovative fiction develops ex-centrically, challenging conventional expectations) and the self-imposed marginality on its author (Sukenick's choice of Boulder, Colorado, on the great Continental Divide, as the site of his artistic resistance).

Sukenick's own fiction follows the subversive principles of an "art brute." Its general disposition is—with a term that the writer actually dislikes—"experimental" but in the sociocultural rather than purely formal sense: "Experimental writing is just writing that breaks out of the modes of literature into something resembling the freaky uniqueness of individual experience beyond the usual brainwash of official culture" (Sukenick, "Introduction: The Dirty Secret" 8). In the "field of the so-called experimental," dissension from the lucrative conventions of "plot, character, verisimilitude, and U.S. Standard English" (*In Form* 55) is intended as a political, reformulative gesture.

2. "CULTIVATE THE UNEXPECTED": THE TECHNIQUES OF IMPROVISATION AND METAMORPHOSIS

Sukenick will probably agree with Iser's caveat that "to conceive of representation not in terms of mimesis but in terms of performance" does not free fiction from the responsibility of its "inventions," but "makes it necessary to dig into the conditions out of which the performative quality arises" (218). In Sukenick's own theory and practice of fiction, the emphasis on improvisation/invention is accompanied by a critical awareness of the limits of an experiential poetics that seeks "the clarity of persistent ellipsis the logic of lacunae the facile discords of discontinuity thinking at the same time the phrase maddeningly local" (*Long Talking Bad Conditions Blues* 9). Sukenick's novels typically move from tentative *project* to multiple *concretizations* and *revisions*. In a significant reversal of focus, the "observation of ragged peripheries" (9) replaces the search for a unifying center, keeping the "coarse fabric" of the story tattered, elliptical. What emerges is a process-oriented narration that emulates the "prosody of experience" with its "irreducible twists and possibilities." Stripped of "inessentials" and the "trappings" of plot, character, and mimetic realism (*In Form* 243), Sukenick's novels have pursued the "white noise of event[s]," trekking their "constantly erased and rewritten" course, "zig of inside through zag of out" (*The Endless Short Story* 47). However, since their job "is to contribute to the ongoing process of cultural building in and through the process of writing itself" (*In Form* 79–80), these narratives have also submitted the superficial data of reality to a continuous process of *(re)articulation* that activates new connections in the flow of experience:

> You see what's happening here you take a few things that interest you and you begin to make connections. The connections are the important thing they don't exist before you make them. This is THE ENDLESS SHORT STORY. It doesn't matter where you start. You must have faith. Life is whole and continuous whatever the appearances. (*The Endless Short Story* 7)

Imaginative *metamorphosis* (idea adapted from Emerson who is quoted with the following line: "Men really have got a new sense, and found within their world, another world, or nest of worlds; for the metamorphosis once seen, we divine that it does not stop" [*In Form* xix]) is Sukenick's response to the crisis of the "post-human" age. Threatened with cancellation by history's "ongoing ego smasher" (*Mosaic Man* 181), the post–Holocaust/Cold War writer resorts to the "metamorphic power of imaginative language," becoming "pretty much anybody [. . .] who gets close enough to allow my antennae to receive their ectoplasmatic transmissions, man, woman, or child" (182). This metamorphic-projective technique is recognizable already in *Up* (1968), a novel whose authorial narrator competes with the character of the inset "Adventures of Strop Banally" in making "up his past as he went along, so that by now it was impossible to separate the truth from the fiction" (4). Hence the improvised aspect of both the frame and the embedded narrative. As a character-reviewer in *Up* complains, this book "is just a collection of disjointed fragments. You don't get anywhere at all. Where's the control, where's the tension?" (222). And yet beyond this surface disorder there is a genuine effort to "work out the essentials of our fate" (223). The frequent shifts in plot, character, and style free experience from formulaic stories, providing a dynamic space wherein character-author and reader can negotiate a "favorable rapport" with reality. The narrative is propelled ahead by a process of invention and revision: "Form is when you look back and you see your footprints in the sand" (*Out* 164).

The impetus behind this narrative dynamics is an "insomniac," "disordered [authorial] imagination," "throwing up its mordant ghosts, its flotsam of desire, the day's remainders, ends of thought, image fragments, lines of verse, phantasmal enactments [...] sterile epiphanies" (*Up* 132). In Sukenick's fiction, authorial figures are often split against themselves. In *Up*, the narrative function is divided between an implied author constructing a picaresque autobiography with "sardonic echoes of Augie March" (40) and a character-author "Ron Sukenick" engaged in writing the "Adventures of Strop Banally." In *98.6*, the character-author "Ron" invents "alter-careers" for "alter-egoes" (80) and enlists the help of his friends in filling the emptiness by "think[ing] and talk[ing] about it [until] it becomes beautiful" (63). In *The Endless Short Story*, "Porfessor Sukenick" (typo intended) spars with his anagrammatic reverse, "Rossefrop Kcinekus," but also with two ironic spin-offs that represent aspects of authorship: Claude Balls with his "pretentious allegories" and "Bitchakokoff" with his "vicious illusions about social reality" (22). The interference between real author and sundry surrogates disrupts both the ontology of fiction and its concept of subjecthood, providing examples of "transworld identity between real and fictional entities" (McHale, *Postmodern Fiction* 203, 205–6). Authors and characters are caught in a metamorphic narrative that explodes identities: the "good bitter brittle glass of self shatters duplicity becoming multiplicity becoming becoming" (*The Endless Short Story* 46).

The tribulations of the author figure come to climax in Sukenick's most recent novel, *Mosaic Man* (1999), which can be read as a summation of the aesthetics and politics of generative fiction. The task of retrieving "a forgotten word, a sign, a sanction, a symbol," a "story" (16) from the dissimulations of "history" (25) is shared by the character-author Ron with a first-person authorial narrator who both supports and critiques Ron's effort to get "straight to the bull's-eye of reality" (22). Ron revisits key sites in Sukenick's own biography (his parents' Brooklyn apartment in the 1940s, Paris in 1958, Venice in 1961 and again thirty years later, Israel after the death of Sukenick's father, Berlin after the collapse of the Wall) to tease out a "real story" (255). Like the authorial narrator who finds that he shares with Maxine Hong Kingston a dislike of "the merely imagined" (196), Ron prefers docunarrative to novels, refusing "imitation" and "[g]raven images" (96), as the Second Commandment teaches. And yet what both of them write "turn[s] into a novel. Behind [their] back" (97). Ron himself is but "an invention. [...] Bar coded, trade marked, copywritten" (255). The "Real Me" converts easily into a "RaMSCaN" (259), a golem controlled by the dominant culture's narratives. Therefore, the authorial narrator intrudes periodically, disrupting the novel with taped conversations, "real time windows" (192), or personal fantasies that rewrite cultural plots, promoting marginal Jewish alter-egos to heroic roles. The story that *Mosaic Man* wants to retrieve is that of the innovative Jewish writer, rendered invisible by history's repressive "writings on the wall," from the Nazi ideology that "exxed" his identity "off the surface of the planet" (181), through the Cold War hysteria that scapegoated "others," to the incomprehensible post–Cold War "babel of symbols, slogans, signs, messages" (184) that—for all its "multicolored multilingual" messages—is still a form of "writing on the wall." Condemned to "progressive invisibility" (182) by official culture, the innovative writer must become metamorphic, a "comic, a crazy" (201), a "Mosaic man" (205) punning his "genetic language" against "the Babel of background DNA" (208–9). The author ends up speaking in the first-person plural, sharing his explorative tasks with several other "personalities [. . .] at least one of which is female" (220).

Character metamorphosis works in similar ways to destabilize the culture's codified narratives and enhance the experiential reach of surfiction. As Sukenick explained in a 1979 interview, his "characters have some very basic minimal identity, but beyond that the changes they go through are enormous, even contradictory. I prefer the characters to be as little consistent with themselves as they can be, so that everything but that tiny, perhaps genetic, trace of identity is canceled out" (*In Form* 133). The writer hopes to reverse the process of cultural reification that converts "Mensch" into "Manikins. Gelded. Gilded. Scripted" (*Mosaic Man* 9), expanding his characters into "multiple possibilities and multiple levels of personality" (interview in Bellamy 63, 65; see also Docherty, *Reading (Absent) Character* 82–84, 262–66). This principle of metamorphic identity that Sukenick has admittedly learned from Emerson, Kafka, Henry Miller, and the psychoanalysis of Wilhelm Reich destabilizes the concept of character. In *Up*, vaguely defined couples trade identities and names, triggering chain reactions of readjustment in the other figures. A leading character in *Out* (1973), referred to as Carl, unexpectedly denies his identity: "Carl isn't my real name. That's because I'm not real. I'm only trying to be real. Like a character in a novel" (26). Yet he credits himself with more "personality" than a reality-based figure like Donald, a Sukenick persona. In this reversed ontology, characters gain more "reality" than their authors once they learn how to reinvent themselves narratively. The implied author's imagination needs to be periodically replenished by resourceful characters who allow "his mind to wander play with possibilities [. . .] inventing stories about himself" (105).

Self–re-creation, predicated on the recognition of alterity in identity, promises a measure of individual freedom. But, as the narrator in *98.6* (1975) warns, this process can also substitute defensive pseudoidentities for the identity one cannot invent: "[I]t's not so easy to find yourself the self beneath the hard crust the hot magma beneath sudden eruptions of feelings the source the base on which everything rests. Not easy to find and impossible to escape. [. . .] The recoil moves in the direction of character. Characterization. Caricature" (80). In *Long Talking Bad Conditions Blues* (1979), interchangeable couples (Carl and Charlene, Drecker and Veronica, Victor and Charlene, Veronica and Bennett) live depersonalized lives, connecting like "peripheries" to other "peripheries" (28). Their relationships obey an "economy of limited resources" (29) in which individuals are mismatched and incomplete versions of one another (Veronica is an either "on" or "off" person, in contrast to Victor who is always "off" and Carl who is "half-on and half-off"). A re-lentless either/or logic that mimics digital syntax condemns people to a "plotless existence, a life beyond judgment, a form of unreflective bliss" (Ruthrof 197). In *Blown Away* (1986), self-contradiction plagues everybody, including the outlandish narrator, Dr. Ccrab, the novel's "Mentist, Astrological Consultant and Syndicated Horoscope Commentator" (26). Dr. Ccrab's ambition is to reduce life to a story line, but his efforts to "plot the future" turn out incoherent because "the future has its own plot and our petty scenarios are often lost" (11).

But imaginative projection can prevail where conventional prediction fails. Dr. Ccrab's redeeming discovery in the novel is a *radical form of narrative metamorphosis* that restructures both self and world. Improvising and projecting (a portion of his nar-rative is told in the future tense), Ccrab manages to upset the fateful order of real-ity that "always leads to doom" (22). Triggered by recurrent seizures, his subversive "intuitions" create a metamorphic counterreality in the "order of fate." His own identity splits into that of character and narrator, enhancing his capacity to inter-vene on both sides of the plot he constructs. Dr. Ccrab causes similar "bifurcations" and "fissions" (to borrow Sukenick's terms from *The Death of the Novel* 163) in the cast of Hollywood characters he associates with. Some of them (vampiric filmmaker Rod Drakenstein, porn producer Strop Banally, demofascist scientist Dr. Frank Stein) migrate into *Mosaic Man* where they participate in a conspiratorial chase of the Golden Calf (Sukenick's metaphor for a materialist, icon-crazed culture), but their "conglomerate personalit[ies]," obtained through "identity raids, borrowing one image after another from junk media sources" (224), are successfully challenged by the "we" narrator who integrates the roles of detective (Ram Shade), reporter (Ronda Fray), reader-writer (Ron, RaMSCaN), and companion (Julia B. Frey). Though his final fate in the hands of the "multinational interplanetary conglomer-ate" (223) remains uncertain, this multiplex narrator manages to intrude a riotous "history rap" of "speculative jocularities" and "stochastic extrapolations" (206) in the grand narratives of Western culture.

Sukenick's narrative program combines thus ontological subversion with cul-tural rearticulation. *Improvisation* and *invention* are used consistently to confer an ex-periential, projective dimension to fiction. The second paragraph of *Blown Away* reads: "Now it will begin. I am the omniscient narrator. But while most tellers tell a tale already set, I tell tales that haven't happened yet. Reading my own mind I, like a crab, think sideways into the future" (9). But as the rest of the novel makes clear, invention is rarely haphazard, starting most often from preexisting situa-tions and opening them up to new experiential possibilities. Likewise, improvisa-tion does not rely on accidental randomness but rather on what Leonard Orr calls "intended randomness," that is, on planned for "accidents" and alterations that are incorporated in the dynamic of the work (105–6). Improvisation inserts an ele-ment of *surprise* in the phenomenology and structure of the narrative world. The quester in *Out* is advised by a palm-reader to "Cultivate the unexpected it's your only chance. [. . .] Hope for surprises welcome the unknown" (34). That advice is reechoed in *The Endless Short Story*: "Think faster feel more go from one thing to an-other take it all in at once" (6). Inspired by the ancient I *Ching* (*Book of Changes*), Sukenick's narrators ascribe therapeutic value to movement and variation: "It's dangerous to stay still to avoid permanent injury move at once" (*Out* 175). The narrator in *98.6* turns these insights into a philosophy of survival against oppressive cultural plots. Through "psychic osmosis," he dreams himself into the role of the Aztec sacrificial victims, confronting the aberrations of history through a "technique of imaginary extending the ordinary to the point of the incredible":

> He decides the best thing would be to play his role through. To resist the torture and keep his mind alive and play his role through in the fullest consciousness. Waiting for the unexpected the aberration the extraordinary event the one chance in a million that will allow him if he's alert enough to slip through. Putting his faith in the unknown. (8)

Ontological subversion acquires a significant function by dint of its capacity to "uncreate the forged consciousness of the mass market," while simultaneously redrawing our cultural space around more rewarding choices. In the spirit of the early twentieth-century avant-garde, *The Endless Short Story* (116) proposes an explicit, albeit self-controverting, poetics of *resistance*:

> the important things is to keep it open
> the important thing is to evade connections
> the important thing is to arrest development
> the important thing is not to look back
> the important thing is to think at the edge
> [. . .] the important thing is to butcher the harmonious
> the important thing is to jerk off good taste
> the important thing is to parenthesize signification
> [. . .] the important thing is to outthink premeditation
> the important thing is to pursue the continuum
> the important thing is a beat
> the important thing is to annihilate the important thing.

In an ironic circularity, Sukenick's avant-garde rejection of traditional frames of thought is counterpointed by a postmodern pursuit of an experiential and cultural "continuum." One side of Sukenick's project is responsive to "the poetry of pure fact," striving to recapture "the cryptic surface of experience that exists despite interpretation and beyond interpretation" (*In Form* 208); the other is reformulative, activating unforeseen connections in experience. As the beginning of *Out* suggests, both tasks present serious challenges to a writer born in the year of Hitler's rise to power (1932), educated during the McCarthy hearings and the Korean War, and defining himself as a novelist in the Vietnam years:

> It all comes together. Don't fall. Each of us carries a stick of dynamite. Concealed on his person. That does several things. One it forms a bond. Two it makes you feel special. Three it's mute articulation of the conditions we live in today [. . .] the *Zeitgeist* you might say if not the human condition itself and keeps you in touch with reality. [. . .] You're either part of the plot or part of the counterplot. (1)

Narration in the Cold War era is a comic-ominous game in which all of us participate, some to relish its violent confrontations, others to pursue the "mute articulation of the conditions we live in." The enigmatic "assignment" received by the narrator of *Out* balances a *subversive* task (suggested by the stick of dynamite he carries around) with an *articulatory* one that tries to make everything "come together."

The authorial narrator of *Mosaic Man* finds that the narratives of the post–Cold War transition are just as violent, connected insidiously to the conflicts of the old world order: "[T]he Kennedy assassinations, the Tri-Lateral Commission, Black Monday, the Unseeables, the War. It all makes sense. From Yalta to Malta, it all makes sense in the name of the Golden Calf" (260). Confronted with the televised images of the Gulf War that turn the other's culture into "a target rich environment," the narrator seeks an alternative reality in which a resurrected Elvis, the homeless black, and the innovative writer come together as versions of archetypal "Jewishness," "advanced case[s] of being human" (16).

Sukenick's own fiction has pursued a similar dialectic, seeking a delicate balance between disruption and rearticulation. In *Out*, the quest plot is interrupted periodically and reframed from surprising new angles. In chapter 8, Harrold awakes from a dream that looks much like the narrative we have been reading, to take a different tack in his quest. Fifty pages later, chapter 6 (chapters are numbered in reverse order) makes a fresh start, taking characters back to their Brooklyn youth. Not only plot and character, but also genre and language are submitted to continuous revision. Sukenick's novels rewrite various genres (existential picaresque, erotic story or film, family chronicle, political thriller, ethnic humor), adjusting them to the demands of experience. Against the negative "dissa" lingo denounced in *The Endless Short Story* ("We discern, we disapprove, we discourse, we dissent, we disdain, we dissemble, we dissipate, we disembody, we disappear. Discontinuity. Dis. Dissolution. Discouraging. Despair" [122]), Sukenick's narrators propose a language unconstrained by axioms of reality, artless and alive like "the secret code on the leopard's fur and the tortoise shell" (*98.6* 4). Sukenick's "sullen rebels against the dictionary" (*The Endless Short Story* 129) want to reform language, "proceeding by puns and triple or quadruple entendre" and establishing "the missing link between mind and body" (*Mosaic Man* 209). The "moment of the STORY gives way to the moment of the DISCOURSE," but not in order to "erase the world" and relieve the novelist of the "responsibilities that pertain to that aesthetic code [of mimesis]," as Carl Darryl Malmgren has argued (186–87), rather in order to realign the discursive space of the novel with social and psychological experience.

According to the introductory note to section 8 of *Out*, "a novel is a flow of energy and feeling as in a piece of music not a subject matter as in a newspaper article. [. . .] In OUT everything flows characters change into one another without warning or change age or grammatical person or become the actual author who lived as a kid on Avenue I in Brooklyn which should help explain the metamorphosis of Harold after he takes the subway" (352). But even as he relishes the flow of narration, Sukenick remains aware of its articulatory function. The effort to guide narration may fail ("Everything's blowing up, falling to pieces. Art dissolves back into life. Chaos." [*The Death of the Novel* 100]); but without that effort, fiction becomes indistinguishable from life's superficia. The condition of Sukenick's prose is not "amorphousness and motion" (Kutnik 87) but provisional articulation. Inasmuch as "communication of our experience to others [remains] the elemental act of civilization," the novelist must reach some "truth" beyond mere contingencies according to a method already described in Sukenick's early stories: "Start with immediate situation. One scene after another, disparate, opaque, absolutely concrete. Later, a fable, a gloss begins to develop, abstraction appear. End with illuminating formulation. Simple, direct utterance" ("What's Your Story," *The Death of the Novel* 154).

3. "STREAM LANGUAGE" VS. MIMETIC LANGUAGE: THE BENEFITS AND FAILINGS OF AN EXPERIENTIAL POETICS

Sukenick's fiction stems from two contradictory propositions that create an incessant tension at the "terminal of all the sense and senses" (*Long Talking Bad Conditions Blues* 10): The weak claim is that surfiction merges art with life ("Life is a lot like a novel you have to make it up" [*98.6* 122]); the strong claim makes of fiction a re-creative *supplement*, "an illuminating addition to its ongoing flow" (*In Form* 6). Sukenick's fiction has tested continually these propositions, pitting an "experiential" approach against a "conceptual" re-creative one. In what follows I will examine each approach as it is rehearsed in Sukenick's novels. This section deals with Sukenick's experiential poetics, the following one will deal with his concept of narrative and cultural rearticulation.

In Sukenick's understanding, fiction is a primarily experiential mode, putting us back in touch with the particulars of "our experience of the world and of ourselves" (*In Form* 209). From this point of view, Sukenick is closer to the experiential direction of surfiction (Kosinski, Hauser, Sorrentino) than to the deconstructionist one (Federman, Katz, Major). Sukenick has made this clear in his conversation with Federman: "You often speak of fiction as being a lie, as lying. I always speak of fiction as telling the truth. [. . .] Fiction, like all other forms of culture, is in process of creating our experience" (Federman and Sukenick 140, 141). For Sukenick, as for Jerzy Kosinski, "There's a place beyond words where experience first occurs to which [they] always want to return" (*The Devil Tree* 32–33). By disrupting the "rationally intelligible" aspect of narrative language (*In Form* 209), both hope to escape—in Paul Bruss's words—their "victimization by [the] language and texts" of their culture, "mak[ing] fresh contact with experience" (15). To be sure, this wistful search for a way out of socialized language is tempered by the awareness that language builds reality so that the novelist's effort should be to reform rather than discard the available narrative language, making it more responsive to the "fragmented, contradictory, anomalous, and progressively dissociating elements of our experience" (*The Death of the Novel* 90).

Sukenick's interest in a poetics of *experiential* and *compositional flow* is evident from both his essays and his novels. As a "theorist," Sukenick has engaged the creative flow of the mind against received ideas, according to a compositional model that he, like William Gass (*Habitations of the Word* 9–49), has derived from Emerson. In Sukenick's historical recapitulation, Emerson inaugurates a great American tradition of literary "resistance" that includes Wallace Stevens, Henry Miller, and Charles Olson—all proponents of a revisionistic *poetics of flow*. As adapted to fiction by Miller, this poetics characteristically relies on "a free-form style of composition whose main technique is improvisation, and the great exemplar of which is jazz. [. . .] In such a style you move from moment to moment and if you make a mistake, have a memory lapse, blow a wrong note, you either build on it anyway or leave it and do it better next time, or you lose the experimental flow" (*In Form* 6–7). By staying with the flow, the novelist can bridge writing and experience. However, staying with the flow does not have to mean passive submission to the "historical process," as George Orwell proposed in his own reading of Miller, which ended with the well-known injunction: "Get inside the whale—or rather admit that you are inside the whale. [. . .] Give yourself over to the world process, stop fighting it or pretending that you control it;

simply accept it, endure it, record it" (526). An experiential poetics such as Miller's is "productive" precisely because it is attuned to the world process, but as a re-creative supplement rather than a substitute for life. Its appeal lies in its ability "to scramble the codes, to cause flows to circulate, to traverse the desert of the body without organs" (Deleuze and Guattari 132–33).

Sukenick's novels have pursued a similar notion of transformative flow, adjusting their compositional dynamics to the flow of experience. What Sukenick's narratives have sought is an ideal "zero gap between what comes in and goes out," between the "fluid flow" of experience and art's "total response" (*Out* 279). *Up* found an early model for this type of experiential integration in Wordsworth's "confessional and auto-therapeutic" poetry: "Unity of experience equals reality of self and that alone is a tremendous source of self-affirmation" (308). *Out* also alluded to a native literary tradition that extolled the flow of things: "I can't explain the universal interest or empathy or whatever it is. From Jonah to Moby Dick. You should hear how quiet they get sort of awed. Some blues can be seen offshore" (262). The Melvillean imagery evoked toward the end of *Out* promises to return us to a metamorphic language whose "calm throbbing poignant" figures participate in a "celebration of life by bulk life."

The yearning for a vibrant language that "will connect body and spirit" (Sukenick, "The Rival Tradition" 8) is ubiquitous in Sukenick's fiction. Against a cultural environment of "constipated [. . .] no whale feeling" and "angst held static by angst" (*98.6* 25), Sukenick's narrators propose to write stories in a "concrete, innocent, beautiful" language of birds (*The Death of the Novel* 157). Ron-Cloud-Sukenick, the character-author of *98.6*, invents his own integrative idiom, "Bjorsq":

> [Bjorsq] is a little like making love a coming together. Of course Bjorsq isn't very exact you couldn't write a textbook in it it expresses nothing with any definition but that's the price you pay. It defines little but it says everything. [. . .] It's not this or that object floating in the stream it's the whole force and direction of the stream the power that moves everything. [. . .] Bjorsq is a vertical language all others are horizontal. Bjorsq is a deep language all others are flat. Bjorsq is a window language all others are mirrors. Bjorsq is rhythmic the rhythm of your pulse the rhythm of the surf. All others are clop-clop. (161–62)

Inspired by the gnomic idiom of the Dead Sea Scrolls, Ron conceives his vertical "stream language" as an alternative to horizontal "mirror languages" (161–62). The rough aurality of this language upsets the repressive silence of symbolic systems: "The important thing is never to stop talking. If you forget the words make sounds make new words. Make words that grunt scream laugh hum sob. The voice is the connection between the body and the head. Silence means you're lost" (*Out* 11–12). Sukenick's work bears out Thomas Docherty's claim that postmodern aurality has subversive, antimimetic potential:

> It attends not to the produced symbolic order of things, but listens to hear the seductive ritualistic voices which mark the labyrinthine meanderings of the process of ritual. It thus combines with the seductive and transgressive aspects of postmodernism. Further, it adds an even more explicit historicity to the postmodern work, for the aural sense is predominantly a temporal, diachronic one, while the visual is primarily spatial and synchronic. (*After Theory* 30)

But how far can a language of "subverbs" and noises take its cultural subversion? If it stays true to itself, Bjorsq can offer little besides dadaist stammerings ("Baba Baba. Baba Baba Baba. Baba Baba Baba Baba. Bjorsq" [*98.6* 152]) or parodic permutations of an obsolete pastoral rhetoric: "Fork jugs vex'd nymph waltz bicq. Nymphs waltz jig fuck vex rod bq. Hymn waltz fuck vex prod big sqj. Hymn waltz fuck sex gip v.d. bjorq. Vex'd nymphs waltz jig fuck bjorq. Futile" (*98.6* 26). Conversely, if it chooses to "express" with any degree of clarity, Bjorsq automatically submits to the repressive will-to-form of ordinary language. Sukenick's fiction is caught in this conflict between an unpremeditated *stream-language*, "responsive to every nuance of mood," and a revamped *representational language* that articulates without forsaking the emotional spontaneity of the former. *98.6* blends "horizontal" narration that skims opaque surfaces with a probing, "vertical" one that enacts unexpected metaphoric connections. The compromise between them reaches poematic heights in *The Endless Short Story*:

> It was admirable [. . .] this flowing through life as if a river totally responsive to the prosody of event you are what you do red white and blue carefree thoughtless ruthless it. He into something heavy she feeling light he groping for poetry of pure consciousness she acting out infolding of his meditation exfoliation of her desire. [. . .] And then the great white fish comes home the power the venom the anger the contempt the fear the dishevelments of Fridays are not enough the vain the vacuum the black suck of nothing dazzling who knocked up the void. (43, 45)

Most often, however, Sukenick maintains a tension between an "unspeakable, unreadable, and unintelligible [. . . w]riting not yet language," and a mimetic language "[m]ultiplied. Viral. [. . .] Sublimated. Assimilated. Dissimulated" (*Mosaic Man* 9). Like other of Sukenick's theoretical propositions, his reflections on the language of fiction are often assigned to a problematic narrator who "thinks he has it all figured out. He won't speak the language that people can understand but he thinks it. He thinks it and that's the trouble. He's so smart" (*98.6* 162). The language of "spermatic consciousness" pursued by the narrator appears problematic for at least two reasons: As a version of the phallogocentric obsession that Sukenick's novels otherwise critique, it reinforces the "paternalistic power structure" (*Long Talking Bad Conditions Blues* 100), reminding us that the 1960s counterculture was "exaggeratedly macho and male dominated," in spite of its alleged "commitment to ideals of common ownership, equitably shared labor, and an egalitarian power structure based on the New Left ideal of participatory democracy" (Dekoven 112). It is also a poor response to the Cold War world, enhancing its confrontational ethos to the point where "noone can any longer distinguish plot from counterplot" (*Out* 94).

The notion of a liberating, spermatic language illustrates a naturalistic mentality that purports to revert us back to the consciousness of animals or prehistoric humans who "couldn't talk about the kinds of things we talk about because their voices weren't connected with their brains they were connected with their bodies" (*98.6* 96). While the return to a simpler bodily semiotic sounds attractive because it retrieves desires and impulses censored by the brain, it admittedly cannot deliver "enough imagination to deal with your particular allotment of biology and chance" (*98.6* 123). Whenever they deprive themselves of their intellection, the dwellers of the experimental commune in *98.6* are reduced to "creatures of biology and chance," "completely at the mercy of what happens" (123). The rhythms of nature are a deceptive barometer of human achievement: Valley miscarries her baby in a black "rain" of blood though her body temperature remains "normal," a constant 98.6. Fiction, as "Ron" gleans at the end of a miscarried psychosocial experiment, cannot merge with the flow of life. Rather than seek the "ambiguous rewards of animality," it must articulate its own concept of reality, "with all the risks of verticality" involved in it (124).

What Sukenick endorses is neither the 1960s notion of spontaneous absorption in the "absolute, self-sufficient performing instant [. . .] unmarked by the shadow of previous knowledge" (Connor, *Postmodernist Culture* 139), nor the controlled autoreflection of the later metafictionists, but a self-conscious, process-oriented fiction capable of making informed decisions "about the medium and its options" (*In Form* 44). Fiction's claim to experiential immediacy is reexamined from novel to novel. *Up* illustrates the seductive side of the experiential approach, but also some of its problems. The vitalistic philosophy that inspired the 1960s concept of "liberated" performance is shown to have questionable cultural outcomes. Not only Strop Banally—"the Albert Einstein of pornographic film" (38) who turns the sex revolution into a lucrative business—but also some of the narrator's "revolutionary" friends allow their youthful existentialism to be exploited and commodified. The narrator's own idea of experiential immediacy is fairly self-serving, reducing the flow of experience to his activities as a writer, "recording whatever happens [to his circle of friends] so they're all characters in his novel including himself" (*98.6* 68). Still, this self-reflexive definition of experience is not left unchallenged: "[F]iction isn't confession. You and I may be interested in your tribulations and so on. But the reader. To the reader this sounds like a maudlin exercise in group therapy" (*Up* 55). Though the character who articulates these objections is a reviewer stuck in a canonical version of modernism ("Something in the first person, as if Bloom were writing Proust in order to recapture the American myth like Faulkner" [57]), his criticism remains partly valid. The questions he raises are later echoed in a debate between the narrator and his friend Otis Slade. Reacting to Ron's Wordsworthian definition of art as "a process of self-creation," Otis urges his friend to "get past the words [he] uses" and take a socially responsible approach to the "brute facts of life." Ron's self-defense sounds lame at first: Like novels, the "brute facts of life" enumerated by Otis (Hitler, Franco, the Siberian Gulag, the Greek prison islands), are "all words and nothing but words. Are we children reading fairy tales or men trying to work out the essentials of our fate?" (223). But he is right in a larger sense: Through his awareness that cultural history unfolds as a succession of repressive "fairy tales," the novelist is better positioned than the fact-finding historian to challenge these "grand narratives" and seek a more honest rapport with experience.

The recurring nature of these auto-disputations (*Up* contains a self-review, literary and aesthetic arguments, passages from the writer's journal) suggests that Sukenick's fiction has been from the beginning engaged in a critical evaluation of its experiential poetics. The autobiographical, self-reflexive focus in *Up* allows two broader explorations to take place: One retraces the effort of a rebellious generation of New York intellectuals, coming of age at the height of the Cold War, to stretch the carefully policed boundaries of cultural discourse; the other foregrounds a writer's struggle to forge a new literary consciousness at a time of existential and cultural stagnation. The concern with the compositional process is explicitly linked to the writer's campaign to salvage experience from "the impositions of official history" and "prefabricated [. . .] Literature" (Sukenick, "Autogyro" 290). *Up* introduces the basics of an experiential poetics but also tests its limits, highlighting the paradoxical mixture of lifelike authenticity and narrative manipulation involved in fiction. The impression of unhampered "flow" is created through lack of punctuation, run-on sentences, character and plot metamorphosis, but these techniques can alternatively suggest experiential immediacy and the artificiality of fiction.

Sukenick's subsequent novels approached the idea of experiential writing with a dialectical understanding. Against an earlier celebration of the spontaneous "surf" of composition, Sukenick's fictions have gradually balanced *chance* and *structuration* as suggested in the following passage from *Out*: "[H]ere we are in the middle of our book speeding along on the breaking crest of the present toward god knows what destination after the first word everything follows anything follows nothing follows the world is pure invention from one minute to the next who said that" (117–18). The balance here is still tipped in favor of ad-lib improvisation. But even as he fantasizes about writing "a book like a cloud that changes as it goes" (136), the character-author has a hunch that improvisation and pattern are intertwined. More self-conscious than previous Sukenick narrators, "Ron" breaks the flow of narration with eye-catching devices (blocks of text with a fixed number of lines, reversed chapter numbering) that foreground the patterned *arbitrariness* of fiction. Only a strategy that balances flow and structure, experience and artifice can get Ron "to the place where the ocean of the unknown begins" (136).

The best illustration of this provisional balance between conflicting narrative approaches is the technique of "psychosynthesis" in *98.6* (180). Based on "the law of mosaics [which is] a way of dealing with parts in the absence of wholes" (122), the technique of psychosynthesis replaces rigid cause-effect enchainments with "luminous moment[s] of coincidence" (180). The whimsical tapestry that Joan pieces together using ten different yarns is a good metaphor for art's delicate balancing act, intermingling experience and ideas "in the nonobjective patterns of weaving." Joan's tapestry follows an unpredictable course "because she feels completely openended" (82). In the long run, her weaving manages to turn the chaos of life into "a vibrant landscape at the end of [which] the sun sets" (105). Sukenick's own novel is closer in spirit to the female art of "piecing and writing" described by Elaine Showalter (226 f) than to any phallogocentric notion of ordering. *98.6* responds to the hedonistic *Zeitgeist* of the 1960s, "racing like a wheel out of contact with the ground" (9–10), with a "patched-up" narrative that seeks provisional points of convergence in contemporary culture. The first part ("Frankenstein"), which has the appearance of a disconnected writer's diary interspersed with journalistic clips, reflects the Frankensteinian disorder of the 1960s "in which eros [...] combined with power to produce a sadistic culture" (Sukenick, "Love Conks Us All" v). This is followed by the fluid, metamorphic narrative of the middle section ("The Children of Frankenstein") that recounts a Californian commune's quest for mind-and-body-harmonizing experiences. After the collapse of this experiment, the third part "Palestine" resorts to "nakedly imaginary" narration (Sukenick, "The Rival Tradition" 6) to solve some of the world's conflicts but also to flag further fissures in the project of the 1960s.

Both *Out* and *98.6* review critically the philosophy of natural flow as part of their reexamination of the cultural myths of the 1960s. After crossing the Continental Divide in *Out*, Roland lands in utopia: A community of scantily clad hippies in a huge natural bowl welcome him as their promised deity. The "mirage" ends soon enough as mounted cowboys charge down the hill to disperse the camp. After an unsuccessful attempt to lead the hippies across the rangeland to safety, Roland exchanges his role of heroic pathfinder for that of an ordinary movie scout, checking out sites for Hollywood. His creative flow is caught in the self-indulgent narcissism of Californian culture. The novel's own flow is undermined by a progressive "thinning" of language to one line paragraphs that dissolve in a final "O," which suggests both an exit hole and an "energy circle" (98.6 76). While subverting sanctioned reality, this technique also has an unsettling effect on the narrative's capacity to create an alternative vision.

98.6 confronts more directly the mythology of flow, opposing two brands of utopian thinking. One introduces the idea of "ancient Caja" (4), an archetypal naturalistic space like a "jungle snowstorm of yellow butterflies" and the "slow throbbing of the fountain of blood" in the body (4); the other intrudes visions of a pre-Columbian civilization of squatting pyramids, gourd-shape women, and orgiastic human sacrifice. The narrator is fascinated with the iconology of the Maya— replicated in the present-day pyramidal skyscrapers—but opposes to its deathly stasis an aspiration for natural movement and survival. The second section of the novel tests the mythology of movement and flow, tracing a Californian commune's "nature trip getting back to earth." Recruited from among the drop-outs of an oppressive urban culture, the settlers of this commune gather around their "Big O" circle, ruled over by a magnanimous earth goddess ("Big Mother") whose "genius is to let things come and to let things go. Let things flow" (123). Looking for a cure to "the particular distortions of [their] psychology" (65), they experiment with fluid states of mind that destabilize referential reality. Their life and work is informed by a concept of flow that emphasizes collective sensual improvisation. Like Ishmael in "A Squeeze of the Hand" chapter of *Moby-Dick*, they become ecstatic in the presence of organic glue (made from bull sperm and yucca sap) that can produce "tissue-like connections."

But 98.6 also interrogates this notion of organic creation. The surf imagery that punctuates the narrative suggests erosion and breakdown more than natural flow:

> It's a beautiful day but under the perfect sky hell is breaking loose along the coast the surf even from the patio sounds like a cave of water collapsing a continuous avalanche the ground shakes when the breakers slam down against the cliffs as if there were an air raid next door. Down in the lagoon the beach is completely washed away and in the cove on the other side of The Tongue *The Wave* is threatening to break up on the rocks. (133–34)

Passive submission to the tides of "cosmic orgasm" erodes the efficacy of an experiential poetics. Before long, the collaborative relations among the settlers are reduced to superficial couplings: Pairs become triangles, quadrangles, or orgiastic "piles," but seldom "magic circles." The collapse of the commune becomes imminent when even its creator-witness Ron submits to the uncontrollable flow, "a stick on the stream a wick on the steam a crick on the scheme" (158).

The failure to rearticulate the "conditions we live in" is epitomized in the outlandish edifice erected by the commune. Built from a hodgepodge of materials, this construction, appropriately called the "Monster," follows a random process of composition: Each member adds something to it at need or whim, turning the "Monster" into a postmodern hybrid, not so much a thing in itself as an *excrescence/comment* on nature and culture. Its very origin is paradoxical: Prevented by the local sanitation inspector to dispose of their waste in the county dump, the settlers decide to use it as building material until the house "begins to look like a collage made from the wreckage of a supermarket" (74). Despite Ron's idealization of the "Monster" as a "furnished womb," this haphazard structure has little to do with nature, springing instead from societal waste and regulations. Its disjointed materials and method of composition deny the organicity of the redwood tree towering above it, "Larger than anything they can invent and against which they are all measured. Not clapped together from the fragments of their world but rooted in it and drawing strength from it " (103).

But the "Monster" is not just a hideous caricature of organicism, it is also a critical revision that turns organic "refuse" into an oddly significant artistic "order." The "Monster" complicates the notion of improvisational form. As a version of process composition, the "Monster" is doubly problematic because its creators believe simultaneously in improvisation and in organic creation. (Halfway through the novel, the settlers adopt naturalistic names like Eucalyptus, Valley, Branch Bud and Blossom, Wind, Cloud, and become engrossed in the idea of a rhythmic continuum.) Constrained to abandon their work, which is finally gutted by fire, the settlers come to realize the limits of their faith in "chance and circumstance." They recognize hidden geological faults underneath the organic order of nature and growing psychological fissures within and among themselves.

The last part of the novel, "Palestine," rescues Ron's philosophy of flow from the "frantic sexuality which act[ed] as a morbid irritant" (173) in the Californian commune, finding new uses for it in the utopian state of Israel. Here phallocentric domination gives way to the postorgasmic phases of Imagination and Illumination (176), in which experience and language are attuned to each other's rhythms. Imagination becomes the big connector in the world ("what is imagination but the waves of the spirit"? [171]), but a connector that often runs wild, following a playful law of improbability. Artists must, therefore, learn to control the flow of their imagination that, like the surf they worship in a kibbutz called The Waves, creates both "stone bleach essentials" and flatness as it "slam[s] down on the landscape without any modulation" (172). What artists need is a language of evaluative "verticality" without which the world "would lose a necessary dimension it would become flat we would all become Nazis" (181). Utopia requires the right intermingling of vertical insight and horizontal flow.

The interplay between experiential fluidity and narrative articulation is explored further in Long Talking Bad Conditions Blues (1979). The imagery of flow continues to play a significant role in this novel, but its definition changes. In lieu of the fast-paced energy of living that looked rather shallow by the end of Out ("Something is moving much faster than we are Roland and Arnie think to themselves something we'll never catch up with so that when we finally get there is always somewhere else or maybe it's something that's catching up with us" [224]), Long Talking Bad Conditions Blues emphasizes the dynamics of thinking/writing. Described by Sukenick as an experiment in the direction of "urban blues, [. . .] sad, but full of energy and wit" (Meyer 142), this novel uses "the symmetries of poetic form, albeit idiosyncratically, within which to improvise the rhythms that most accurately express the unpredictable flow of experience" ("Autogyro" 294). But this flow is disrupted by incremental spaces between sentences and paragraphs that thin out the text until we reach a blank on page 57. These gaps have "physical correspondences" in the dilapidated cityscapes evoked by the novel; they also translate metaphorically the "shatter" of postindustrial culture. The novel's characters cannot partake of the experiential flow due to various blockages ("clots of knowledge") and verbal "holes" (*Long Talking Bad Conditions Blues* 12, 32). The women submit unconditionally to the repetitive surges of man's desire (65). The men try to ride the flow, imposing rhythmic "units" on the "new conditions." Most often they fail, dropping back into the "flow without pattern" (109).

In their quest for a harmonious rhythm in the midst of the general cultural dispersion, Sukenick's young rebels exile themselves on an unnamed city-island that gives them the thrilling sense of living in a space of unrealized possibilities. Several of them (Carl, Victor, Drecker) treat the island as a huge laboratory where the "new conditions" of postindustrial culture can be studied in their "subtle transitions from nothing to something in the empty air" (21). The island teaches Sukenick's field researcher and writer respect for process as against "final fact" (26). But the "new situation" they pursue turns out to be "nothing more than a crystallization of the agitated stasis of the old situation that is politics degree zero weird lapse of civic consciousness" (17). Controlled by "international superfinancial intelligence powers," the hip culture of the island converts the flow of experience into a cultural commodity.

The "new religion" that Sukenick's islanders invent contributes to this process of commodification, promoting a self-indulgent a fantasy of "liberated" desire (37) instead of true communitarianism. As Horst Ruthrof argues in a discussion of contemporary digitalized culture, "Seriality, replaceability and iterability become the mantras of postindustrial society" (195). In Sukenick's dystopia, an elaborate system of rituals preempts the freedoms of the counterculture: The islanders' couplings are regulated by marriage bureaus according to "sympathetic" genes; their sexuality is "liberated" through "death vaccines"; improvisation is stimulated through "neoindustrial calibrations"; tactical squads of comedians and peasant gurus push collective optimism; "achronic terrorists" perform stunts that block time and progress. The island's paternalistic power co-opts the "dropouts critics mystics counter culture crazies rasputins of the marginal" by instilling emotional dependency in them (90).

By the end of *Long Talking Bad Conditions Blues*, several characters attain a more critical view of the "flow without pattern" that lacks "the headwind of mind" (109–10). Unstructured flow, whether biologic or informational, "sooner or later lead[s] to total blockage" (11). Sukenick's more creative characters try to oppose a Rimbauldian notion of "decontrol[ling]" poetic flow (105) to these forms of entropic breakdown. But, on the whole, the philosophy of experiential flow fails to provide a satisfactory answer to the "new conditions" of the postindustrial age that are "fluid atraditional and constantly changing" anyway (32), increasing the embattled insularity of urban culture. *Long Talking Bad Conditions Blues* ends much like *98.6*, with the desertion of Sukenick's experimenters, their dream of "essential togetherness" thwarted by a "mop up" party of parents.

By the time we reach *Blown Away* (1986), the only defenders of the karmic philosophy of flow are bumbling prophets like Dr. Boris O. Ccrab, the "greying, balding, bellying, middle-aging" narrator who crosses the "Golden State" in a run-down whale-shape car, accompanied by his clairvoyant cat, Leo. The imagery of the flow has more negative connotations here, associated with the stream of tinsel simulacra emanating from Shaky City (Los Angeles). There is no direction or coherence to this flow: The culture is reduced to a broken rhythm of incongruous screen fantasies that even Ccrab's metamorphic imagination cannot always redeem. *Interruption* is also the dominant motif in *The Endless Short Story* (1986), affecting every level of narration. The compositional flow is fractured in various ways: One section is presented as an "interrupted manuscript," its last pages missing and the penciled scribbling smudged beyond legibility. Several other sections move out of focus, ending in a "swarm of possibilities." The authorial agency is split into conflicting personas at various removes from the site of writing (authorial surrogate, narrating character, editor, author of a "Ronald Sukenick" monograph, and so on). The world these authorial avatars construct is a paradoxical mixture of anarchic flow and disruption, as epitomized by the character of Ricardo, an anarchist who can only sustain a murderous flow of action. Other characters with comic names and pseudorevolutionary biographies (Zatzat, Jungle Jim, Dickie Dick, Kewpie Slitz, Jim Slitz Nutz) contribute their share to the general "song of slow disintegration ending with the ambiguous blessings of entropy" (109). For Sukenick the worship of the "vacuums of pure energy" is "just another Am/erican story [of] the past thrown o/verboard" (110), a desire to elude the "fatal prosody of events" by situating oneself outside the constraints of history.

But if submission to the flow of experience fails to turn the world's "insupportable [. . .] inertia" into "an expectancy" (18), Sukenick's novels beginning with 98.6 suggest a different approach that valorizes the transformative flow of narration. In the author's own account, 98.6 moves from "documentary to fantasy," resolving "the conflicts that have developed in the book" through a bold exercise of narrative imagination ("Love Conks Us All" v). The protagonists of Long Talking Bad Conditions Blues, Car and Victor, try to fill the "gaping holes" (22) of their existence with extemporaneous speech ("keep talking [. . .] if you stop talking you're in trouble" [104]), but they soon adopt a more complex strategy that detaches and reformulates experience through narration. The poetics of the novel undergoes thus a significant shift from the rhythms of speech to the complex arrangements of writing: Long Talking Bad Conditions Blues begins by emphasizing the "natural" flow of speech, replacing punctuation with rhythmic spacing and complex syntax with free association. This emphasis, which seems to return us to Plato's distinction between diegesis and mimesis ("speaking voice" and dramatic "impersonation") is subsequently reversed. The flow of language is disrupted increasingly until the narrative voice is wiped out entirely midway in the novel, as the characters undergo an experiential crisis atop Mount Medwick. Then the printed text returns, expanding slowly from half columns to full-page blocks in direct relation to Carl's growing awareness of the world-building role of narration. Writing is reaffirmed as a more accurate barometer of "the rhythms of the mind itself" (In Form 89) than speech.

Thematically and compositionally, Long Talking Bad Conditions Blues epitomizes Sukenick's doubly constituted poetics, emphasizing simultaneously flow and pattern, experience and re-creation. The roots of this poetics can be traced back to "The New Tradition" essay (1972) that prescribed a singular blend of antimimetic procedures, abstraction, and improvisation: "As abstraction frees fiction from the representational and the need to imitate some version of reality other than its own, so improvisation liberates it from any a priori order and allows it to discover new sequences and interconnections in the flow of experience" (In Form 211). The poetics of improvisation, inherited from the historical avant-garde and jazz (Sukenick's interest in such experimental techniques as improvisation and collage linkage is well documented in the "Art and the Underground" introduction to In Form xiii-xxii), accommodates "the vagaries of experience, its randomness, its arbitrariness" (In Form 19). But the compositional flow is also interrupted by arbitrary typographic and thematic arrangements that introduce the element of resistance on which rhythm is based: "Pure experience" or "pure thought" has "no sound no rhythm no meaning only movement through time the rest comes in as interference as resistance as assertion of being the first consequence is rhythm you have this stream of pure nothing interrupted by intermittent beeps of pure something" (*The Endless Short Story* 103).

By itself neither abstraction nor improvisation can resolve the "schizoid" split between art and experience. Abstract "conceptual art" is useful in challenging the "idea of the novel as a consumer item," teaching us that "fiction is an event in the field of experience that, like many other events, has the power to alter that field in significant ways." However, this mode also conduces to "false reduction/delusive essentials non-existent separation" (*Out* 127), moving "us away from that union of sense and concept, matter and spirit, ego and world" (*In Form* 42). The "experiential digression," on the other hand, claims "the continuity of poetry with experience as against its discontinuity in terms of technique, abstraction, or content" (43). While revitalizing art, this approach reduces it to a nonreflexive focus on "'real life' also known as 'fact'" (44). What surfiction needs is to balance its own nostalgia for experiential immersion with a self-critical focus on narrative and cultural articulation, pitting the phenomenological approach against the formalistic one.

Sukenick's fiction has successfully exploited the dissonance between the desired immediacy of experience and the technological artifice of art. His early attempts to take "sonic snapshots" of reality in *The Death of the Novel and Other Stories* (1969) revealed both the strengths and the pitfalls of the experiential mode. By recording unedited language events, stories like "Roast Beef: A Slice of Life" and "Momentum" promise a first-order, unretouched "reality." But a transcribed tape does not guarantee experiential authenticity: The real-life conversations included in Sukenick's early narratives were interspersed with linguistic and behavioral stereotypes that made them predictable enough. Other Sukenick narratives worked from the opposite direction, using conceptual and typographic "abstraction" to dispel the realistic illusion. Rigid numerological schemes determine the order of narration in *Out* (the book opens with chapter "$1\frac{0}{0}$"followed by chapters numbered in descending order from 9 to 0, each with a corresponding number of lines per paragraph), *98.6*, and *Long Talking Bad Conditions Blues*. Interruptions are also created by arbitrary permutational idioms like Bjorsq. Together these procedures violate the semantic cohesion of the narrative and the stability of the novel as a cultural object. But abstraction can also become an obstacle in the writer's effort "to rescue experience from any system [. . .] that threatens to devitalize or manipulate experience" (*In Form* 11). Sukenick's early work viewed abstraction as a necessary evil. In the "Brief Erotic Autobiography" section of *Up*, the narrator extolled the virtues of an experiential focus "free of the banality of the mechanical, of, literally, the cliché" (174); but then he proceeded to reconstruct his erotic "autobiography" from the only source he had anyway—a few "stilted snapshots" that belie the "truth of our experience."

More recent Sukenick narratives have enrolled both compositional modes in the service of a radically transformative poetics. Experience and articulation, improvisation and abstraction are seen now as the two interrelated sides of the narrative process. The "duck tape" section of *The Endless Short Story*—a rewrite of the older "Roast Beef" sonic "snapshot"—unfolds as a complex narrative performance in which invention and transcription mix freely in "one big, endless, bottomless story." The narrator and his female collaborator preplan, talk about, and then perform events from the story, so that it becomes increasingly difficult to tell on what diegetic level we find ourselves. The very definition of Sukenick's improvisational poetics is revised in *The Endless Short Story*. Employed in the "Bush Fever" section to "sustain the morale" of an imprisoned writer, improvisation produces no experiential immediacy but only puzzling discontinuities that can be interpreted variously as coded autobiography, metaphoric reflections, or parodies of traditional literature. Refusing to take part in the mimetic illusion, improvisation functions as a bridge to the *possible*: "the real is a matter of attention of focus   this story pays no attention has no focus   this story is everything all the time   are you grown up enough to step beyond reality" (95). Dissonant and coiled "like a great mamba in the bush" (101), improvisation articulates the "uncorked stories," the "absent unsung song," releasing us from the "tedium of prefabrication" (97). *The Endless Short Story* suggests that narration can produce a subtle "rhythm" that is "a way of paying attention without paying attention" to the "vital and irrelevant system of daily things" (99). The connection between the "rhythm of your sentences" and the "rhythm of your life" remains oblique, re-creative rather than mimetic. By contrast to *history*, *story* pursues a self-defining rhythm, a "tuning of the nerves to the score of life in all its irreducible twists and possibilities contained in the prophetic multiplicity of language in the wisdom of its evasions" (102).

Mosaic Man is likewise concerned with rescuing story from history and the "iconoclastic" "generative" word from the "viral" language of imitation that proliferates only "simulacra" (9). Resorting to a number of well-tested experimental techniques (taped conversations, mixing of documentary and fiction, improvisation in "real-time windows"), the narrator rushes headlong into an encounter with "reality at last" (33). Like "Dr. Frankenstein or the Rabbi of Prague and his Golem" (198), the Sukenick narrator wants to create rather than imitate life. But his self-acknowledged "fetish of transparence" (199) is put to test by the historical nature of his work. The "reality" he encounters in his journey through the second half of the twentieth century is not some unformed flow of experience but the "nightmare of history" itself, which threatens to write "the final chapter to [his] dreams" (68). History catches up with the narrator in two ways: as human drama ("But at our backs we always fear the growl of history slouching near," writes the narrator during the Gulf War [249]); and as "writing on the wall" (ideologies and conventions that alienate the innovative writer from experience).

Ironically, the narrator's own documentary fetish allows history to overtake and dead-end the personal story. As the narrator quips, the tape or "reel is unreal [. . .] for the true story go to the teller not the tale—the writer not the writ" (36). After some hesitations, the narrator takes his own advice, turning his novel into a "history rap" that mixes reconstruction with "[prophetic] riffs" (206). While still refusing to "make it up," the Sukenick narrator applies his critical imagination to the past, using dream, surreal fantasy, and rewriting to open up the "petrified palimpsest" of history to new possibilities (182). What these possibilities are becomes clear from three key sections of *Mosaic Man*: an early Captain Midnight fantasy with Jewish Ron in the role of the traditional cartoon avenger, casting bombs with "dream vaccines" (50) on a crypto-fascist Paris; a fantastic reverie in Venice during which a disenchanted Ron meets an earlier version of himself, experiencing a vision of the lost "sophistication of innocence" in the "riptide of history" (178, 180); and a comic-ominous rewriting of the "Raiders of the Lost Calf" at the end of the novel, which pits Ron's search for the sacred "book of life" (196) against the pursuit of the Golden Calf by an assortment of American, Israeli, and Arab agents. Common to all three sections is Ron's effort to reconstruct a world of "Inbetween[s]" (239) as a response to the polarizing "white way" that "would rescue a man from woman, rescue the world from nature, rescue civilization from negation with the explosive high-velocity-projectiles of supertechnology" (238). In a bold and befitting gesture, the authorial narrator claims a place among such "Peasant Crazies" as "Yorick, Sterne, Schweick, Charlie, Groucho, Harpo, Costello, Abbott" (224), the inspired squatters in the world's interstices, who "jack into the outlaw frequencies of the dark matter, the ninety to ninety-nine percent of matter that has been erased from the universe" (232).

4. "THE MUTE ARTICULATION OF THE CONDITIONS WE LIVE IN"

Sukenick's struggle against conventional realism does not preclude a strong interest in the "conditions we live in" (Out 1). By the writer's own admission, the innovative push of fiction should "always be in the direction of what we sense as real. [. . .] The novelist accommodates to the ongoing flow of experience, smashing anything that impedes his sense of it, even if it happens to be the novel. Especially if it happens to be the novel" [In Form 206]). To suggest, as Alan Wilde does in Middle Grounds, that reflexive fiction "resists reality" (20) is largely incorrect. What Sukenick resists is not the existential substratum of reality but our conventional representations of it. The writer's ultimate task "is to break through the literary, though usually the only way to do that is by means of the literary itself: form against form, idea against idea, choice of subject against choice of subject" ("Introduction: The Dirty Secret" 8). Sukenick's fictions are neither self-absorbed "textual games" (Couturier 215) nor attempts to "incorporate the events of his life seemingly without the mediation of fiction" (Trachtenberg 57–58), but complex negotiations of the relationship between "story" and "history."

The characters entrusted with the task of reformulating the rapport between fiction and history are not only authorial surrogates, as in the early works, but also sociologists (Carl in Long Talking Bad Conditions Blues), amateur architects (Rodia in The Endless Short Story), makers of culture high and low (a former 1960s activist turned film producer in Blown Away or a resurrected Elvis in Mosaic Man). The latter characters continue to share Sukenick's interest in improvisation and self–re-creation, but they are also more solidly anchored in the cultural praxis, in "the complications of the new conditions." The aesthetic concerns of Sukenick's early fictions are supplemented with a sharper sociocultural focus beginning with 98.6, a novel that foregrounds the significant role that narrative imagination plays in the construction of "reality." The later works are also more aware of the manipulative side of experimentation and of its periodic co-option by the popular market: "mumbling became the style intentional incoherence discontinuity ending in a virtual celebration of autism whole populations were held incommunicado agents planted in the media discouraged reading encouraged self expression" [Long Talking Bad Condition Blues 18]). If Long Talking Bad Condition Blues is concerned with the "cybernetization" of the mind, Blown Away and Mosaic Man examine the abuses of narrative imagination in contemporary culture, from high-brow literature to popular entertainment.

Attentive readers have recognized a revisionist mode of realism in Sukenick's fiction that "finds reality not by projecting subjective ideals, but by 'discovering significant relations' with what is really out there" (Klinkowitz, *Literary Disruptions* 128). To Christine Brooke-Rose "Sukenick seems almost neo-realist in contrast [to Ishmael Reed], but again in such a stylized way that the effect can be unreal" (*A Rhetoric of the Unreal* 380). In other words, Sukenick's "realism" is not putative but re-creative and "experimental" (Klinkowitz, *Literary Disruptions* 64), engaged in several articulatory tasks.

a. Phenomenological Articulation: The Plot of the "Missing Lunk"

On a phenomenological level, Sukenick attempts to reconnect us with a more "in-clusive 'reality'" beyond our "fossilized formulas of discourse" ("Autogyro" 294). His novels have been from the beginning concerned with "a question of being totally here a question of response concentration the feel of things like the landscape from a car like continual improvisation" (*Up* 314). Periodically, Sukenick's prose reaches moments of quasi-epiphanic "presence": "I've never been so totally here warmth of the sun warmth of coffee the feel of table" (314). The "simple thereness" of details, scrupulously recorded, promises to make experience leap into significance. But all attempts at transcribing events and "luminous thoughts" even as they "happen" (324) remain problematic, defeated by a haunting circularity. Consider this example from *Up*: "Sukenick" leaves his East Side apartment where he has just had a night-mare, to sit at a bar whose "solidity" reassures him ("In being so absolutely *there*, the bar communicated to me an absolute sense of my own presence" [160]). But his grasp on reality is denied by everything that follows. First, the narrator confesses that photographs of himself give him the same feeling of self-presence. Then he in-volves himself in a "regular masquerade party," interacting with people who play roles and spin stories, until he lands on his back, asking candidly: "Who am I?" (169).

Especially difficult for Sukenick are his incursions into personal history, from his childhood spent in Brooklyn's "Great Gobi Desert" (*Up* 52) to his youth caught in frazzled idealisms and "half-assed ideas about life." Sukenick's literary autobiogra-phy has emerged slowly and fitfully from *Up* to *Mosaic Man*, making rewriting its chief narrative tool. As the narrator of *Up* already knows, writing autobiography remains "unreal without an effort of recreation, an effort like that of a diver moving painfully in his leaden suit across the ocean floor" (*Up* 9). Familiar figures rise from the past "with the slight discontinuity of a home movie" (58), remaining "spooky" until the narrator manages to capture them in comic-metaphoric detail. The narrator himself passes through a painful process of redefinition, as he struggles to emerge "from the stunted shadow world of [his] Brooklyn childhood" (60).

Sukenick's narratives often bounce off the opaque surfaces of phenomenological reality. Their refusal of "transcendent" depth has not gone unnoticed: One function of surfiction, according to Richard Pearce, is to "evoke a world that is all surface, or all surf. [. . .] Everything is there on the page, where the author, his characters, his story, their stories, fragments of the contemporary world, figments of the imagina-tion, and the text itself [. . .] create the surf and ride it" (*The Novel in Motion* 123, 125). According to Sukenick's own essays, the novel's function is not to penetrate a preexisting structure of reality but to create an experiential obverse that makes pos-sible the "kind of knowledge one could never know directly but only through using wisdom winking from the other side of the wind" (*In Form* 212). This revisionist narrative hermeneutic is well illustrated in Sukenick's early novels. *Up* and *Out* tease the reader with a conjectural type of narrative that promises significant revelations. But the hermeneutic level does not offer satisfactory answers to the puzzle of post-war culture characterized by growing divisiveness and irrationality. As "speed increases [blank] space expands" (*Out* illustrates this also visually), "data accumulates obscurity persists" (*Out* 104, 163). In the words of the comic-enigmatic "Porfessor," all is finally "mazol. Mazol is as mazol does. When you find something that's mazol not the thing but the finding. And sometimes it's losing but not often it's often hard to say" (70). This kind of babble can be read as a failure on the part of Sukenick's "urban, literary, Jewish Don Quixote[s]" (Sokolov 34) to decipher postwar culture, but also as a deliberate "stubbornness to make sense [. . .] out of itself, and out of the past [. . .]. OUT is the most obstinate, the most perverse, the most unreasonable story I've ever read" (Federman, "In" 137).

Reread from the perspective of Sukenick's most recent novel that demonstrates the subversive potential of a narrative structure "mosaicked together" from conflicting materials (*Mosaic Man* 218), the early works bear out Federman's comment about the "stubbornness to make sense" as a form of literary *resistance*. Sukenick's fiction refuses to submit to a logic that is not inherent to the narrative itself. The accrual process that establishes meaning in traditional fiction is all but blocked:

> OUT stubbornly goes OUT of itself. Empties itself rather than fulfills itself (like most novels). Rather than augmenting as it progresses, rather than establishing its purpose with each new sentence, each new paragraph, each new page, and with each additional complication, [. . .] OUT juggles away its purpose and its complications. (introductory note to section 8 of *Out* 353)

The refusal of conventional "sense" is not a metafictional gimmick, but an earnest engagement with questions of reality and power. The "commission" that Sukenick's characters receive in *Out* involves a search for a power center (the "Admiral," the "Commissioner"). This quest unfolds contradictorily, taking characters from their narrowly plotted lives in New York to the open spaces of western America. As they move from one enigmatic "meet" to the next, characters learn to read the world in "psychic and visionary" ways that detour messages from their original meaning. They literally internalize the quest, ingesting and reorganizing the signs of reality in order to produce liberating messages. (Carl and Donald rearrange the letters in their bowls of "alphabet soup" to spell "The Commissioner diing.") Reality is expanded creatively through experiential rather than conceptual gap filling. The distinction is important: Abstract ideas land the main character in a ditch, feeling "a nonentity a vacuum a campground of conflicting tribes in uneasy truce" (90). Only experiential responses and internalized ideas ("Words come out of the body into the air. Then they have to come out the air back into the body" [26]) can meaningfully expand the ontological and cultural landscape.

What Sukenick's novels thus propose is not an all-out deconstruction but a redefinition of meaning as experiential and relational. During his trip west, Carl stumbles on an alternative hermeneutic. After crossing the Great Divide, he discovers (through Empty Fox, an Indian from the Black Hills of North Dakota) "one of the magical places where everything comes together" (*Out* 140–41). This omphalic zone stimulates sensual experiencing and re-creative imagination. The first approach provides "living" connections with physical reality; the second, imagined connections that transcend boundaries of time and space. The latter are especially important because they teach Carl to "think sideways" (107), in explorative images that expand reality. Halfway through the novel, the dutiful quester Carl metamorphoses into an imaginative re-creator of reality now called "Ronald Sukenick."

In order to regain the visionary dimension of fiction, the Sukenick persona must follow Empty Fox's advice and learn the art of "seeing with the mind," sharing "visions" with others (136). Against the traditional hermeneutics defended by Skuul (school/skull), which relies on "carefully sifting of data painful word by word exegesis of evidence till the theory of it all starts to crystallize that will en-//lighten us as to cause and effect account for deviation rationalize history and explain the future" (126), "Sukenick" proposes an experiential hermeneutics that connects with "rhythmic" contingencies:

> [. . .] you practice a discipline of abstraction I practice a discipline of inclusion. You practice a discipline of reduction I of addition. You pursue essentials I ride with the random. You cultivate separation I union. You struggle toward stillness I rest in movement. [. . .] I like to put things together. I don't think it matters much what they are. Connections develop meaning falls away. (127, 128)

Sukenick's narrators often seek but only occasionally achieve that interconnectedness with the sensible world refused to Sartre's Roquentin: "I look around finally pick up a large jagged pink stone glittering many-faceted with glassy fractures orange layers deeply translucent faulted to its heart I feel an indefinable connection with it I hold it up to Empty Fox good this is your stone it's good luck" (146). The theme of "knotting and unknotting" is central to this narrative phenomenology: As an allegorical Sailor explains at the end of *Out*, "a knot is a connection. [. . .] The way you tie it is first you have to know the right connection then you have to know the knot for it then you have to make it with the right feeling" (265–66). The "Sailor" reappears in *98.6* to serve as guide to "Frankenstein," entrusting the narrator with a "message" whose decoding generates creative connections with reality. Through the middle section of the novel, the "children of Frankenstein" search for a "missing Lunk," a quasi-mythic representation of their need for intersubjective connections. The quest for a missing "other" makes Sukenick's characters more involved and experimental, replacing their "straight-line" mentality with "zigzaggy reasons for doing something" (144). While for Lance, the "missing Lunk" is literal and threatening (a suspected intruder into the commune), Ron understands it as a revealing linguistic clue: "The Missing Lunk [. . .] I didn't see it I heard it. I heard it come out of your mouth" (158).

Ideally, a "good knot is not": Right connections are slippery, embodied in circumstance. *The Endless Short Story* debunks the "petit bourgeois" propensity for tracing "trivial continuities among grandiose disruptions" (48). The kind of connections "Porfessor Sukenick" proposes are metamorphic, unpredictable, and "non-narrative," strictly speaking. They do not stem from the logic of plot, character motivation, or causal relating, but from "collage linkage." They follow an aleatoric compositional principle that Roland Barthes called *constellation* (*Sollers écrivain* 40). As Philitis explains in *Blown Away*, these unforeseen connections create a "dense psychic time" that renders everything co-present. The cause-and-effect enchainment is replaced by a "sideways" thinking that, like prophecy, can sweep time and space, constructing "a vast coincidence" (20–21).

Compositionally, Sukenick's novels reassert the author's configurational function but also prevent it from rigidifying the experiential flow of fiction. *Up,* for example, reconstructs—in tentative, discontinuous vignettes—the narrator's Brooklyn childhood. Through association and "modulation" (*Up* 70), which replays a motif in a different key, the narrator achieves some narrative control, but his recounting of the past remains "experimental" and contradictory. In *Out,* the character-author alternates between "dreamy" improvisations and a more conventional questing that betrays Empty Fox's existential emphasis without giving the protagonist more narrative control. The authorial figure retreats behind aliases such as Roland Sycamore, "peeled off from the Sukenick character" (164). *Out* also hints at the agonistic aspects of narration. In a burlesque replay of the Scheherazade plot, the character-author writes furiously with electrodes fastened to his tongue and penis. Whenever he pauses, electricity is discharged through his body so that every lapse from writing is punctuated with a shriek. In the end, the author is revived by Doctor Frank Stein and taught to improvise his way out of painful gaps. Writing starts as a reflex response to an existential crisis; but as long as it obeys a panic hermeneutics, it cannot participate in a significant restructuring of reality. The character-author must learn to take control of his situation, expanding the plots available to him through invention and rewriting.

Sukenick's subsequent novels have mapped more clearly the range of possibilities between an experiential phenomenology and cultural rewriting. The movement from instantaneous recording to cultural metaphor is negotiated in stages, with the sense detail triggering a thematic transposition that never loses concreteness. Sukenick's conflicting foci are held together by a free "sliding." The metaphor that best describes this provisional balance in *Up* is kite flying. As their kite soars over Brooklyn Bridge, Ron and his friends pass incredulously the 1,200-foot line from hand to hand: "And we went off together in the chill of the late afternoon, projecting huge, beautiful homemade kites, hypothesizing an impossible, ultimate kite, heading for a local bar, happy and nothing solved" (274). In *Out,* a limping Kent eulogizes the "cultivated balance" of skating (42). Later in the book, Empty Fox initiates Ron in his philosophy of sliding-and-connecting. Ron's visionary flights are at first comically Melvillean:

> Soon I'm on the wing I'm in the ocean sky sunny whales are spouting their heads break water they slide slowly to the surface spout their long backs glide roll under with a slow sweep of the flukes one especially surfacing next to me a great dappled grey whale very old his immense back // mossy and barnacled I can see the bristly folds of his blow hole a navel that connects him with his mother the air his flukes as he sounds waving high over me like a gesture of goodbye follow me [. . .]. (157)

But the ensuing events recall the narrator back to earth, anchoring his vision equidistantly between "fact," "feeling," and "vision." The best fiction emerges from an imaginative straddling of literal and symbolical divides, focusing the "visionary moment" on the contingencies of life (see also Tatham, "Mythotherapy and Postmodern Fiction"; Maltby, "Postmodern Thoughts on the Visionary Moment").

The technique of *sliding* and *connecting* ascribes a visionary quality to Sukenick's fiction. Drawing on Carlos Castaneda's utopian psychology (see Sukenick's "Castaneda: Upward and Juanward," *In Form* 214–25, and *Down and In* 253–55), Sukenick's fiction suggests that "narrative is or can be a mediumistic form, rather than the empirical form positivism has delineated for it" ("Autogyro" 194). While the Sukenick narrators inevitably fail to discover the "omphalic" center of existence, they slowly acquire what Empty Fox—the Castaneda-type philosopher in *Out*—calls a visionary "second sight." Dr. Ccrab in *Blown Away* is an offspring not only of Castaneda, but also of "our tradition's most eminent literary wizard," Prospero. The role of the typical Sukenick narrator is to "fill what he called the verbal hole with airy provocations" (*Long Talking Bad Conditions Blues* 4), releasing the "darker powers of consciousness that energize art." He trains his "double vision" on "whales" that spout connections with "their mother air," on sliding kites that defy the laws of gravitation, or on the great interface between regional American cultures. From his strategic location in Boulder, Colorado, close to the Continental Divide, the authorial persona learns to read simultaneously the cultures of the American east and west and to feel "connected to other connections" (*The Endless Short Story* 34).

At the same time, Sukenick's fiction has emphasized with increasing clarity the difficulties encountered by a visionary narrative imagination in a period dominated by modes of cultural reproduction. In *Up*, the narrator's effort to turn his "practice work" (55) into an effective cultural response is undercut by his realization that all contemporary fiction is "a tissue of petty lies" (53). *Out* acquaints the character-author with Empty Fox's notion of visionary imagination only to jolt him back to a more pragmatic (less "clairvoyant") notion of fiction. The narrator in *98.6* extols the "powers" of narration to extend "the ordinary to the point of the incredible," but his rewriting cannot always circumvent "the negative hallucinations of our culture" (11). The characters of *The Endless Short Story* strive to come out of their "boxed" plots (53), connecting with a larger narrative. But the "big picture" they discover "is that there is no big picture" (39), or that the "big picture" is history itself that—as in *Mosaic Man*—links the fate of the postwar writer to the grand plots of the Cold War and the materialist pursuit of the Golden Calf. In order to free individual story from repressive history, Sukenick (whose name recalls the Polish *sukiennik*, "cloth weaver" [*Mosaic Man* 114]) must rearticulate both individual experience and the cultural "big picture," "opening doors" in dead-ended history (170).

b. Cultural Rearticulation: The Uses and Abuses of Narrative Power

In its attempt to bring "experience and reflection unified in charting flow" (*The Endless Short Story* 58–59), Sukenick's fiction has to confront the tension between the fluidity of phenomenological reality and the rigidity of the ensuing narrative plots: "Evidence is ubiquitous suspicion is pervasive fantasy hardens reality sweeps away. / I have an overwhelming intuition of a plot one more clue one more theory and everything falls into place. / Will fall. / Must fall into place. Oof" (*Out* 115–16). Sukenick's hesitations dramatize a genuine suspicion of narrative conventions. In a vicious circularity, the "official version of life [is] compiled by some bored clerk or statistician roughly on the basis of crude figures highly modified by political necessity and mostly conditioned by remnants of Victorian novels floating through his underdeveloped imagination" (*Long Talking Bad Conditions Blues* 7). Traditional literature participates in the process of cultural construction both as a symbolic technology and an ideology of ordering. Its reifying effects are well suggested in the image of the infernal autoclave that devours placentas and stillbirths in the maternity ward of *Up*. The product of this universal swallower is a test tube reality, calculated to render incontrovertibly certain essential facts. Here there is no evasion, no waste, every event is measured, every response is meaningful. An aseptic world with no tolerance for the fiction people whisper to themselves. It's a re-lief. Still it depresses me. It's not human. (69)

Intimidated by this "huge sterilizer" at the center of his adolescent world, the narrator welcomes the first "unaccountable spot of blood" (68), hoping that this "incontrovertible fact" of life will undercut the intolerant order of abstraction. In his own budding work, however, he learns to play the terms of this opposition ("essential fact" vs. "fiction," phenomenology vs. "test tube reality") against each other. In Sukenick's own poetics, experience and rearticulation collaborate to create a rich "reality" from the "vital interchange" of "mentation" and "observation" (*Long Talking Bad Conditions Blues* 5–6).

The prevailing narrative representations create a screen of "negative hallucinations" (*98.6* 11), distancing the writer from immediate experience. Therefore, in Sukenick's fiction the pursuit of experiential freedom is inextricably bound to cultural and linguistic revision. Sukenick's novels reconstruct the biography of an experimental generation growing up during the 1950s, at a time of "suffocating restrictions" (*Up* 9). The story of a generation's revolt against "the perpetual conspiracy of inappropriate institutions" (225) is told against the backdrop of a "collective narrative about the rise, decline and future prospects" of artistic and cultural innovation (*Down and In* 5). This twofold plot is easier to recognize today, after Sukenick has commented on it in his "docunovel" *Down and In: Life in the Underground* (1987). His main contention is that "underground innovation" (a term he prefers to that of the "avant-garde") has endured "because it is not the result of a willed strategy, but responds to an unchanging antagonism between the way of life imposed by our pragmatic business society and the humanistic values by which our culture has taught us to experience and judge the quality of our individual and collective lives" (5–6). Sukenick's own "descent into the underground" (the revolutionary bohemia of Greenwich Village in the late 1940s; the postmodern East Village culture of the 1960s) has been politically and aesthetically motivated, pursuing "freedom from money," the liberation of desire and thought, and the performative intertwining of life and art (6). But, as *Down and In* warns, a "too-literal take" (170) on these pursuits erodes the efficacy of adversary culture. Sukenick's own relation to the underground has never been "literal" or one-sided. Culturally, he has cleverly drawn on the resources of both bohemia and the marketplace, defining new publishing alternatives at their interface (Fiction Collective, *The American Book Review,* the *Black Ice* magazine and collections). Aesthetically, Sukenick's "campaign against Literature" ("Autogyro" 290) has been fought from within literature, in an effort to redeem the innovative aspects of writing. As the title of *Down and In* makes clear, Sukenick is concerned with the conflicting roles played by the artistic underground (adversarial outsider vs. hip insider), praising its cultural politics of resistance against the "idolater of Things, [the] consumer at the feet of the Golden Calf" (12), while also exposing "its darker, unredeemed" aspects: "the cult of failure" (245ff), sexploitation (177–78), anti-intellectualism, and cultural anarchy.

The narrative of cultural and artistic resistance retraced in Sukenick's fiction is itself complex, "desentimentalizing" the underground. Emerging from its cocoon at the end of the 1950s, Sukenick's "brave new generation" of Brooklyn intellectuals felt "trapped in a mediocre, unreal kind of life" (*Up* 225). With little to do, they "all put [them]selves on ice, some in the army, some on Mad Avenue, some in grad school, some disapproving in the anonymous slums of the cities" (175). Their social disobedience took most often (self-)destructive forms such as the cultural purgation advocated by Yissis ("We must function as social laxatives. Elimination is our duty" [112]) or the "elimination" of one's own work welcomed by the amateur painter Otis Slade ("It frees you from your possessions, it tends to flush you out of your hole" [222]). Another character touts deconstructive parody: "Since you can't take what you're doing seriously you turn it into parody. It's a doctrine of no-think. Nobody with any intelligence believes anything any more" (95). These superficial forms of revolt conceal a narcissistic preoccupation with power images. As Ernie Slade urges, "It's time to drop our incognitos and take a stand. You've got to identify yourself. Things are getting polarized. There are no neutrals. Looking nondescript just makes you everybody's enemy" (208). But taking sides in the polarized world of the Cold War 1950s and the anarchistic 1960s can lead to the betrayal of one's personal beliefs. Tired of the "old alienation crap" (51), several of Sukenick's rebels trade their radicalism on the "fringes of established institutions" for a lucrative form of cultural "salesmanship" on the side of power (285).

The Sukenick character-author fits in this generation of "left-handed" geniuses with appetence for the oblique and the contrary. His self-portrait is replete with ironic contrasts: "Ronald Sukenick: exploiter of women, sadist, guru, quick-change artist, juggler, one-man band, master of the shell game, satanic creator who thrives on destruction and deconstruction, descendant of the trickster god Hermes, fleet-footed surfer" (Pearce, *The Novel in Motion* 124). Son of a Brooklyn immigrant, "Ronnie Suchanitch Sukanitch Subanitch [. . .] Sukenick" (97) feels at home in his East Side bachelor apartment surrounded by "images of heroic isolation, types of exile and self-exile, Kafka, Joyce, Lawrence, Melville" (*Up* 265). An underground bohemian rather than a revolutionary (he loathes the "chalk slogans and bumper stickers" of political activism), he acquires nonetheless a subversive renown through his writing. In the eyes of a government agent, his first novel shows "what we call tendencies, Ronnie. Definite tendencies" (37). Drawing on his "schizoid genius for lucky blunders, convenient ignorance, accidental last minute narrow escapes, and a knack for doing everything ass-backwards" (105), Ronnie's fiction challenges sanctioned reality through imaginative invention and "rebellious farce" (38).

For the Sukenick persona fiction is not escapist but reformulative, engaging us in a creative dialogue with history: "*Liberation*, maybe. *Escape* never. There's a big difference. It's a question of synthesis rather than a negation" (*Up* 184). Narrative synthesis, "incorporat[ing] negation in the affirmation," is what surfiction seeks. But its acts of imaginative rewriting are constantly undermined by a cultural space of simulations. Sukenick's books are underwritten by a tension between two seemingly irreconcilable modes of narration, fictional and visual-electronic. As understood by Sukenick, the fictional mode is improvisational, transgressive, allowing a certain degree of artistic intervention in reality. By contrast, the visual-electronic discourse is reproductive and uncritical, treating everything as a skeletal story ("TV is total Aristotle, the story at all costs" [Charles Newman 130]). While innovative fiction emphasizes the transformative flow of composition, cinematic or electronic discourse exploits fetishistic appearances: "[T]his is a Huge Production. // It's big. Very big. It's got everything. [. . .] You're in it. I'm in it. We're being filmed right now the cameras are rolling all the time we have set them up all over the world" (*Out* 209). The job of the innovative novelist is to disrupt/transform this "culture of 'monstra-tion'" and "scenic hallucination" (Baudrillard, *Forget Foucault* 22, 40).

Like the personal, recollective narratives in Pynchon's *Vineland,* Sukenick's fictions strive to break through conventional discursive screens to the repressed data of expe-rience: "Our lives are an appalling slapstick. On the other side of the TV screen real blood flows, I've got to get out of this" (*Out* 115). But they are also troubled by the re-alization that the two narrative modes, one reproductive-manipulative, the other re-creative, are facets of the same process, and that innovative fiction cannot entirely escape the mastery system of representation. Sukenick's novels often illustrate a "messy" poetics, alternating imaginative narration with "scenic hallucinations." In *Out* a prototypic couple experiences a sequence of erotic encounters and squabbles, only the names of the characters change (Rex and Ova, Harrold and Trixie, etc.). By con-trast, when the storyteller succeeds in overcoming his sense of "unreality" through imaginative narration, his language becomes "dangerous but exhilarating always tit-tering on the edge of control" (*98.6* 27). The Sukenick narrators find it difficult to sus-tain their imaginative effort for more than a few sections. Halfway through *Out,* "Roland" trades the role of path-breaking novelist for that of movie scout. Hugh Der-rekker's "epic-mythic-comic spectacular," with the South American god Titicaca in the guise of a gang-raped Iowan woman, "so you get the Precolumbian elements side by side with a kind of peasant populism" (204), badly needs a "finder" to explain "what the hell is going on in the movie." The traditional task of the path-breaking novelist was to provide a lead through a plethora of details, making order out of chaos. Sukenick's explorer-novelist fails to do that because he is too immersed in a pre-scripted "celluloid unreality" (*98.6* 11); also because he views ambivalently the medium of writing that combines "acquisitive behavior" with "nonacquisitive acts of intellec-tual energy, possession and [imaginative] assimilation" (Weiman 183).

Sukenick's fiction illustrates both aspects, exposing "imperial"-"acquisitive" ap-propriation and practicing "imaginative assimilation" (re-creation). Like Jerzy Kosinski, Sukenick demystifies the thin dividing line between these two aspects of the narrative production. His narrators confess to a culpable weakness for manipu-lation. Their "inventions" often echo the sadomasochistic fantasies of popular cul-ture. In *Up,* Ronnie's erotic autobiography is intersected with the fictional adventures of Strop Banally, a "ferocious All-American rascal," "the Albert Einstein of the pornographic film" (38). This insouciant character steps outside the bound-aries of the novel Ronnie is writing, intruding in his creator's autobiographical nar-rative, abusing Ronnie's former girlfriends and teaching his author the "art of rape": "It's simple Pavlov, reward and punishment. Work on their reflexes. Once you culti-vate that craving you got them hooked" (299). Pornography, as Baudrillard under-stands it, is a lucrative institution that has turned an art of seduction based on secrecy and withholding into a frenzied production (*Forget Foucault* 21–22). Through ritualized repetition (all participants have to rehearse their rape scene in front of Strop Banally's film camera), human subjects are reduced to "celluloid unreality" (170) exploited socially and culturally.

Strop's cynical advice to Ronnie to "spice up" his narratives with a few rape scenes, "a little satire, a typographical trick or two, and call it avant-garde" (*Up* 300), points self-ironically to the prominent place occupied by macho sexuality in the early surfiction of Federman, Katz, Kosinski, Major, and Sukenick. As Sukenick claimed in a 1972 interview, "There is an apparent authenticity [in these scenes of sexual violence], however desperate, because it is out of our conceptual control, out of our cultural control, out of our conventional control" (Bellamy 67). His early novels tried to support this claim by allowing characters to "escape into the present" of feeling and "open-endedness." Freed from inhibitions, Sukenick's couples interact with seeming ease, expanding the boundaries of acceptable culture. But their rediscovered libido also fills them with an "immense instability": "A new beat a new rhythm is starting and nobody yet knows how they are going to answer it what kind of music it will make" (98.6 87). The gift of this "emotional flood" is ambiguous, unsettling especially for women.

Sukenick's fiction has not been oblivious to this exploitative side of eroticism, focusing critically on the "sado-masochistic explosion" in contemporary culture that, as the author put it in his 1981 interview with McCaffery, confuses "power and sex" (In Form 113). The typical Sukenick male tries to overcome his sense of social deficiency by "club[bing his] way out with [his] cock" (104). The Sukenick woman is portrayed as a self-deluded victim of an ongoing mental and physical rape: "[S]he loves the feeling of being overwhelmed she loves the repetitiveness of it and she loves loving it. [. . .] All men are rapists at heart just like all women love a fascist right. Right" (98.6 19). Clover Bottom, the wide-eyed porn goddess of Blown Away, is a composite of the sexually victimized and culturally manipulated women in Sukenick's previous fiction. Her response to the "pervasive sexual aggression" is a "streak of in-turned violence" (Blown Away 47), a "passionate submission" to the "negative hallucinations" of popular culture. Through her complicity, sexuality is co-opted in a system of simulacra that enhances the "celluloid quality of the emotion here—more Hollywood than Hollywood" (Up 170). The masters of this art of sexploitation are Drackenstein, Bottom's movie director, and Strop Banally, the "boy-genius of the communication industry" in Sukenick's first novel. In their theater of sado-masochism, sexuality can be experienced only as a "hyperreal" simulacrum (Kroker, Kroker, and Cook 203), a "hysteric reproduction" of an encounter emptied of reality (Baudrillard, Simulations 2, 44).

The fictionalized author Ronnie, responsible for creating Strop Banally, cannot draw a reassuring line between his character's pornographic productions and his own lurid fantasies that avenge his "youth of lust and misery" on the withholding "demi-vierges of Brooklyn" (93). In seeking a "prostitution of himself and the rape of what he desires" (Up 41), Ronnie comes closer than he thinks to a cultural predator like Strop. Finally, the same ritual scene of rape is replayed with variations in Sukenick's frame-novel, in the embedded Strop Banally story, and in Banally's porn films. As the novel shrewdly suggests, Ronnie's effort to control the story is itself a disguised form of cultural rape. "You don't want friends," another character chides him, "you want slaves. You're afraid to let things happen. [. . .] You treat people as comic characters" (294). Mosaic Man makes this point again, bringing the character-author Ron in contact not only with Strop Banally, but also with other cryptofascist creators, real or invented (Céline, filmmaker Drackenstein, Dr. Frank Stein, and Mr. Huge himself, the eccentric ruler of the phallic "multinational interplanetary conglomerate" [223]). Drawn to the Parisian bohemia of the late 1950s, Ron is initiated into the "Esthetic of Evil" (84) that mixes decadent aestheticism with sexism and antisemitism. Though Ron remains mostly a spectator through these "swine event[s]," his guilty voyeurism makes him feel queasy and weak, as if he's fighting off an infection" (84). The virus he is succumbing to has many insidious forms in the "thinky and kinky" (85) postwar avant-garde, from sexual violence against women to the "bracing bitterness" of Céline's texts (92).

Critics of postmodernism would probably interpret this admission of contamination as a proof that innovative fiction ends by duplicating the cultural mechanisms it critiques: "After all, to 'rape out of boredom' is a perfect metaphor for the aesthetic which attempts to destroy the clichés of life by infibulating them with the clichés of art" (Charles Newman 143). Yet Sukenick's approach is not very different from that of postmodern feminism (Joanna Russ, Kathy Acker), engaging the political aspects of sexuality and denouncing the "intersections of rape and representation, [. . .] their inseparability from questions of subjectivity, authority, meaning, power and voice" (Higgins and Silver 1). Sukenick's focus is on both the impact of sexual violence on female and male identity and the insidious analogy between literary representation and rape. Already *The Death of the Novel* (1969) exposed the complicity between sexual and narrative manipulation. The narrator of that early story recounts a spicy sexual encounter between himself and an underage girl only to recant it subsequently as a ploy designed to increase the story's seduction: "How's that? Not bad? A little sex? Okay. Now let's do a retake of that, with a little more accuracy this time" (*The Death of the Novel* 51).

Sukenick's novels have also illustrated and denounced this libidinal economy of narration. The "story" in *Out* is propelled ahead by a pulsating movement: "There's been an escalation of the latest deescalation intensifying protest increasing backlash" (94). These cycles of rising and falling action, which suggest the charged atmosphere of the Vietnam era, entice the reader's hermeneutic interest. The usurpation of the real by a seductive fictional economy is very much at issue also in *Up*. While the character-author Ronnie is not always aware of his own acts of mystification, he is quick to denounce them in others. When Nancy, the ex-prude turned nude model, praises striptease as an epiphanic production that makes her truer self appear in response to the "intensity of [man's] desire," the narrator counters by calling it a contribution to the institution of "make-believe. The dream trade. [. . .] It sounds to me like a bad parody of socialism" (187, 254). Instead of defining a "truer" femininity, striptease functions as a "professional writing" for the benefit of the male gaze, always present just outside the frame. Within the generational history that Sukenick reconstructs for us, Nancy and Strop Banally epitomize the culture's growing fascination with a seductive economy of signs that bears little relation to reality. Sukenick's generation experiences simulation in both its repressive-deluded, and its compensatory–re-creative forms suggested by Baudrillard (*Selected Writings* 114). As "kids [they] already knew how to use conformity as a mask, [. . .] how to camouflage themselves in their own innocence" (*Up* 209). After they emerge from their 1950s "trance of adolescent ignorance," Sukenick's characters embrace insidious forms of make-believe or cultural "salesmanship" (Nancy's pornographic "dream trade," Finch's dog food advertisement, Otis's job as orderly among the rich "barbarians"). The Sukenick persona shares with them a somewhat self-indulging desire to "make up [his] own version of everything as [he] goes along" (210). But he has the candor to expose his manipulations and the courage to unlock the uncontrollable potential of his "inventions": "[O]f course all this probably sounds wacky to you. That's because none of it is true. The dynamite stick's a dud. Light the fuse and you will see. Or maybe you better not. Maybe it'll blow your head off." (*Out* 2).

In *98.6*, seduction, simulation, and the fantasy of control are intimately connected. This novel probes the roots of our postmodern condition, exposing the "grotesqueries [...] of the vast cultural failures of the recent generation" (32): invasive technology, sexual and political violence, cultural simulation. (Sukenick quotes Andy Warhol's "a pop person is a vacuum that eats up everything" [45].) Rape becomes a literal nexus between sexual and cultural domination. This transformation is essential for a successful rereading of cultural violence: The "Insistence on taking rape literally" allows a "conscious critical act of reading the violence and sexuality back into texts where it has been deflected, either by the text itself or by the critics: where it has been turned into a metaphor or a symbol or represented rhetorically as titillation, persuasion, ravishment, seduction, or desire" (Higgins and Silver 4). Rape is in *98.6* a ubiquitous paradigm that connects the ritual sacrifices of the Aztec priests, who submit captive virgins to the deathly embrace of "promised gods" (8), to the modern forms of "mind rape nature manipulation exploitation control" (176) summed up in the "Frankenstein" trope. The other literal-metaphoric term evoked in *98.6*, "colonization," makes the connection between cultural violence (the bloody conquest of the Aztec Empire) and anal rape even clearer.

The novel's first section ("Frankenstein") abounds in images of captivity, torture, and rape, with the men playing "god the father" and the women sacrificial roles both in the past (the Aztec death rituals reimagined by the narrator) and in the present (Sharon Tate's murder evoked in press clippings, the relationship between the unnamed diary-keeper of the section and his beautiful "witch" [13]). Sukenick's female characters define their identity in love by getting "colonized," "torn to pieces" in "a long ongoing rape" (19). Even the liberation pursued by them in section two, "Children of Frankenstein," ends by reinforcing the fantasy of control and submission. The Californian commune collapses under the pressure of the mysterious "Slaughter that has started again" (140), pitting males against females, insiders against outsiders, and "gooks" against "natives." Sukenick refrains from direct commentary, letting the contrapuntal energy of his quotes and illustrations do the work. Thus a Joanna Russ ironic comment from *Village Voice* ("leave sado-masochism to the gifted few who choose it voluntarily and can make something of it what most of us do is just dreary" [48]) is surrounded by descriptions of gang rapes, sacrificial murders, and genocide. Similarly, a quote from Ed Sanders's book on Charles Manson, which suggests that the role of sexual orgies is to tear "down the mind through pain persuasion drugs and repetitive weirdness—just like a magnet erases recording tape—and rebuilding the mind according to the desires of the cult" (49–50), is juxtaposed to images of mass sacrifice during the Spanish conquest. Such juxtapositions bear out the feminist argument that "in androcracy making love *is* making war," a "means of asserting male dominance" (Eisler 15).

In *Long Talking Bad Conditions Blues*, the system of male power that makes sexual violence necessary comes under questioning, and so does its matriarchal counterpart. Carl writes a report that disparages maternity, urging women to adopt the "path of power," paternity. But his own relationships suggest a sulking dependency on women interrupted by occasional fits of adolescent rebellion. Some of the women struggle to develop an alternative to the pervasive system of male rationality, emphasizing analogy over polarity, fluidity over boundaries, and responsiveness in place of phallic aggression. But as "feelers," women continue to function as metaphors of unrepresentability, as "dull null[s]" (58) that drive their men into a frenzy of invective whenever they feel trapped in "phantoms ambiguities contradictions paradoxes un-certainties the maybe in her yesno system" (41). Their relationship to power remains problematic: Veronica is uncomfortable with political power, pursuing solely the power of orgasm, even though this leaves her "an empty hole in the nameless abyss" (102). Charlene submits to sadistic abuse on the false pretense that her work as an escort gives her an identity ("I'm loved therefore I exist" [69]), helping her become a "factor" in the "world of endless possibilities" (64).

Long Talking Bad Conditions Blues employs the island metaphor to dramatize the dera-cinated insularity of postmodern culture, but also to suggest an alternative to its "ac-celerated shatter": an active intermingling of human "units," each capable of "self-invention" while also of giving itself to a reciprocal relation. Blown Away focuses on the two narrative vehicles of contemporary culture, film and fiction, to denounce their participation in the institution of "make-believe" but also to assess their refor-mulative potential. Hollywood is driven by "masters of the [fabricated] obvious": Dr. Boris O. Ccrab, "foul-smelling, middle-aged [. . .] mentist" (28) and subject-finder, Rod Drackenstein, producer-director of hip porn films, and Victor Plotz, the "hack screenwriter" who always wanted to be a novelist but has to content himself with "moving plots" (144). The three compete to gain control over a powerful medium of mystification ("what you see with your own eyes is fake" [28]). In order to corrobo-rate his gift for prediction and avenge himself on his unscrupulous friend, Ccrab se-duces a beautiful UCLA student and sells her to Drackenstein as "Clover Bottom," a new porn starlet. He subsequently "predicts" her fate, giving her a Marilyn Monroe identity and weaving a twisted plot around her and his other friends. Drackenstein collaborates in this makeover, casting Clover Bottom in a crude form of "ciné verité" that mixes improvisation, parody, and a trite story about a teenage prostitute inviting gang sex and audience voyeurism. Conceived as a loose parody of The Tempest, Drack-enstein's movie claims to be culturally "liberating." What it delivers, after the revi-sions operated by Victor Plotz and the studio director O. U. Miracle, is orgiastic sex mixed with "Shakespeare in La La Land" (77). Though their initial motivations may be different, Ccrab, Drackenstein, and Plotz collaborate toward creating what Robert D'Amico calls a "pornotopia," a "male universe [. . .] occupied by continually orgas-mic women who inexhaustibly entertain continually potent men" (87). In this an-drocentric universe that demands and punishes female submission, the object of manipulation is Cathy June alias Clover Bottom. "Numbed by simplicity, stunned with easy pleasure, [. . .] blank look on face at words of more than two syllables" (Blown Away 133), Clover Bottom allows herself to be used in a degrading game of sex-ual provocation. As a "sexploitation starlet" (151) whose only "reality" is that of the movies, she is destined to end cracked up from a "pharmacopoeia" of drugs (153) or in a plane crash. The only plot revision that the Drackenstein team can offer after her death is one in which Clover Bottom is "resurrected" as an icon for the "Consumers World of Zomboid Shoppers" (162).

Not only Clover Bottom, but also her male handlers are caught in a loop of nar-rative fabrication that begins with a "prenovelization" (the screen plot), goes through a movie production, and ends with a "renovelization" of the screen story. To emphasize this self-reflexive circularity even further, Plotz's "renovelization" is called "Blown Away, a Californian Novel," just like Sukenick's own book. Blown Away warns us that innovative fiction is not impervious to manipulation by popular cul-ture. The three framers of the Clover Bottom story use improvisation, temporal disruption, and narrative prophecy as trendy techniques that substitute illusion for reality. When Victor Plotz leaves both the filmscript and his renovelization unfinished, dying in a car accident, a "cockamamie avant-garde writer" is hired to finish it as "experimental fiction. But experimental fiction that everybody can understand" (138). Genuine innovation is excluded from this market-driven emplotment that keeps "everything at the moral level of a Walt Disney animation" (139).

But *Blown Away* also illustrates the revisionary power of narration. Another fortuneteller, Madame Lazonga, urges Ccrab to take control of his predictive plots: "Use your own power. We all make up our own scenarios, that's what it's all about" (58). Under her influence, Ccrab's definition of "fortune telling" changes, becoming attuned to the voices of alterity. His job now is "to listen to all of them, to listen carefully and to hear the ones most people can't hear because they don't listen. And to always believe them" (60). Uttering his "predictions" from a double diegetic position, as character "Crab" and narrator "Ccrab," he is able to intervene more meaningfully in "the occult crosscurrents of experience" (128). Since his visions as narrator and character tend to diverge, the narrative is pluralized. In Ccrab's rewriting, each character develops a creative side, becoming more human. Clover herself enters a phase of self-awareness, disrupting the script of her male handlers with her need to be more than a "fake, a fraud, a cartoon" (142). Drackenstein is also beset by strange dreams in which his twin brother advocates an art of silent "unwriting" that poses a challenge to his own pursuit of cheap visual thrills: "I unwrite the cultural destiny that's been written and over written. I convert the stain of the visible to the pure radiance of the invisible. I repatriate the spirit from the realm of determinism to that of potential" (114). At the end of the novel, several characters converge on this concept of creative unwriting/rewriting: Victor Plotz plans to convert his "renovelization" into an original narrative about filmmaking that intends to let "some stain of truth [. . .] seep through on the stripped stage" (144–45). After his death, Drackenstein imagines himself keeping Plotz's appointment with Henry Miller and Anaïs Nin who teach him that life and novels are messy, open-ended. Ccrab himself thinks more and more like Sukenick, trading fortune telling for an art of narrative reformulation.

5. *"Repatriating" Fiction from the "Realm of Determinism to that of Potential": A Dialectic of Unwriting/Rewriting*

As the ending of *Blown Away* makes clear, truth begins to "seep through on the stripped stage" (145) only after the Drackenstein team renounces its fantasy of control over the Clover Bottom production, opening both her narrative and their own stories to the play of possibility. *Blown Away* thus reconfirms a basic dialectic in Sukenick's fiction that combines the effort to "unwrite" conventional representations with an imaginative rewriting that converts the "petty iconography of the quotidian" into a comprehensive act of "invention and prophecy" (*Long Talking Bad Conditions Blues* 5). But there is some risk involved in this process of revision. The Aunt Theldas among us will want recognizable and resolved stories: "Why don't you write a story like Arthur Miller? It's all true. I knew the family just put in a lot of sex they all do you can make a lot of money on one story" (*Up* 231). Sukenick's fiction teases the reader to expect predictable narratives but disrupts/rewrites them beyond

recognition. Through Thor Hamstrung, the blue-eyed janitor who is busily composing a modern-day saga about a Viking in the Pacific campaign, *Up* treats us to a
savory parody of the heroic-epic tradition, with echoes from Jack London and the
Hollywood theme of "misunderstood giants." *Out* mocks in similar ways the expectations of a Great American Novel (a heroic character, a grand narrative of exploration, symbolic significance, readability). Sukenick's fiction also parodies/rewrites
the existential picaresque, the quest narrative, the gauche-pornographic novel, the
family chronicle, and even the grand narratives of the Hebrew Bible in *Mosaic Man.*
Readers may find disconcerting this dislocation of narrative tradition. But they can
take heart after reading the exchange at the end of *Up,* between Ron's real and fictitious friends who meet in his apartment to celebrate the completion of his book.
When Ron reassures his wife, "It's *only a novel* don't forget," Lynn replies: "I understand that very well, but don't say only. It's been three years of our lives" (327). Experimental fiction is not inconsequential to the lives of those who write and read it,
ascribing them the dimension of becoming.

Rewriting also entails a degree of manipulation of experience that Sukenick's fiction carefully highlights. While rejecting "a pure form of mastery" equated with the
imagery of rape, Sukenick endorses an "interventive" form of narrative capable of reclaiming excluded experiences:

> To write out of the uncensored data of the self is to break the rules—as you note—&
> feminists are in a good position to understand that. [. . .] Anyway, the big issue for me
> [. . .] is the issue of authority—to which that of form & "reality" are secondary. Our big
> push [. . .] is to include more of the currently censored data of experience—& it is our
> ongoing struggle: as soon as a block of data is decensored, you go to the next block cry-
> ing out to be liberated into the personal, & then the general consciousness. [. . .] The
> issue of authority will finally be decided on who manages to salvage the most impor-
> tant of the repressed data of experience. (January 17, 1989, letter to the author; qtd. by
> permission)

But the surfictionist mode of articulation must remain "subversively personal, [. . .]
unruly [and] unpredictable" enough to threaten "a world that pushes constantly in
the direction of the impersonal and systematic" (*In Form* 33). Its oblique strategies
seek not meanings, but "angles and velocities and effects of effects of effects" (*Out*
90). The writer is only partly in control of the process that reperforms reality. This
fact does not release him from narrative responsibility but does allow him to share a
"quirk of ignorance [. . .] with history," not knowing what is going to happen next and
why: "Stories don't have reasons. Or if they have them they have them after the fact
like the weather. Then the reasons become part of the story" (*The Endless Short Story* 3).

Sukenick's imaginative mode of unwriting/rewriting is fully displayed in *98.6.* The
first section of the novel undertakes a deconstructive rereading of the "Frankensteinian" features of contemporary society (cultural violence, sexual domination, psychological stagnation), allowing an alternative perspective to slowly seep through:

> [E]verything is different that is it's the same but as if he's seeing it from the other side
> as if he's passed through a door or a mirror what seemed the reflection is now the real-
> ity or maybe there are two realities each the reflection of the other and he's aware or
> both. Everything is more intense imagine looking at everything through one eye all
> your life and suddenly you open the other. (52–53)

The second part of the novel tests the strengths of this alternative perspective, rearticulating the cultural scene around new narrative propositions. Ron, the intradiegetic character-author, initiates a social experiment (a commune in California) in order to provide his novel-in-progress with a "real-life" theme: "One of the main things Ron had in mind when he thought up the idea for the settlement besides getting back to earth was to write a novel about it. He feels that novels should be about real life so instead of making up some story he gets a cast of characters and invents a situation for them and he simply writes what happens" (68). It soon becomes clear, however, that the relationship between social and narrative experimentation is more complex than Ron's initial idea suggested. The book he ends up writing records the failure of a communal project; therefore, it is not celebratory but ironic of the Rousseauistic idealism that infuses both the social experiment and the narrator's effort to turn his novel into "something organic, some addition to nature" (70).

The last part of the book, "Palestine," picks up again the themes of cultural and narrative reformulation, taking a more pragmatic view of them. Sukenick's narrative imagination revises both collective history and individual autobiography, promising "composure grown out of ongoing decomposition" (167). By way of a few chance "connections" (such as Sukenick's identity of name with the discoverer of the Dead Sea Scrolls), the fictionalized author finds himself in a land of new possibilities, one of those postmodern "zones" constructed through imaginative "misattribution" that McHale describes in *Postmodern Fiction* (45–58). "[I]mprobabilities of the unknown" (*98.6* 170) expand the real, rewriting the political and cultural scene to make it more acceptable: Robert Kennedy has survived his assassination attempt, Jews and Arabs live peacefully together in "Palestine," the geography of this region features mountains, jungle, monorails, canals, perfect weather, and zebras instead of cars. Artists and scientists are acknowledged here as creators of reality, living in a kibbutz called The Wave that reenacts a ritual return to the Mother Sea. Their cultural identity is appropriately hybrid as suggested by the name of the astronomer Yitzak Fawzi who teaches Ron that "the waves are the fingerprints of the spirit on the blank page of matter. [. . .] They fill the gaps. They are the missing dimension" (170). The place itself is multidimensional, combining the "certain" with the "improbable." Pictured as an alternative zone "in which certain things that have happened have not. At the same time that they have" (180), Sukenick's metamorphic Israel fills the "blank space the clean slate [. . .] where terror is" with "moments of Luminous Coincidence between an inner and an outer necessity" (171).

But the last part also foregrounds the dramatic tensions between historical reality and utopia. The voice that articulates the utopian vision is self-controverting, split into several identities one of which is again Dr. Frankenstein. This emanation of the Age of Positivism and the Cold War (Dr. Frankenstein's nails are "claw-like and painted in V shaped stripes of red white and blue" [172]), remains a controversial speaker for the new age of postmodernity: an "illogical positivist [. . .] that says all those strange things that we do not [. . .] like to think about" and an inspired prophet who "has taught us how to create ourselves in finer and finer harmony with the rhythms of the cosmos" (172). While the Sukenick narrator is seduced by Dr. Frankenstein's hip discourse of harmonization, he carefully avoids the impression of a "seamless perfect" (188) in his own work, allowing everything to both "come together" and "unravel" (187). Through a contradictory mode of writing that redefines the novel as "bungled fragments stitched together" (188), the narrator achieves a progressive "loosening of tongues" (*Long Talking Bad Conditions Blues* 29), "decontrolling" and interplaying his narrative messages.

The struggle to rearticulate the culture's discourse has found its wittiest "interventionist" expression in *Doggy Bag* (1994), a collection of "hyperfictions" that critique and revise narrative practices from the Cold War to the hyperconsumerist 1990s. Conceived in parodic response to Eliot's *The Waste Land,* Sukenick's *Doggy Bag* pits European high culture pieties against American pop culture irreverence, recycling the expectations of both. The cover design—representing a wolf dog holding a green bag with a skull-and-crossbones drawing—suggests that recycling within and between cultures resembles a process of cannibalization. The closing piece, "Death on the Supply Side," makes the metaphor explicit, attributing Western culture "a history of voracious looting and scavenging, culture feeding on itself in a progressive comedy of transformation, the spoils of conquest ornamenting the Roman Empire, Roman columns used to build Christian churches, Romanesque frescoes ripped off for baroque buildings, the Pantheon robbed to decorate St. Peter's, antique monuments as marble quarries for newer palazzi" (141). The story contrasts two approaches to consumption and death: European high culture decadence and American pop culture obsolesce. Both have the ability to survive by cannibalizing their own histories and each other.

If in "Death on the Supply Side" American pop culture compares unfavorably with "Rome's cannibalistic jungle of stone" (147) that lives confidently off its own "detritus of dead civilizations" (145), other *Doggy Bag* stories redress the balance, adopting the irreverent approach of Mark Twain's *Innocents Abroad* toward all things European. Sukenick's hyperfictions "cannibalize" European styles, from the Jamesian intellectual travelogue with its tripartite cast of roles (the American "reflector," his "companion," and his "interlocutor" [16]), to Beckett's metaphysical quests for elusive "Gott Knots" (51). The semiautobiographical narrators mock Europe's institutionalized "high culture," "detached from any communal matrix" (141). In the "Name of the Dog," the narrator recalls his first journey to Paris at the age of twenty-five, reevaluating it with the help of his philosophic bird companion, Edgar Allen Crow. If for the young graduate student "Europe was the solution" (23) to the "ordinary, mediocre, and meaningless life" of America in the 1950s, the adult writer views Europe as a postcolonial myth created by "the illusion of the colonials who, without realizing, had outgrown their colonizers" (24). As evidence, the narrator recalls his own chance encounter with Federico Fellini, which at the time amounted to little more than an overheard conversation about the name of Disney's "mad dog" (28), but which he reinterprets now as a comic prophecy of Disney's carnivalization of Western Europe and of his own successful retranslation of Fellini's carnival to American fiction.

"A Mummy's Curse" denounces Europe's "classy [cultural] conspiracy" (72) that made possible the usurpation of native tongues by Latinate languages. Echoing the search of previous Sukenick characters for a redeeming experiential language, the unnamed hero peruses cryptic inscriptions in search of the displaced "lingua franca inseparable from the body and its physical existence" (72). What he stumbles over is an idiosyncratic, unstructured language that defies understanding much like Bjorsq. (Sukenick illustrates this with sequences of random typographic signs and passages of alliterative language that bridge individual stories.) Opposed to this "cryptic language we don't need to understand until it interrupts, a language that understands us" (77) are the frozen icons of established culture, ruled over by the

Sphinx (a riddling "sphincter, holding things back" [75]) and by the dog-faced god of war, Set. In a parodic rewriting of the archetypal descent plot, the hero ends up in an Egyptian temple where he is about to be sacrificed to the gods of "Total Control" (76). A narrator steps in to rescue him but the latter soon finds that the entire world is a (post)colonial outpost of the international death cult of the Sphinx. As he follows his dismembered hero (by now clearly identified with Osiris) to the "Netherlands," the narrator records other aspects of the "constipated Nazi Sphinx stasis" that has taken hold of Europe. Mummified European culture still has one redeeming value: the power of eccentric perception that the narrator finds exemplified in Vermeer's puzzling *View of Delft*. The paradoxical distribution of light in this painting (sunlight touching only a few barely visible houses in the background) teaches the narrator a form of perception that foregrounds the invisible backside of the real.

This lesson in denaturalized "seeing" ("See sight. Only the invisible makes visible" [86]) is also applied to various aspects of American culture. Sukenick's "hypertexts" parody not only European traditions but also the American pop narratives that cannibalize them: spy adventures among ancient ruins ("A Mummy's Curse"), interactive porn with a classical gothic theme ("The Burial of Count Orgasm"), vampire narratives in cyberpunk key ("Doggy Bag"). Sukenick's own fiction is cannibalized when the familiar authorial surrogate, Ronald Sycamore, announces in "Doggy Bag" that the world is "infiltrated" by the monstrous creations of "idea doctors" like Dr. Frank Stein of *98.6* and "guard dogs of the mind" (49) like movie producer Rod Drackenstein of *Blown Away*. Sukenick's alter ego spots "Frankensteinian creations" (43) everywhere, disguised as TV zombies and members of the "Get Set" (45). Ronald's mission in "Doggy Bag" is to find an antidote against "the white voodoo mind control" (34) of mass consumption. Through a Kissinger look-alike, he obtains a blood-draining substance called Thot after the "ibis-headed" god of "all things written" (41), but this antidote fails in the United States because it is turned into a "tepid mush" (46) for yuppie consumption. The last section of "Doggy Bag" depicts the dystopian society of generalized white noise that emerges after Zombies take control of society. But it also prescribes a stronger form of Thot as remedy to America's "white death" (70): a "wolfish" language of self-expression that—in a parodic revision of Eliot's famous homily at the end of *The Waste Land*—pursues simultaneously the "datta" (the "leftovers of reality"), "yatta" (uncontrollable desire), and the punning "intelligence of music."

Three other stories illustrate this triple focus, deconstructing, punning, and transforming recognizable plots in order to make room for new "datta." "50,010,008" begins with the voice of an archeologist ("anthroapologist," 90) whose ambition is to find a continuous evolutionary plot in human history from the ice age to the year 50,010,008. Ironically, his discourse of continuity is interrupted by recurring textual gaps, until he is entirely displaced by an investigative writer who announces that the "anthroapologist" has mysteriously vanished. The writer's perspective is also divided, defending the archeologist's notion of a single evolutionary plot while also promoting the idea of an alternative order made up of infinite chromatic patterning. Trying to verify this intuition, the investigative writer undertakes his own exploration of European cultural sites, only to have his own text soon scattered. The story is then picked up by a narratee who announces that the writer-detective has been infected with a "metamorphic cybervirus" that creates memory gaps. The ending of this narratological fable, told in the voice of yet another narratee-reader, suggests that the

virus that "infects and revises" (101) may have been present in Europe's earliest inscriptions, preventing any rigorous interpretation or transmission of knowledge.
Therefore, the role of readers is not to decipher the story through a unified
hermeneutic but to retell/revision it, filling its inhibiting gaps with their own imaginative discourse.

"The Wondering Jew and the Black Widow Murders, or the Return of the Planet
of the Apes" plays the game of re-visioning skillfully, shifting the identity of the narrator and rewriting several plots simultaneously. The first layer of cannibalized plots
comes from American popular culture, and it ranges from Western fiction to pornographic romance and time travel. This layer is embedded in recognizable European
plots (the Wandering Jew, the search for cultural origins, sexual and racial travesties)
that force further revisions of the American material. The story begins with an
American narrator who experiences Europe "at a remove," like a ghost returning to
a place from which he had been transplanted (102). This alienated perspective is
subsequently transposed to America as seen through the eyes of a renowned anthropology professor—"occasionally Belge," "occasionally Polish," and "sometimes
Jewish" (106). In savory broken English, the old professor tells a number of semial-
legorical stories, beginning with his experience in the American academe, learning
to cope with "exclusive racisms and classicisms" (107). His education in American
mores looks benign by comparison to that of a middle-age Jewish lady whose story
he retells/invents. Gang-raped in the barracks of the Women's Army Auxiliary
Corps, Anita decapitates her subsequent lovers in a series of Black Widow murders
that make her famous. The film that she and her first lover watch suggests other
racial and gender tensions embedded in American culture. In this parody of a Western, the "bad guys" are Jewish Indians who compete with the Christian settlers in establishing their own commerce. The last story told by the professor features him in
the role of cultural detective, trying to decode the cryptic messages of the Middle
East. As his own example suggests, the enigmatic messages of history can be understood only in imaginative acts of retelling that cannibalize known plots.

Though "The Burial of Count Orgasm" pursues a simpler narrative agenda, engaging critically only one popular genre, this story spans—much like Coover's *Spanking the Maid*—a range of permutations within pornographic fiction. The difference is
that Sukenick's text is composed entirely of interrupted sentences that need the
reader's participation to be completed. If we accept this task, we can flesh out a
number of stock pornographic scenes. In a crescendo of sexual violence, Sukenick's
story parodies pornographic discourse, exposing its mindless repetitiveness ("as-
Randyagain sheagain thenRam then Randy as she again with Randy Ram jammed"
[123]). Yet like all parodies this story also reinforces the model, suggesting that
pornography draws readers into an interactive complicity.

Sukenick's ironic recycling of popular forms of fiction has been related to the
trend of avant-pop. But the same way his postmodernist "riffs" undo T. S. Eliot's
modernist "runes" (52), Sukenick's "hyperfictions" appropriate "mass-market modes
to undercut the mass market" (Rooney 10). Instead of happily embracing the "Hypertextual Consciousness of the Net," whose "virtual ubiquity" may appear radically
liberating to his younger technophile colleagues (Amerika 1), Sukenick continues to
write from a skeptical position that suspects that the Net may be an insidious replica
of the System and that avant-pop, insofar as it fails to distinguish "the real thing
from the holywood version [. . .] is liable to lead us back into the shopping mall"

("aVANT-pOP" 49). In answer to a criticism—heard all too often—that he will not "give the audience a break and sell more books by using the kind of plot and character narrative they're used to" (*Doggy Bag* 149), Sukenick has periodically reasserted his right to produce "half-baked and [. . .] interrupted" stories (Rooney 8) that challenge narrative and cultural expectations.

His most recent novel restates this argument with an urgency that turns Sukenick's "règles du jeu" (*Mosaic Man* 11) into operating principles for the post-Holocaust/post–Cold War writer. Unwriting and rewriting function in *Mosaic Man* as strategies of reformulation against a "bankrupt modern" world (188). They submit the author's heritage (European, Jewish, and American) to a thorough reexamination, discarding one-sided definitions in favor of cross-cultural interactions. Since, as the authorial narrator states in a section written after the fall of the Berlin Wall, "we're all post humous as Europeans, including the Europeans themselves. Post human" (197), Europe becomes Sukenick's chief critical focus through the early pages of *Mosaic Man*. The book revisits significant sites in Sukenick's European education—Paris in 1958, Venice in the early 1960s, Poland at the time of the Solidarity movement, and Berlin after 1989—mainly to conclude that the "great tradition of Humanism wasn't worth shit in chateau, a fox hole, a death camp, a stately state palace" (197). Ron finds postwar Europe infested with more insidious versions of the "virus" that tore the continent apart during the war: from the neofascist attacks on "Blacks, Arabs, and other 'foreigners' in the metros of Paris" (12), to "anti-semitism without Jews" (76) in postcommunist Poland, or a more general sense of "dead-ended" history experienced in the "ghost ghettoes" (189) of Europe. European artistic culture is likewise contaminated, offering the American novice varieties of decadence.

But Europe can still surprise the authorial narrator with a few "luminous coincidences" that reinforce his own iconoclastic art. In Poland, Ron recognizes a "Siamese twin" in Marta—writer, translator, and editor of an underground publication—with whom he shares the foolishness of believing in "an impossible promise to meet the impossibility of the situation" (112). In Italy, Ron recaptures briefly the brash vision of his own youthful self and is encouraged to continue his pursuit of "an opening to the unworldly in a world that can't honor it" (180). Finally, post-1989 Berlin inspires a meditation on walls and ideas frozen in concrete ("All ideas are wrong. Especially the right ones" [189]). The mutilated Berlin Wall suggests to the narrator a collapsed ideological discourse, "all hacked and pocked and pitted," written over with "multicolored multilingual writing [. . .] epigraphs, pictograms, hieroglyphs, riddles, runes, alphabetic jocularities, plays on words and words playing on themselves" (184). Though this "concrete poetry" cannot blot out the wall's "skeleton of reinforcement rods" (184), its hybridity manages to challenge the monologic discourse of the last half century.

American culture offers Ron the possibility to stake out a new cultural-aesthetic space beyond the "bankrupt modern." The authorial narrator begins by submitting the defining words/icons of American culture and of the Jewish subnarrative within it, to a reevaluation in the first three sections, "Genes," "Ex/Ode," and "Umbilicus." Echoing Burroughs's metaphor of the virus as "a very small unit of word and image" (3), Ron regards traditional representations as "viral" (9), disseminating empty imitations. Against their infectious reproducibility, section 6 of "Numbers" proposes a (re)generative concept of writing that mixes the Judaic topos of the sacred book of life with Sukenick's iconoclastic emphasis on what "is true beyond illusion" (106) and his postmodern visionarism, "a kind of praying that resembles a kind of wishing that resembles a kind of dreaming" (105). Ron seeks models for his projected "Wholly Book" (105) in alternative American culture. Its liminal spaces such as the Colorado frontier absorb Jews like Ron who have had their "ties cut" but who are willing to "extend [them]selves to re-establish connections as soon as possible" (190). Its "Peasant Crazies," from "Charlie Groucho, Harpo, Costello, Abbott" (225) to a reinvented Jewish Elvis, teach Ron the art of speaking in "conundrums, riddles, nonsense, stupidity, slapstick, wordplay, clowning, quips" (224). American popular culture also inspires Ron's Captain Midnight expeditions against the neofascists of the world and his witty avant-pop rewritings ("Great Expectorations" and "Raiders of the Lost Calf"). While postmodern America is more attractive to the innovative writer than "bankrupt modern" Europe, it is by no means impervious to the "virus." Both Ron's Brooklyn boyhood and his struggling youth in New York are marked by traumatic encounters with anti-Semitism, sexism, social discrimination. The post–Cold War America of the 1990s is the choice playfield of the idolatrous, money-driven "Raiders of the Lost Calf" and the instigator of new racial and economic wars. (Ron is especially concerned with the Gulf War and the situation of the homeless.) But unlike Europe, Ron's America still hopes to be saved by the "collective conscious" (105) of its eccentric dreamers.

The utopian-regenerative promise is even stronger in the "Jerusalem" section (118–59) that re-creates Sukenick's journey to Israel after his father's death. In this imaginative travelogue, Israel functions as a metaphor of paradoxical inclusiveness. Here "everything is happening at the same time [. . .] because in Israel the past is the present" (136). This historical copresence gives the country a "hard definition, redefinition, over-definition" (146), but also a sense of self-contradiction. The three major religions resemble the "jewel-like birds that inhabit its parks—bulbul, hopoe, sunbird—each monopolizing the attention in turn with its dazzling, singular and dominating brilliance" (146). The intercultural competition gives the country "a certain state of intensity" (122) that Ron's writer friend Shmulie associates interestingly with "the States during Vietnam which was something that was happening and not happening for a long time. Until it happened" (136).

Ron gets a taste of this "state of intensity" during his trip to the Hezekiah tunnel. Caught in the multilingual throng of people, Ron experiences a claustrophobic sense of the "Babel" of tongues and desires that reduce each other to an indistinguishable "babble" (148). He also has a revelation of the infrastructure of his own culture divided between "two tracks, the Mosaic and the Aaronian, [. . .] the Rabbi and the Mogul" (248). Both tracks are dangerous when absolutized. The "Extreme Rabbi" track leads to new prejudice: "The Israelis are now in the business of telling others they aren't really Jewish, the Orthodox tell it to the Reformed, the Hassidim tell it to the Orthodox, the white Jews tell it to the black Jews" (151–52). The "Mogul track leads [. . .] to Mr. Huge['s]" mindless corporate consumerism. Deep in the womb of the Hezekiah tunnel, Ron feels threatened by "brutal chtonic powers," "the Minotaur, Zeus as bull, the Golden Calf [. . .] demanding blood sacrifice" (155). His quest for his spiritual roots is continually detoured by surreptitious messages that send him on a goose chase for the Golden Calf idol. But the Middle East also provides Ron with a fraternity of like-minded writers: Ron's Jewish friend, Shmulie, who wants to publish a book about the humanity of the Arabs; an innovative Israeli novelist who writes in Ron's familiar self-canceling style, subverting "the virile thrust of his native tongue with a kind of semiotic hysteria" (149); and the Arab writer Anton Shammas whose "soft irony" is "more Jewish than the aggressive assertiveness of the Israelis" (149). These writers reconfirm for him the idea that imaginative literature can cross ethnic boundaries, providing the necessary antidote to the European virus of prejudice, the Middle Eastern "fratricidal Babel" (148), and the "American disease" of not thinking, just buying (151).

Ron's own option is the "generative" word (9), writing as a "nest of possibilities" (142). Revisionist writing turns Ron's search for his Jewish roots into a re-creative experience, making a new "bible" out of the "babble" (148) of historical interests. *Mosaic Man* rewrites the titles of the Books of Moses (Genesis becomes "Genes," Exodus "Ex/Ode," Leviticus "Umbilicus," and Deuteronomy "Autonomy") and some of the plots of the Old and New Testament from an iconoclastic perspective that emphasizes the progress from "Testimony," Commandments, and "Numbers," to "Writing." In similar ways, the self-indulgent babble of contemporary culture is opened up through rewriting to excluded voices: Holocaust survivors, "unmentionable" Arabs, Mosaic dreamers, homeless blacks. The narrator feels solidarity with all those "burned in the belly of the whale" (253), adopting the first-person plural in the last section of his narrative. Feeling "unhomed" not only in the "post-human" Europe, but also in the fratricidal Middle East and the American "Space Bubble" (220), the narrator makes a home for himself in the "collective conscious" of history's excluded and in the "stupid [b]ut beautiful" (261) art of rewriting that makes the "literally true [. . .] fundamentally mysterious" (105).

CHAPTER 5

Narrative (Dis-)Articulation in the "Shadowbox" of History

RAYMOND FEDERMAN'S EXPLORATORY SURFICTION

[T]he contradictions encountered in the attempt to understand and present the self in all its truth provide a powerful narrative machine. Any time one goes over a moment of the past, the machine can be relied on to produce more narrative—not only differing stories of the past, but future scenarios and narratives of writing itself.

—Peter Brooks, *Reading for the Plot* (33)

While pretending to be telling the story of his life, or the story of any life, the fiction writer can at the same time tell the story of the story he is telling, the story of the language he is manipulating, the story of the methods he is using, the story of the pencil or the typewriter he is using to write his story, the story of the fiction he is inventing, and even the story of the anguish (or joy, or disgust, or exhilaration) he is feeling while telling his story.

—Raymond Federman, "Surfiction—Four Propositions" (12)

I. FICTION UNDER THE SIGN OF SATURN: REIMAGINED TRUTH VS. FACTUAL VERISIMILITUDE

Writing for the protean narrators in Raymond Federman's fiction is like falling "into a white precipice," a heroic subjection to a "lovely lonely season in hell" (*To Whom It May Concern* 30). Stranded in "the elasticity of time," these narrators face countless writing decisions:

> So many unexpected problems. A smart guy would simply
>
> > give up
> > the whole thing before starting
> > before starting
> > to get involved with such a story

                                        before falling into
                                              the
                                             hole
                                        (feet first)
                                   (*Double or Nothing* 22)

Despite repeated failures to "stumble on the right design" (*To Whom It May Concern*
103), the narrators of Federman's successive novels keep renewing their commitment to writing, which for these "word-addicted idiots" (34) is the only sane response to the "abyss of darkness" (129) we call history. Short of taking Primo Levi's
suicidal leap into nothingness (129), the dramatized author of Federman's most recent novel relearns to speak from the "basement of [his] own despair" (30). Exchanging his nihilistic peace of mind for a harrowing sense of hope, he "once again
confront[s] that avalanche of simultaneous events we call life" (31).

Even though, as Federman has freely admitted, "writing is such an inhuman thing
to be doing, so brutally asocial, unnatural. So much against nature" ("A Version of
My Life" 64), for him writing has been a matter of necessity rather than one of
choice. Federman fits the pattern of the exilic, post-Holocaust writer. He stumbled
over writing after living the first two decades of his life "oblivious to [himself] and
to the sordid affairs of the world around [him], unaware that the experiences [he]
was [. . .] enduring would someday make of [him] a writer." But once discovered,
writing became for him an alternative to self-annihilation. From the start, Federman's fiction has been "on the verge of a great tale," circling "something really big
and profound" (*To Whom It May Concern* 82); but just as stubbornly, it has
deferred/detoured that "great tale" with its irreverent approach to storytelling. Both
attitudes, serious and playful, articulative and disruptive, are justified biographically:
"My life began in incoherence and discontinuity, and my work has undoubtedly been
marked by this. Perhaps that is why it has been called experimental" ("A Version of
My Life" 64). Federman's fiction—like that of Beckett, his acknowledged mentor in
the art of discontinuity and survival—has been about the "perception of chaos, the
survey of chaos, the immersion into chaos" (Strauss 505). A historical chaos, primarily, that could be broached only through the endless detours of a
postmodern/post-Holocaust writer: "It is as if the experience of the Holocaust is
more than language can comprehend or communicate—except, perhaps, by a denial
of language" (Sukenick, "Refugee from the Holocaust" 40). But this denial has been
attended by a reconstructive effort, recalling Vonnegut's own attempts to deal with
historical cataclysms in *Slaughterhouse Five, Slapstick,* or *Galápagos.* According to Federman's own explanation, postmodernism was invented to deal with the catastrophes
of World War II: "The prewar split between form and content was incapable of
dealing with the moral crisis provoked by the Holocaust, and therefore writers like
Beckett, Walter Abish, Ronald Sukenick, Primo Levi, Raymond Federman, Jerzy
Kosinski, and many others, invented Postmodernism to search among the dead
[. . .] in order to re-animate wasted blood and wasted tears [. . .] or perhaps simply
in order *to create something more interesting than death*" (*Critifiction* 122).

Faced with what Philippe Sollers has called an "experience of limits" (185)—the
Holocaust, exile and transplantation on the New Continent—Federman has pushed
narration to the edge, breaking open its traditional frameworks in order to accommodate a more honest response to history. His revisionist narratives begin by dis-

rupting conventional representations of history but simultaneously propose an imaginative rewriting that interplays "reality" and "potentiality." Federman's novels usually unfold in a "conditional" space: "Most of our life is spent in the if of our desires and regrets, and I am as iffy as the next guy [. . .]. But who knows, this conditional incapacity is perhaps destined to have its place in the work process, assigned to it by a furor devoid of desires and regrets" (*To Whom It May Concern* 128). What begins as an "iffy" narrative, concerned with the many gaps in the author's own existence, gradually achieves some structure through projection/re-creation:

> Ahead of him like a huge hole (an enormous hollow sphere) lies his future. A kind of void. Emptiness. America. The hollow of his fate. His future in America still unrealized. Of course his imagination (or mine working for his) fills the void with all sorts of notions. (*Double or Nothing* 156)

The multiplication of voices and narrative frames in Federman's fiction testifies to the difficulty of bridging the central gap "between the function of memory and that of thinking (between being-then and being-now or if you prefer being and non-being)" (*Take It or Leave It* 335; my pagination). A gush of words and typographic designs spans that "hollow" space for us, trying to convert it into a meaningful textual surface. But, as Federman's narrators soon find out, narration cannot fill the existential and historical gaps with strong performative speech: Most often it tenders a voice-in-the closet, "whirl[ing] me in a verbal vacuum" (*The Voice in the Closet* 13; my pagination).

Federman's novels illustrate the paradoxical nature of narrative articulation. Every narrative act aspires to be final and all-inclusive, putting the other words to rest; but in the same breath it must acknowledge its provisional, fallible status: "[E]ach word must be written [. . .] as though it were the last one, yes the last gasp, standing on the edge of a precipice leaning against the wind, and the one word that happens to have no successor can only be the last one for just a moment" (*The Twofold Vibration* 117). The *said* remains to a great extent unsaid as long as, in Robert Pinget's words quoted in the epigraph of *Double or Nothing,* "on peut le dire autrement." Federman's dynamic of articulation can be best understood in terms of Lyotard's agonistic theory of representation. Reality for Lyotard is not a "given" but "a state of the referent (that about which one speaks)" (*The Differend* 4). History's complex referents can be negotiated only through partial testimonies and questionable cognitive appropriations (9). Lyotard's postulation of an incomplete and contentious encoding of reality undermines traditional hermeneutic expectations. Readers are asked to identify not only the explicit but also the implicit statements that a text makes with its "silences." They are encouraged to recognize that the

> silence of the book is not a lack to be remedied, an inadequacy to be made up for. It is not a temporary silence that could be finally abolished. We must distinguish the necessity of this silence. For example, it can be shown that it is the juxtaposition and conflict of several meanings that produces the radical otherness that shapes the work: this conflict is not resolved or absorbed, but simply *displayed.* (Macherey 84)

The narrative hermeneutics that an innovative novelist like Raymond Federman pursues refuses totalization, remaining—in Macherey's terms—"incomplete" or rather "interminable." It is also true that by way of its incompleteness, this type of narration "*manifests,* uncovers, what it cannot say. This silence gives it life" (83, 84). The "holes, the gaps, the voids, the empty spaces, the blank pages" point to what is unspeakable in Federman's fiction (the Holocaust, the trauma of his uprooting, etc.); but also to what must be spoken through "cunning and devious stratagems" (*Critifiction* 86).

The theoretical rationale for Federman's project is clarified in *To Whom It May Concern:*

> [. . .] I'm convinced that we must now move beyond mere fables, beyond the neatly packaged stories which provide a chain of terminal satisfactions from predictable beginnings to foreshadowed endings. We have come so far in the long journey of literature that all the stories whisper the same old thing to us in the same cracked voice. And so we must dig in to see where the raw words and fundamental sounds are buried so that the great silence within can finally be decoded. (86)

Like other contemporary innovators, Federman confronts here a twofold problem: Aesthetically, he has to ward off a postmodern feeling of belatedness and exhaustion of possibilities; ideologically, he must recover a story silenced or distorted by misrepresentation. Refusing his culture's prepackaged narratives with their "neat beginning, middle and end," the innovative novelist redeems Diderot's method of digressive detouring: "[T]o get on with the story one must avoid precision. One must digress. Skip around. Improvise. Leave blank what cannot be filled in. Offer multiple choices. Deviate from the facts, from the where and the when, in order to reach the truth" (104).

Federman·thus places the genre of the novel under "the sign of Saturn, the planet of detours and delays" (19); but also in the more meaningful mode of revisionary writing. As the authorial narrator of *To Whom It May Concern* gleans, our "access to an event of the past is never unmediated, [. . .] it is always manipulated by false restitutions" (183–84). Therefore, the Federman narrator rejects "the paralyzing holiness" of conventional historical representation (106), pursuing emotional and intellectual *truth* in lieu of superficial *verisimilitude:*

> We're not dealing with credibility here, but with the truth. That's not the same. Certain truths do not need the specificity of time and place to be asserted. A war is a war, doesn't matter where and when it happened. And suffering is timeless. (39)

Rather than "paint" a story thickly with "layers of reality," Federman has tried to decertainize time and place references that give "a semblance of stability and continuity," opening history to new narrative possibilities. The section of *To Whom It May Concern* published in *Formations* as "From the Book of Sarah" still provided explicit time and space specifications as well as biographical detail to contextualize the story of Sarah and her cousin. The final version of the novel no longer names places, hinting only vaguely at particular cultural backgrounds and documentary sources. The narrator resists even the temptation of giving "Sarah's cousin" a name, arguing that this character may be "too close to me to foist a false name on him. Perhaps eventually the necessity of a name will become apparent" (*To Whom It May Concern* 39).

By eschewing the "tricks" of specificity, Federman has kept at bay what he calls the "imposture of realism, that ugly beast that stands [. . .] ready to leap in the moment you begin scribbling your fiction" (106). The history he has been rehearsing is too brittle and complex to withstand the "banality of realism" behind which always lurks "a catastrophe or a bad joke" (107). Official history unfolds in Federman's view as a series of betrayals and substitutions: The spot in Paris where the apartment building of a number of Holocaust victims once stood has been usurped by an art museum that stands there as a "scandalous substitution. The immorality of history replaced by the playfulness of modern art" (107). Federman would probably agree with the director of *Shoah* that conventional representations of traumatic history are obscene, a "way of escaping," "a way not to face the horror" (Lanzmann 481). That is why "it is essential to avoid the specificity of time and place, even at the risk of skirting allegory. History is a joke whose punch line is always messed up in advance" (*To Whom It May Concern* 107–8). There is little one can say about the scandalous disappearance of Federman's own family in the extermination camps. The chapter that ponders "the correctness or incorrectness of the unforgivable enormity" (99–100) is only a half-page long, interrupted by silence. This "absolute erasure" can be represented only through a sequence of Xs ("X-X-X-X"). As Thomas Hartl comments, the "nakedness, the symmetry of enforced exposure" that these symbols transmit mirror better than any language "the cold figure of the cross, [. . .] those two quick scraps of pen on the paper of death"; in the face of a "loss of words, of an absence of language," the "symbolic irreality of the Xs explodes any attempted assertion of sense, causing the finger-pointers to become conscious" (McCaffery, Hartl, and Rice 2).

Since recent history is to Federman a "story of erasures, [. . .] why not erase all traces of pretense, and have a story that empties itself of references" (*To Whom It May Concern* 168)? This proposition has tempted Federman's narrators; still, what they have sought is not the erasure of history but—in Shoshana Felman's terminology—a "performative" reenactment in which events are not "simply relayed, repeated or reported" but reimagined (Felman and Laub 3, 5). Resisting the "arrogance of story-telling" (*To Whom It May Concern* 169) that turns historical fiascoes into successful narratives, Federman's narrators weave "exploratory or better yet extemporaneous" stories that approach self and humanity "from a potential point of view" (*The Twofold Vibration* 1–2). The historical experience they re-create is left deliberately open-ended, open to further revisions. To borrow Lyotard's distinctions, Federman's fiction opposes an endless process of anamnesis to official "history" that forgets by memorializing: since "one forgets as soon as one believes, draws conclusions, and holds for certain" (*Heidegger and "the Jews"* 10), the task of the fiction writer is to puncture the "protective shield" of official history (8), confronting it with the unexplainable, the unrepresentable, the silent.

The "great silence within" (*To Whom It May Concern* 86) that Federman's fiction has struggled to decode is largely autobiographical. There are at least three ruptures in Federman's life story that the writer has tried to come to terms with, "straddling two languages, two continents and two lives, two cultures also" (*Take It or Leave It* 100; my pagination). The first and most forbidding involves Federman's "closet" experience, his survival as a thirteen-year-old boy from the Nazi genocide, hidden by his mother in a closet. As Federman has suggested, this is his "primal scene," his "real birthday, for that day [he] was given an excess of life"; also his great narrative challenge, the "necessity and impossibility of expressing the erasure of [his] family" and understanding "the darkness into which I was plunged that day" ("A Version of My Life" 65). Federman's entire "pre-closet" period has remained a blur of disjointed remembrances, only recently recuperated/reimagined in his French-language novel, *La*

fourrure de ma tante Rachel: Roman improvisé en fourire (Aunt Rachel's Fur: Novel Improvised in Mad Laughter, 1996). Robbed not only of his childhood, but also of the normal transfer of familial narratives from parents to their progenies, the writer has had
to "spin out [. . .] infinite stories," speaking "both the truth and the lie of [his] condition" as a Holocaust escapee (*The Voice in the Closet* 11, 20; my pagination). As trauma
psychologists argue, "trauma is not simply an effect of destruction but also, fundamentally, an enigma of survival. It is only by recognizing traumatic experience as a
paradoxical relation between destruction and survival that [the author] can also recognize the incomprehensibility at the heart of catastrophic experience" (Caruth, *Unclaimed Experience* 58).

The second rupture concerns Federman's later "exilic" leap across the Atlantic,
from native France to the New World; the third, his educational wanderings
through another "kind of void [. . .] America." Across these three basic gaps, Federman's fiction has moved digressively toward some form of personal enlightenment. This enlightenment is conditional and incomplete since the writer has had
to negotiate one gap in terms of the others (in *La fourrure de ma tante Rachel,* the narrator returns to Paris to finish his first novel about the American experiences of a
young Jewish refugee whose family was exterminated in the Holocaust, but gets
sidetracked into memories about prewar and wartime France, superposing several
temporal levels in a desperate effort to reestablish the coherence of his life). The
process of continuous approximation and revision involved in articulating his
"story" is well suggested in *The Voice in the Closet:*

> begins again closet confined as selectricstud resumes movement among empty skins im-
> ages crumble through distortions spins out lies into a false version leapfrogs infinite sto-
> ries falling silently into abyss to be replaced retold confusion foretelling subsequent
> enlightenment (20; my pagination)

A positive quest is involved in Federman's treatment of each traumatic gap in his
autobiography: the search for a meaningful, morally justifiable way out of the "primordial closet" that claimed Federman's family; an exile's pursuit of the elusive
"terre mater-Amer Eldorado," a foreigner's emotional quest for integration ("penis-
tration") of America. All three quests are highly dramatic and "tellable." A successful romance can be woven around the partly autobiographical episode in which an
alluring middle-class American girl smiles on an "unemployed, unskilled" foreigner,
"just returned from army duty in Korea." After all, "many great loves of the past were
initiated by little more than a conniving smile" (*Smiles on Washington Square* 66). The
"big crossing" of America can likewise inspire literature because America itself is a
kind of "fictitious discourse" wherein "everything can happen" (*Double or Nothing* 114).
Even the "unthinkable, the unspeakable, the unbearable" event of the genocide can
induce narration along predictable lines:

> Genocide is [. . .] a machinery which fabricates death on a large scale. Though we are
> disturbed, troubled, outraged, horrified by the final product (the final solution!) of
> genocide, when talking or writing about it, it is usually the machine, the mechanism
> which becomes the central topic of our discourse rather than the tragic fate of the vic-
> tims. [. . .] Whether it is to destroy a family, a tribe, a race, or a whole nation of peo-
> ple, genocide functions like a gigantic machine which once set in motion cannot stop
> its work. [. . .] The genocide-machine thrives on deception. After all it is a form of the-

ater. Like all great tragedies, it purges, it offers catharsis, but in the reverse direction."
("The Art of Genocide" 1)

It is this machinery of deception, the triumphant "mise-en-scène" of disasters that Federman has resisted in his own writing. With the exception of his unpublished autobiographical novel, *And I Followed My Shadow,* Federman has not been interested in pursuing a conventional quest narrative or a Holocaust novel, which would have misrepresented the distinctive quality of events with "disaster words [. . .] unqualifiable babble" (*The Voice in the Closet* 17; my pagination). Against this fraudulent economy of representation that "always tells the truth while cheating the original experience" (8–9), Federman has had to mobilize a rhetoric of "unsaying," spinning out self-controverting stories whose "questions and answers [. . .] annul each other" (*To Whom It May Concern* 106). Federman's narrators do not as a rule allow their stories to find their "final resting place," keeping them "temporary," "just trying out things" (136). In his effort to write "something simple about two people tormented by their past, and anguished about the state of their present life" (83), the character-author of *To Whom It May Concern* undercuts the plot every time it lapses into "pure solid naturalism" (135) or melodrama.

But can melodrama be kept out of a narrative about "people [who] are dragged into history in spite of themselves" (128)? The innovative novelist is caught between "the story with its [sordid] details that demands to be told" and his awareness that to "tell a story [unfractured from the cozy realm of conventional practices] is to discredit it" (136). His story cannot escape a historical logic that generates endless plots of disaster: "To each his own disaster. The disaster of the street accident, the disaster of the plane crash, the disaster of racial or political assassination, the disaster of the terrorist attack, the disaster of famine, the disaster of natural cataclysm" (17). Because they "take care of everything," such narratives of disaster absorb us in a "heedless unlimited" (Blanchot, *The Writing of the Disaster* 2–3), dispensing with the need for careful thought and writing. For that reason, the novelist's job is to disrupt history's settled plot that can only cause a "gnawing feeling of frustration." What Federman has been interested to write is not a historical recapitulation of "what happened and how it was or was not resolved" (*To Whom It May Concern* 108), but rather an explorative narrative of "consequences" and "refractions," restoring history the dimension of becoming

Federman's commitment to a deconstructive–re-creative response to history has something of the uncompromising integrity of Beckett, even though Federman has replaced the latter's "literary monasticism" (Howe 26) with a more participatory attitude that draws on the conflictive energies of contemporary culture. Federman's works abound in fond gestures toward Beckett, his precursor in the teetering land of exile and existential or political chaos, whose death in 1989 marked for Federman the final point of postmodernism itself (*Critifiction* 105–6). As Melvin J. Friedman has suggested, one can liken Federman's apprenticeship to Beckett with Beckett's own creative exposure to Joyce and Proust: "In short, the early Beckett of *Murphy, More Pricks than Kicks,* and *Echo's Bones and Other Precipitates* has left its stamp on the critifiction phase of Federman's work. [. . .] The later Beckett, with his spare, accentless prose, clearly helped form the brief, unparagraphed, unpunctuated, bilingual *The Voice in the Closet*" (138). Beckett's spiritual and stylistic presence is recognizable also in Federman's two most exploratory novels to date,

The Twofold Vibration and *To Whom It May Concern.* But these novels that use the "pla(y)giaristic" talents of their protagonists (great "thieves of language" both) to rewrite personal and collective history, suggest that Beckett has provided Federman with more than a few thematic-linguistic echoes. Following Beckett's example, Federman's fiction has struggled with versions of the "unnamable," replacing "plot" with a conditional "story" that allows for rupture but also for some development, at least of the circular kind—the "spherical [purgatory which] excludes culmination" that Beckett attributed to *Finnegans Wake* ("Dante . . . Bruno, Vico . . . Joyce" 21–22). Federman's own version of narrative purgatory is (with a word purloined from Joyce) a "Bethickett," a Beckett-made thicket: "all good story tellers go to BETHICKETT on the way to Heaven and that is why perhaps they are so long in reaching their destination" (*Take It or Leave It* 176; my pagination).

Federman's early narratives dramatized the writer's painful struggle to articulate a past of unthinkable events, moving Beckett-like "by affirmations and negations invalidated as uttered" (*Critifiction* 120). Two impetuses competed in these works: "a force committed to delirium, disorder, **laughter**, tears, fantastic postures and impostures; in short a force familiar with and expressive of the multiple"; and another "afraid of delirium and abandon, requiring control" (Tatham review rpt. in McCaffery, Hartl, and Rice 88). Federman's prose seemed thus to distrust "control nearly as much as abandon" (88). His more recent novels have been more willing to speak the unspeakable, tilting the balance toward rearticulation at the risk of submitting experience to what Tatham calls "order words" (88). But they have also tried to retain the open-endedness of the process of historical interpretation, mixing retrospection with projection: "In order for the story to be told, to be written in fact, the narrative must leap forward into the future before it can turn towards the past to recapture it" (*Critifiction* 102). Approaching the "central unspeakable event" of the Holocaust with a "devenant" rather than "revenant" approach (50), *The Twofold Vibration* (1982) allows its characters to survive by reinventing themselves and their place in history. *Smiles on Washington Square* (1985) and *To Whom It May Concern* (1990) move beyond the earlier plots of (self-)cancellation, accommodating "good lousy stories" of encounter (*Smiles on Washington Square* 67): the immigrant's "love story of sorts" with an American girl in the former; the reunion and partial reconciliation with history of two Holocaust survivors in the latter. Their relative thematic and syntactic coherence should not deceive us: Under the guise of a modest task (to write "something simple about two people tormented by their past" [*To Whom It May Concern* 83]), the narrator of Federman's most recent novel remains committed to a re-creative approach to twentieth century history.

2. HOW TO RE-PLACE LIFE: REVISIONISTIC STRATEGIES OF STORYTELLING

The question most frequently asked by Federman has been "how to replace a life in its context when in most cases one has forgotten or falsified the original text?" (*Critifiction* 87). In other words, Federman's challenge has not been that of inventing a story from scratch, but rather that of submitting an already existing historical script to a process of unraveling/revision, foregrounding the gaps or unfulfilled possibilities within it. Federman has been blessed (or cursed) with enough biography for several epic cycles. In his unpublished *And I Followed My Shadow*, this autobiographical material yielded not one but several conflicting "storifications" separated by substantial geographic, temporal, and cultural gaps: "the story of the closet [in war-torn Paris], but also the story of the raw potatoes on the train, the story of the farm [in southern France], the story of the journey to America, of the factory in Detroit, the story of Charlie Parker and the tenor saxophone, and of the Buick Special, and all the other stories he has told" (*Critifiction* 99). Federman could hardly contain the onrush of words, producing between 1955 and 1958 a 600-page novel manuscript in addition to scores of poems and stories. He shifted from first to third person and from authorial to figural narrative, hoping to find a controlling frame for his life saga. The thinly disguised autobiographical narrator had no qualms about "making up all sorts of stories" when his memory failed him, or emphasizing a pattern of self-enlightenment, the "meaning of life and of myself moving into life."

Federman's life story lends itself to this kind of treatment, unfolding like a grand plot packed with uncanny or tragic twists: a surrealistic painter and inveterate dreamer with Trotskyite sympathies for a father; father, mother, and two sisters erased in the gas chambers of Auschwitz, while Raymond escaped the Nazi roundup hidden in a closet; surviving the war on a farm in southern France; returning to Paris in 1945 atop an American tank; brought to America in 1947 by a surviving paternal uncle, after the council of rich maternal aunts could not decide into whose care to entrust him; beginning his American apprenticeship working the night shift on a Chrysler assembly line while attending a predominantly black Detroit high school during the day; playing tenor saxophone with three of his schoolmates (once jamming with Charlie Parker at the Blue Bird club); moving to New York in 1950 to work briefly in a lampshade factory; drafted into the U.S. Army in 1951, initiated into combat and racial prejudice by his fellow paratroopers at Fort Bennington, Georgia; shipped to Korea in January 1952, made a U.S. citizen one year later in Japan while serving as an interpreter for Military Intelligence; writing his first poems about hustlers and black marketers in Japan; admitted to Columbia University on a G.I. Bill in 1954; discovering Beckett in 1956, the only writer whose work he has never stopped reading and whose literary executor he has become; returning to Paris in 1958 to make it as a writer, but finding only disenchantment there; writing poetry in both French and English, gambling, and playing in golf tournaments while completing his Ph.D. in comparative literature at UCLA; moving to SUNY Buffalo in 1964 to write, teach, and become an international literary citizen.

The life story is all there, rich in *peripety*: "All this sounds so much like the script of bad movie. But I suppose, in retrospect, one's life always becomes a series of clichés. Even the most horrendous moments appear banal" ("A Version of My Life" 70). Consequently, Federman's effort after *And I Followed My Shadow* has been to upset this predictable life script. "To create fiction," Federman proposes, "is [. . .] a way to abolish reality, and especially to abolish the notion that reality is truth" ("Surfiction—Four Propositions" 8). But this surfictionist claim is accompanied by the realization that even the most disruptive novel cannot escape realism:

> This mortgage weighs upon it since its origin, since the period when for justifying itself of the suspicion of frivolity, it had to present itself as a means of knowledge—and not only since the 19th century. The history of the novel is [. . .] nothing else but the succession of its efforts to "appresent" a reality which always evades, always substitutes

for vulgar mirrors finer mirrors, more selective mirrors. But, in another sense, the novel
is nothing else but a denunciation, by its very reality, of the illusion which animates it.
(*Double or Nothing*, unnumbered pages between 146–47)

Against the "probabilistic" march of mimetic fiction (Nash 46), Federman has
employed a host of "illusion-breaking" devices: self-controverting narrators, improvised or "borrowed" characters, metaleptic switches between story planes, narrative
"erasure," concrete prose ideograms. The narratological significance of these procedures is well established in Federman criticism; but Federman's fiction has always
counterbalanced disruption with reconfiguration, opening narration up to a "renewed and renewing content" (*To Whom It May Concern* 157). As explained by Federman in his programmatic essays, his procedures (derived mostly from Beckett) take
a narrative attached to a particular linguistic and ontological order and
"unwrite/rewrite" it: destabilizing and relocating it into another frame (*displacement*);
removing or complicating the meaning of words by "double exposure" or the mixing
of voices/languages/media (*cancellation*); decomposing the formal structure and syntax of texts through oral or visual dislocation (*pulverization*); repeating text by oral and
visual overlapping, causing variation and distortion in it (*repetition*); pluralizing static
texts through "pla(y)giarizing [. . .] voices within voices" (*revision*) ("Federman:
Voices within Voices" 159–61). Federman's essays have argued the ontological value
of these procedures that unsettle and pluralize a given story. What they have not emphasized enough, at least until recently, is the cultural significance of these strategies
that allow the writer's "imagination to supersede memory" (*Critifiction* 102), moving
from a constrictive past to an explorative present. Freed from the compulsion to repeat "who or what he was or what he did," the writer can "invent who or what he
wants to be or have been or have done. [. . .] In this sense it is no longer memory
which controls the movement of the narrative, but imagination" (101, 102).

Unlike his literary master Beckett who knowingly pursued "the ultimate reduction and deprivation of fictional, human and linguistic possibilities" (Federman,
"The Impossibility of Saying the Same Old Thing the Same Old Way 38–39), Federman has steered an elastic course between deconstruction and rearticulation. His
early books (*Double or Nothing, Take It or Leave It, The Voice in the Closet*) treated the space
of the page as a compositional field, deploying across it disruptive typographic designs. But the loquacious narrators of these and the following books still made an effort to extricate a "real story" from their "edifice of words" (*The Voice in the Closet* 18;
my pagination). At its best, Federman's fiction foregrounds the deconstructive impulse within narration itself, exploiting comically the "pitfalls" of realism enumerated by J. P. Stern: the "overburdening of intimation" (overuse of allusions and
assumptions); "the descent into banality"; "inadvertent" change of mode and style;
"the break in continuity [. . .] and coherence"; the "turn to *kitsch*" (77–81). His self-
problematizing narratives discredit conventional representation, revealing its failures and errors. By the same token, they make us aware of alternative modes of
narrative articulation and intelligibility.

Errors play an important part in Federman's process of narration: The monumental preparations for writing made by the intradiegetic narrator in *Double or Nothing* are
foiled by the contingencies of life and by the unpredictability of the creative process
itself. In *Take It or Leave It*, an initial "little error" perpetrated by his army captain prevents "Frenchy" from undertaking his quest across America, in the tradition of Céline

(*Voyage au bout de la nuit*). Both his journey and his retelling of it are blocked, detoured. Echoing Beckett's *Molloy*, where the journey and its narration end in a ditch, Federman's story takes its hero to the edge of a natural and symbolic precipice. By only a hairbreadth the protagonist misses self-cancellation, the final "error." But that makes the whole difference: The Beckett motif of cancellation through error is converted into an acceptance of error as constitutive of the narrative process. Federman's candid heroes and "second-hand" tellers outlive their errors that decry their effort to rationalize life. Thus, while the "discovery novel" is thwarted ironically, something does emerge from this discourse of errors: a writer's imaginative journey through a vast *textual America* that other novelists before him, from Melville to Henry Miller and Michel Butor, have tried to chart.

The concept of authorship is submitted to a similar deconstruction/revision. While insisting that the novel "invents its own reality," Federman does not "recirculate the rhetoric of the artist's supreme and unconditional artistic freedom" (Connor, *Postmodernist Culture* 116). For Federman the writer's freedom to invent is framed by historical and narrative constraints. The author is allowed to "come back in the Text" mostly as a "guest," in Barthes's sense of the word, "inscribed in the novel like one of the characters [. . .]; no longer privileged, paternal, aletheological" (*Image-Music-Text* 161). He is often split into conflicting selves or reduced to a "mere assembler of various bits of language and culture into writings that [are] no longer works of art but simply cultural collages or *texts*" (Federman, *Critifiction* 117–18). The author's control over his life story is subverted by the proliferation of "versions" (*Critifiction* 91). What Federman writes is not autobiographical fiction, in the classic sense, but a series of self-recreations that remind us that "the presentation of self through language is necessarily a selection, a fragment, a fiction" (Bailey 80). Federman's work fits Peter Bailey's category of the "novel-as-autobiography," reinventing autobiography as a self-referential reality (81) while simultaneously "conduct[ing] us back toward the gratuitous complexities of existence rather than toward the artfully contrived, all-encompassing aesthetic resolutions" of traditional fiction (80). Federman's own term for this speculative and interrogative self-narration is "avant-garde" or "experimental" autobiographical fiction (*Critifiction* 102). This term flags the paradoxical status of the genre that, like Jerzy Kosinski's "autofiction," is concerned less with testimonial accuracy and more with "reaching a deeper explanatory level by rendering the broader sense of the events" (Ornatowski and Durczak 12). What both Federman's and Kosinski's fiction asks, at the risk of undermining its immediate historical credibility, is "whether any human memory or even perception can be literally 'true,' rather than constituting an interpretation that aims at recovering (or making up?) some broader sense for ourselves and for others" (Ornatowski and Durczac 12). Federman's "avant-garde autobiographical fiction" delights in that contradiction, blurring the lines between "life and fiction, fact and fiction, language and fiction, [. . .] history and story" (*Critifiction* 89). Thus, the character-narrator of *The Twofold Vibration* can state simultaneously: "I have no country, my life is the story," and "my story is my life" (159).

This paradox, however, is not insurmountable. The two competing propositions can be mediated through the "futuristic, explorative" writing (*Critifiction* 102) proposed by Federman that unfolds as a journey of disruption and transformation: "My stories are usually based on a journey of some sort. [. . .] And whenever there is movement in the story, there is also displacement, discovery, loss, and mystery" (interview in LeClair and McCaffery 129).

3. EXTRICATING A "REAL-FICTITIOUS" STORY
FROM THE JUMBLE OF POSTWAR HISTORY:
FEDERMAN'S EXISTENTIAL AND CULTURAL EXPLORATIONS

Double or Nothing, Take It or Leave It, and *The Voice in the Closet* demonstrate the length to which Federman has been willing to carry the process of "destorification." Described as stories "that cancel [themselves] as they go," these experimental novels continually disrupt the narrative line with digressions on a variety of subjects (the traumatic legacy of the Holocaust, the challenges of contemporary history, the fraudulent devices of realism, the uses of English and French in fiction, and so on). Still, Federman's "critifictional" digressions (as chapter 21 of *Take It or Leave It* calls them) are not entirely "storyless." They allow a phenomenological story to emerge, based on a loose concatenation of "real-fictitious" anecdotes, accompanied by a "literary story" of sorts, made up of polemical references to the novelistic tradition. As the writer conceded in his exchange with Ronald Sukenick, he still found use for the "series of little anecdotes strung together to make a sequence, although the sequence is no longer that of a plot [. . .]. Your idea of incident, movement, action, is what keeps the story going. My sequential movement works the same way" (Federman and Sukenick 144). If I understand it correctly, Federman's distinction between "plot" and sequential "story" is one of static structure vs. dynamic structuration: Plot remains a rigid, "fraudulent way of holding things together," while "story" allows for surprise and transformation, "unload[ing] in all directions without respect for logic and with rather crooked means" (*Take It or Leave It* 292–93; my pagination).

Federman's succession of books promises a "real thing, a search for the truth" (*The Twofold Vibration* 122). Their narrators renew periodically their pledge:

> This time we dive in
> we plunge in
> we move forward
> we progress rather than digress
> we push on
> we organize rather than disorganize
> we recite correctly
> we keep it going neatly
> we don't fuck around anymore
> we don't try to be fancy
> to say more than we know
> we do it right
> we pull no punches
> we pile up the stuff in a straight line along a linear course of action
> we tell it as he told it
> we try not to exaggerate too much
> we say it as we heard it! (*Take It or Leave It* 124; my pagination)

The irony behind this dutiful recitation of best intentions is that no amount of "correct," straight-line narration can guarantee "truth." Were he able to follow his conscientious literary program, the character-author would still offer little more than "an exaggerated second-hand tale to be read aloud either standing or sitting," as the subtitle of *Take It or Leave It* quips. Narration can only duplicate "the carbon design of my life" (*The Voice in the Closet* 9; my pagination), retranslating in countless versions an "original" story that is itself but an imperfect translation of an elusive truth. Federman's fiction makes this vertigo of *translation* immediately apparent: Most of his books have French versions or pages with bilingual columns of text, his narrative persona is defined as a "schizotype" (186) who straddles "two languages, two continents [. . .] spread-eagle over the Atlantic" (*Take It or Leave It* 100; my pagination), and each novel spins out "entre croisée" versions of the same sequence of "autobiographical" events. Narration becomes a "big crossing" of texts for the "mad acrobat of fiction" who seeks enlightenment in the "accumulation of facts," "signs," and languages (*Take It or Leave It* 297).

In a "lecture/demonstration" on bilingualism, Federman takes this idea further, constructing (on Beckett's model) a literary epistemology and personal myth around the notion of self-translation. Autobiographically, the author is a "bilingual being, a double-headed mumbler, [. . .] and as such also a bicultural being" ("A Voice within a Voice," *Critifiction* 76). Federman's narrative poetics also relies heavily on "self-translating" (76), in both a literal and a metaphoric sense. Self-translation yields contradictory outcomes: On one hand, it "results in a **loss**, in a betrayal and weakening of the original work" (80). The intersecting voices/languages threaten to displace each other. On the other hand, translation offers the writer an invaluable second chance for "correcting the errors of the original text":

> Usually when I finish a novel [. . .] I am immediately tempted to write (rewrite, adapt, transform, transact, transcreate—I'm not sure what term I should use here, but certainly not translate) the original into the other language. [. . .] My feeling here is that the original text is not complete until there is an equivalent version in French or in English. [. . .] [S]ince we know that language is what gets us where we want to go but at the same time prevents us from getting there (I am paraphrasing Samuel Beckett here), then by using [. . .] the other language in us, we may have a better chance of getting where we want to go, a better chance of saying what we wanted to say, or at least we have a second chance of succeeding. (79, 80–81)

In terms that recall the poststructuralist philosophy of translation developed by Jacques Derrida ("Des Tours de Babel") and Paul de Man ("Conclusions") in their respective commentaries on Walter Benjamin, Federman views translation as a supplementing act, enriching both the "original" and the "ambivalent (ambidextrous) psyche" of the writer (77). Any narration remains incomplete, suspended in its "ignorance"; but while de Man considered this aporetic "blindness" insurmountable, Federman suggests that narration can move toward partial "knowledge" by multiplying its languages and versions:

> The original creative act [. . .] always proceeds in the DARK—**in the dark, in ignorance and error.** Though the act of translating (and especially self-translating) is also a creative act, nevertheless it is performed in the LIGHT (**in the light** of the original text), it is performed in KNOWLEDGE (**in the knowledge** of the existing text), and therefore it is performed **without error**, at least at the start. (81)

Ideally, narration becomes a "double-headed mumbler," translating itself back and forth, probing simultaneously the horizontal of articulation and the vertical of displacement (merging). A perfect example of this can be found in Beckett's "twin-texts—whether

French/English or English/French—[that] are not to be read as translations or substitutes for another. They are always complementary to one another" (78). But what Federman has pursued in his own bilingual fiction is something less than perfect—and therefore more tensional, dialogic—reflecting his divided literary biography. While his characteristic protagonist remains "a Frenchman in exile," his fiction always has an "implicit, active interlocutor/listener present in the text [. . . who] is of the English and not the French language" (82). This creates a "dual internal (one should almost say infernal) dialogue" (*Critifiction* 81) in Federman's books as his narrators try to negotiate the "walls of antithesis" (Caramello, "On Styles of Postmodern Writing" 230): French and English, voice and "mute language," story and history.

Federman's books have gradually moved toward this dialogic understanding of narration as an interplay of incomplete articulations and self-translations. His early books gambled with impossible narrative and formal obstacles, with "ways to cancel my life digressively each space relating to nothing other than itself" (*The Voice in the Closet* 5; my pagination). *Double or Nothing* (1971; winner of the Frances Steloff Fiction Award and the Panache Experimental Fiction Prize) placed the struggling Federman narrator in a characteristic "intra-mural set-up" (*Double or Nothing* 00000). His story never quite gets under way, framed as it is by two enormous "holes": the real-symbolic "closet" from which the protagonist emerges in his early youth, at one end; and the prescience of a similarly uncharted void, at the other end of his adult, American experience. The self-controverting narrator (split into a matter-of-fact and an emotional-experimental voice) struggles not only with the "facts" of life but also with the preliminaries of storytelling. He tries to balance means and ends, preparing himself for the task of writing the way others would prepare for a year spent on the Moon. Not only does Federman's novel invent a limit situation for its writer, it also becomes a "limit text" in Sollers's sense of the word, involved in an "open-ended experience that necessarily put[s the writer's] life fundamentally at risk, [. . .] inverting that perspective of the world in which [he] found [himself], concretely touching for [himself] its limits" (198–99). The limits tested in this first novel are at once textual and existential. The "boxes of noodles," initially destined to ensure the narrator's survival through a year's hard work on his novel, are used by him to map out his writerly time. (The narrator sets up an ingenious noodle calendar to keep track of the days spent in seclusion.) He also uses word noodles to structure the blank space of the page with iconic designs and to "noodle" out a text about a "noodle reality" (110), employing an improvisational technique that recalls the "noodling around" ("casual, aimless playing") in jazz (Wielgosz 104).

The first edition of the novel, photographed from Federman's meticulously prepared typescript, made the restructuring effort of writing "visible" at every step, suggesting more than a casual relationship between the topics and the topology of each page (see Federman's page-by-page "Summary of the Discourse" at the end of the book). The novelist's desire to gain self-determination of movement is pursued more systematically in the third edition of the novel (1992), redesigned on the computer. The increased range of options offered by electronic publishing allows a more coherent interplay between the visual and the verbal. But some of the anxious effort involved in the original wager made by the novelist with his medium seems lost in the revised edition.

Take It or Leave It (1976) splits narration further into competing diegetic levels: teller, reteller, listeners. The composite narrator still tries to eschew the "tidal events of his narration," destabilizing its emerging plots through digression, "word-designs" and "double-talk" (179; my pagination). After successfully deferring his unspeakable story as a Holocaust survivor, the narrator broaches digressively his equally problematic present in an America "hard to take to swallow hard to conquer" (114). *Take It or Leave It* evokes for the reader the "classic American initiation story, in which the young initiate will discover America while he engages himself in a process of self-discovery" (McCaffery, "New Rules of the Game" 145). But the contemplated transcontinental journey never takes place, slipping "from its ontological status of anticipated fact into the limbo of the merely hypothetical" (McHale, *Postmodern Fiction* 50). The novel ends with the cancellation of the young foreigner's trip west, his hopes for discovering America (before sailing overseas to fight in Korea) shattered. Still, although this version of the discovery novel "cancels itself out not only as it progresses, but also in advance" (to quote the book's "Pretext"), *Take It or Leave It* manages to take the reexamination of his American experience a step further than the previous novel.

The visual and dramaturgic resources of narration (both in the sense of a narrative "mise en scène" and a polemical confrontation between teller, reteller, and listener) are tapped again in *The Voice in the Closet/La voix dans le cabinet de débarras* (1979). This remarkable experiment in graphic and intertextual composition allows several kinds of "voices" to intersect: human and mechanical (the IBM "selectricstud"), narrative and textual, "past-self" and "present-self." They are all locked inside rigidly justified boxes of text—square in the English text, rectangular in the French version—that evoke the closet that saved young Federman from deportation to Auschwitz. The text "tiptoes" ahead through association, translation, and echo. (Federman's French version of the text, as well as Maurice Roche's response essay entitled "Échoes à Raymond Federman," make "echoing" even more of a thematic obsession.) This plotless one-sentence "story" that stretches for twenty pages reminds us that "reality" is a matter of *literation*, an "edifice of words" displayed across a typographic surface.

While the two previous novels circled indecisively Federman's "closet" experience, this book confronts it more directly in an act of anamnesis that tries to bring guilt-ridden history out of the "box-room." This process of recovery cannot entirely succeed: The same techniques (digression, repetition, mixing of voices and languages, rewriting) that enable the gradual encircling of the past also undermine its ontological status, turning it into an ambiguous "real-fictitious" discourse. The problematic nature of historical writing is played out especially at the level of narrative voice: "The voice in the closet becomes a voice in the typewriter, resisting the typist, resisting the reader, [. . .] resisting the naming perhaps maiming processes of words, figuring forth for us [. . .] the unreality of our world and of ourselves as we have invented them and allowed others to invent them for us" (Quartermain 73–74).

After reaching peaks of polyphonic performance, Federman's *voice in the closet* becomes disarticulated again in the end, returning (much like the earlier books) to paradoxical "mute speech/sign of my presence" (*The Voice in the Closet* 20; my pagination). By contrast, Federman's more recent novels seem eager to break out of the author's "primordial closet," projecting a *narrative of possibilities* rather than one of "cancellation," as before. *The Twofold Vibration* (1982) recovers more of the "unspeakable" story than any of Federman's previous books: the "great round-up" of the French Jews (July 16, 1942), the "closet" episode, the hero's frantic wandering around Paris, his raw potato binge on the escape freight train to southern France, his adult visit to Dachau are all featured here in rich, readable detail. Still they are sub-sumed to an allegorical theme, with the protagonist in the role of a "Jewish space hero" (6) awaiting—on New Year's Eve, 1999—deportation "to the space colonies" (43) together with other cultural "undesirables." As usual, the Old Man's story is framed by two voids: the "primordial closet of his strange birth" and the "ultimate metallic box" that awaits him at the other end. The Old Man is ultimately spared de-portation, remaining all alone in the middle of the huge spaceport, a baffling signi-fier for us to interpret. We are aided in the process by a triple-headed narrator: a speaker-scribe "Federman" and his two informants, Moinous and Namredef. These "three Musketeers of Surfiction" are happier spinning stories about the old man than studying him in any orderly fashion. But a form of enlightenment does emerge, a collective catharsis in which all of us participate as interpreters of history in a postapocalyptic world.

In a 1979 interview, Federman called attention to the shift from "typographical play" to "syntactical" experimentation in The Twofold Vibration, arguing that this was part of his attempt to reconfigure fiction at a more basic level ("Inside the Thing" 90–94). Borrowed from Beckett's Le Depeupleur (The Lost Ones, 1971), the novel's title also testifies to an effort of historical rearticulation. Beckett's short prose text de-picts a postapocalyptic world trapped in a cylinder fifty meters in diameter and sub-mitted to a "twofold vibration" of light and temperature. Unlike Beckett's closed world, which experiences only mechanical changes of rhythm leading to entropic stasis, Federman's post-Holocaust world is more responsive to the "twofold vibra-tion" of history, allowing characters to appropriate a space for their imaginative sur-vival. As a comic-heroic offspin of Beckett's Malone Dies, the old man "somehow manages to outwit and outlive his own death by being reborn" in his own fiction (Critifiction 113). This great "thief of language" (The Twofold Vibration 9) borrows freely from modern existentialist masters (Céline, Jabès, Sartre, Beckett) in order to fill in the gaps of his personal narrative.

Smiles on Washington Square (1985; winner of the American Book Award) emphasized even further the rearticulative side of fiction, accommodating—part seriously, part in jest—our appetite for a story line. A passing exchange of smiles between Moinous and Sucette at an anti-McCarthy rally in Washington Square Park is "good enough to get things started, especially since they are both lonely that day and in need of human fel-lowship" (3). As long as the two protagonists "are in a state of emotional availability within the confines of their loneliness, anything can happen" (37), even "a love story of sorts" (the novel's subtitle) engaging critically the conventions of the genre. In our unabashed craving for "nutritional literature," we may miss through first reading the ironic flavor of this experimentation in "love-storyism." It may well be that, as Umberto Eco argues in "Reflections on The Name of the Rose," the "postmodern reply to the modern consists of recognizing that the past, since it cannot be destroyed, because its destruction leads to silence, must be revisited: but with irony, not innocently" (17). Federman couples this notion of "irony"—somewhat devalued by now—with a criti-cal mode of rewriting that reconfigures personal story and history. History becomes a multilevel narrative performance (the narrator shares his tasks with Moinous and Soucette), a true "polylogic" discourse.

Federman's most recent English novel, To Whom It May Concern (1990), illustrates the high stakes involved in rearticulating history. The narrator of this "winter-book" (35) that reconstructs the post-Holocaust destinies of Sarah and her American cousin is helped in his task by the two protagonists who, in anticipation of their reunion in Israel, seek an imaginative convergence between their individual stories. The narrative is as usual pulled in conflicting directions, mimicking the dynamic of separation-reunion in the lives of the two cousins who, thirty-five years ago, "set out in opposite directions. He to achieve his artistic vocation, and she to help shape a sterile piece of desert into a serene yet ruthless country" (14). By making the cousin a famous sculptor, this narrative allegorizes the motif of difficult articulation: "The figures in his sculptures barely emerge from the raw material. They seem either to be struggling to come out and become or else receding into a condition of non-being" (15–16).

The character of the artist shares the essential data of Federman's formative biography as well as a common anxiety over the fragility of artistic creation. As the authorial narrator suggests, this anxiety has autobiographical roots, linked to Federman's traumatic loss of parental heritage (his painter father disappeared in the gas chambers, leaving no trace of his work) and to the threat of cancellation hanging over his own work (as the narrator's daughter comments with unintended cruelty, his epitaph should be "OUT OF PRINT!" [33]). The "book of Sarah & her Cousin" thematizes the drama of *interrupted and revised creation*. The authorial narrator must first "visualize the whole thing, hear the voices, draw the geometry before I can get going" (38–39). The emerging narrative remains hesitant, but it manages to expand the narrator's understanding of twentieth-century history and its representations.

While careful to "verify the details of [their] moment in history," the narrator frees Sarah and her cousin from the naturalistic "specificity of time and place," resituating them in "a place of perfect certainty where something fundamental can be said about them" (39). As members of "a race doomed to survival by its impatience and inability to remain still," the two characters acquire multidimensionality, their lives spanning three geocultural zones (the United States, Israel, and France of their childhood):

> On one side, a land of misrepresentation where Sarah's cousin has been living for the past thirty-five years. On the other, far away, across the ocean, a land of false promises, a piece of desert full of mirages, where Sarah has been living her own exile for as many years. And bracketed in between, the country where the two cousins were born, and where an unforgivable enormity was committed during the war. That place will remain parenthetical. It will linger in the depth of the cousins' background. (10)

The insomniac narrator reaches moments of epiphanic understanding as he chases their story, a "tiny black spot, like a spider," across this triple world. His imagination finds its own three-dimensionality, taking together with his characters a leap out of linear historical time into multilevel temporality. *To Whom It May Concern* makes clear that only by resituating imaginatively the "pure chaos" of history (102) can the writer achieve enlightenment. The tasks of articulating a complex narrative fable in a responsive language are interdependent: If too absorbed in the meanders of plot, narration finds "no shape, no music to the words," its language falling "flat," "empty of resonance" (102); if, on the contrary, narration gets waylaid in endless speculations and procrastinations, no "tale" will take shape.

We need to keep this interplay of tasks in mind as we reassess Federman's fiction. Criticism has traditionally faulted (or praised) Federman for finessing the relationship between "authorial self and worldly history" to the point where the second term almost disappears in "dazzlingly inventive, elegantly written [. . .] small master-piece[s]" that do not "strive deeply into human experience" (Dienstfrey 147). By contrast, David Dowling has urged us to see Federman's narrative experimentation as "metonymically" emerging from American experience:

> [A]n approach which sees Federman's fictional experience as embedded in the American experience rather than metaphoric of it—does fuller justice to his sophisticated sense of America as a semiotic system and to the interplay in his work between biography and history, voice and narrative, text and context. His texts arise out of a specifically American predicament and simultaneously critique it. An apparent flight from political engagement turns out to be a thoroughly engaged polemic. (348)

Dowling's approach helps reconnect the experimental and experiential sides of Federman's fiction, but their interplay need not be reduced to a metonymic relation. The disjointed, multivoiced structure of Federman's books does not simply "imitate" the form of American experience. A substantial amount of restructuring takes place at the discursive level, disarticulating those narrative conventions that perpetuate "manipulations and lies" and "purifying" language "so that it can no longer structure or even enslave the individual into a sociohistorical scenario prepared in advance and replayed by the official discourse on television, in mass-media, in the political arena, and in literature" (33). A similar restructuring goes on at the level of the "story." Federman may well be "one the most intensely personal and autobiographical of the American postmodernists" (Dowling 349), but his approach is always revisionistic, an imaginative rewriting of self and world. As the narrator of *Take It or Leave It* confesses in verse: "I undouble / I multiply / I play hide-and-seek with myself / I subdivide / I cry and decry in two languages / [. . .] I cut and recut myself / [. . .] I disperse / [. . .]I singularize / I pluralize also / I decenter / [. . .] I schizophrenize / [. . .] I double up and undouble again / I redouble or nothing / I multiply by two and demultiply by four" (271; my pagination). Federman's fiction submits personal and collective history to multiple storifications that resist a reductive reading. Writing becomes a "form of liberation" (Federman and Sukenick 134), in both an existential and a cultural sense.

Dowling's analysis focuses primarily on the cultural aspect of Federman's "search for freedom," his "trauma of escaping from totalitarianism and encountering American liberalism" in a form that resisted "Jewish anger, guilt, and sentimentality" (353). Federman's fictional autobiography revolves around "warring opposites": past self, "totally defined by his Jewishness," and present, immigrant self, "defined by his lack of definition" (356). The two selves try eagerly to connect in a "bifurcated sense of history" (356) caught between the traumatic void created by World War II and a metamorphic postwar America that appears to the immigrant as another "kind of void" (*Double or Nothing* 157). This bifurcation remains largely unsettled, as the antithetic titles of Federman's narratives ("double or nothing," "take it or leave it," "the twofold vibration," "amer eldorado") aptly suggest. Federman's work is underwritten by an existentialist tension that criticism, busy praising or questioning his textual devices, has often overlooked. In *The Twofold Vibration*, Federman quotes "his old Rumanian friend [Emil] Cioran" (86) who taught the existential value of laughter. Other existentialist citations include Dostoevsky, Rimbaud of "Une Saison en Enfer," Nietzsche, Céline, and Edmond Jabès; also a number of unnamed sources like Sartre's

Being and Nothingness from which the "old man" in the story steals freely. Even the love-struck characters of *Smiles on Washington Square* adopt existentialist poses:

> As Moinous sits there, crushed by his own superfluous presence and the lurking inse-curities of his being, he has a sudden urge to scream, to jump up and down and scream in the middle of that politely hushed clank of the serving dishes and that sweet buzzing of conversation. He feels like shouting vile obscenities. (33)

Moinous may well be another Roquentin "who wallows in the disorder of his ob-sessive imagination. For him the chaos of imagination always supersedes the order and veracity of facts. That is the basic condition of his existence" (12). But Feder-man's existentialism remains ambiguous, both "felt" and borrowed from literary tra-dition, playing the role of ironic commentary on a bifurcated personal and historical existence. Moinous is rather incongruently paired with the daughter of a rich Bostonian family who reads French, takes writing classes at Columbia, and puts Moinous through a forty-two-day ordeal, refusing to "surrender to the physical plea-sures of love" until she has found a proper ending for her native-meets-foreigner love story. After their exchange of glances at an anti-McCarthy rally, Federman's novel grants one more meeting opportunity to its protagonists, at the Librairie Française where "Moinous happens to be looking at a mystery novel of the Série Noire, and Sucette is inquiring about the latest novel by Simone de Beauvoir" (15). Even after this intertextual help from French existentialism, the fate of the two play-ers is left hanging in incertitude.

The Federman hero is allowed to consummate love only with smothering mother figures or with vampirical but prudish virgins, flaunting their "crooked morality and twisted oedipal complex" (*Take It or Leave It* 309; my pagination). Both are predictable clichés produced by the excitable imagination of a "foreigner." Looming over them all is the mysteriously inviting image of America as a "big fat broad that one must seize with one's arms squeeze passionately a big sexy bitch [. . .] and if you want to possess to explore to search that magnificent bit of geography you have to go a long way and have the desire and courage not only to speak about it but do something about it" (*Take It or Leave It* 114; my pagination). The Federman persona seems to mimic here the stereotypical struggle of the American male to define a "transcen-dent" self "through conflict with and by achieving power over a female antagonist, who [. . .] symbolically represents a vulgar, unrealized culture—the whore America" (Fetterley xx). But this symbolic plot, which Judith Fetterley associates with the classical American novel, is appropriated ironically by a "foreigner" and converted into a narrative of praise for, rather than mastery of, the "other." The scene of the narrator's furtive exchange of glances with a black subway girl, retold in different versions through Federman's books, suggests a transformative encounter with the rich "other" side of the New Continent, a descent into the "belly of America source of life and death" (*Double or Nothing* 164). Every new experience becomes a "discourse of opportunity," every chance missed haunts the protagonist with the awareness of his failed initiation into America.

Implied in these frustrated erotic experiences is the discovery of a "flawed par-adise," an imperfect New World. But there is a flaw also in the ill-adjusted foreigner Boris/Frenchy/Moinous/Dominique, who behaves like a cultural innocent unversed in the American way of life:

His first two years in America, before he was drafted into the army, had been disastrous on all counts. Poverty, humiliation, hunger, unemployment, loneliness. Moinous endured all that during those two years. Plus, of course, the chaotic burden of his obsessive memories. But this time it would be different. [. . .] A new life, yes, a fresh start. (*Smiles on Washington Square* 57–58)

The Federman hero is always en route, a *Homo duplex* (an exile) suspended between two cultures. He has "abandoned the rationality of Europe, however unreal and obsolete it may be, for the temptation of trying to achieve comfort, and perhaps even success, in such a disjointed reality" (21). While avoiding the conventional "good lousy story" of the American immigrant, Federman explores facets of the "crucial confrontation" with a new culture in *Take It or Leave It, The Twofold Vibration,* and *Smiles on Washington Square*. This confrontation involves the narrator in a rethinking not only of his cultural commitments, but also of his notions of language and narration. As Federman suggested in the French version of his autobiographical narrative, the "errance et vertige américain" disrupt conventional narration, creating a vertiginous "errance de l'écriture" (*Amer Eldorado* 2) that both asserts and calls into question the possibilities of fiction. Consider the grand finale in *Take It or Leave It*:

Finished . . . it's finished . . . foutu . . . america . . . the great journey . . . the great discovery . . . the great plains . . . north south . . . east west . . . through the middle . . . the rockies and the plateaux . . . the mississippi . . . the grand canyon . . . rattlesnakes in the desert [. . .] las vegas and the crap games . . . indians with feathers . . . and the naked movie stars in hollywood [. . .] up and down and across the middle [. . .] all the beautiful dreams . . . finished (417; my pagination)

And so he folded himself upon himself like an old wrinkled piece of yellow paper there on that hospital bed as I took leave of him [. . .] closed himself like a used torn book [. . .] as he thought of the trip the big beautiful journey he could have made cross.country coast to coast and which someday he could have told like a beautiful story or retold with all the exciting details to a friend or to some gathering of interested listeners with all the passion necessary to tell such a story directly or indirectly but now it was finished cancelled cancelled and so empty of his last drop of courage and the last words of his story which is now cancelled cancelled since they were shipping him back to where it all started he said sadly to himself
no need trying to go on
no need . . . .
but perhaps the next time . . .
yes . . . the next time . . . (418)

To our surprise the cancellation of the immediate "story" (a projected trip across America) has triggered a vivid discourse about cancellation, a metanarrative thematization (note the book metaphors in the second excerpt) that in many ways is more interesting qua story than the "real-life" narrative.

While the existentialist-erotic plots are most often denied consummation, they occasion important reflections on the nature and ideology of fiction. The shift from an existential to a cultural focus can be illustrated with Federman's "first encounter" motif. Several of his books tease us with the promise of a subway or "boat romance" that is never fulfilled. These unrealized encounters prod the narrators to find expression even if "it is very difficult if not impossible to relate exactly [. . .] what goes through a guy's mind while sitting in a subway (or in front of a sheet of paper) looking between a girl's legs while on his way to the end of his story the first day he arrived in America" (*Double or Nothing* 179). Upon closer scrutiny, these "first encounter" scenes are as much about the capabilities of fiction as they are about erotic desire. While mapping a range of erotic possibilities, Federman's novels also test the limits of narrative creativity, moving from tentative "noodling" in words to seductive emplotments and from the postponed apocalyptic ending of *The Twofold Vibration* to ambiguous attainment in *Smiles on Washington Square*.

Federman's confrontation with his other existential-cultural theme (his "closet" experience and survival from the Holocaust) has made his recent fiction more aware of the tensions between an innovative writing project and the constraints of history. As Bolling wrote in a review of *The Twofold Vibration*, "the boldness (and the value for contemporary experimental fiction) lies in the confrontation of the postmodern aesthetic of disavowal and an event in recent history which demands a passionate response" (7). If Federman's earlier books emphasized primarily an "aesthetic of disavowal," *The Twofold Vibration* and *To Whom It May Concern* articulate a forceful, polemical response to history. But in spite of their different approaches, all of Federman's novels have been concerned with the success or failure of language to rearticulate history. Syntax promises to fix events "into a place, a space, prescribes an order to them, it prevents them from wondering" (Federman, "Why Maurice Roche?" 132). But narration must also remain open to what is nonsystematizable in history. Aware of both demands, *The Twofold Vibration* sets out to "explore the future retrospectively," calling attention to "our bondage to an uninhabitable past and our need to create a future" (Bolling 7). The old man finds himself repeating recognizable literary situations (such as the exploits of Dostoevsky's "gambler") or events that belong to Federman's own history. In order to escape the scripted plots of history, the old man must turn from "observing," "rationalizing and explaining," to "imagining and projecting" (163). His imaginative rearticulation is not the work of a naive *Homo aestheticus*, camped outside of history, but rather that of a *Homo politicus*, engaged like Federman in the transformations of his time.

The challenges of this theoretical and narrative rethinking of history are dramatized in *To Whom It May Concern*. History frames everything in this book, superposing personal and world events. (December 7 marks in the narrator's manuscript the beginning of another chapter/letter, the birth of the narrator's daughter, Pearl Harbor Day, and the day of the extermination of Jews at Chelmno [79].) Sarah and her cousin are equally burdened with history. Both have experienced a traumatic dissociation in their youth, their existence "stalled," resigned to "a condition of temporariness" (46). The adult reunion of the two cousins can take place only after both uncover their short-circuited histories and begin to negotiate their many gaps through letters, conversations, and imaginative "reinvention."

To Whom It May Concern retraces the implacable order of Federman's own autobiographical narrative, but creates interesting variations within it by distributing its events to two different characters, both entrusted with the significant task of bringing history out of the closet. The male protagonist inherits the author's "closet experience" ("Should he call it a birth? A salvation? Or should he name it the beginning of a long absence from himself? [. . .] a false resurrection"? [142]), translating it into a complex metaphor for the re-presentation of history in art. Emerging out of his childhood closet like an Orpheus "who carved his way out of the stone block into which he was buried alive" (142), the cousin rediscovers his Orphic vocation as a young sculptor: "[I]n the space a statue occupies it gives form to an absence. His whole life he had been obsessed with absence, and now he had found a way to render that absence present" (118). But to what extent can art's "form[s] without substance" (117) fill the existential void? As he prepares to accompany a personal exhibition to the Middle Eastern country of his estranged cousin, the sculptor undergoes a predictable crisis of faith, suspecting that his sculptures do not "reflect reality but the crumbling of reality in the mind" (92). In response to the violence of history, the sculptor has been practicing a violent art: Rather than seek "an easy accommodation with the material; he invades the material to destroy it" (92). His savage (de)constructive poetics, forcing together disparate materials, bespeaks deeply unsettled anxieties.

Although the sculptor undertakes his journey to Israel without much enthusiasm, unsure not only of the value of his work but also of the "contradictory politics" and "obtuse views on religion" (91) of Sarah's country, he benefits enormously from the process of narrative recollection that his journey triggers. In the course of their final conversation, the sculptor gains a more accurate understanding of his lifelong artistic struggle. His reunion with Sarah allows him to understand that the historical reconciliation he failed to achieve in the medium of sculpture has better chances of coming about through the mediation of narrative. He collaborates toward that purpose with Sarah and the extradiegetic narrator, bringing a provisional resolution to the mess of history. Sarah's own contribution to this effort is essential. Though psychologically as divided as her cousin, she manages to provide a richer human story, retold without complications in the past tense. Her recollections bridge important gaps in the cousin's own narrative, bearing on other victims and survivors of the war years, whose "entanglements" give purpose to Sarah's own life (137). Her experiences in a country whose very survival continues to be threatened provide her with a philosophic understanding of historical contingencies:

> In this country everything is in a constant state of transition and erosion. The past does not belong to anyone, and nor will the future. When we first came here this place looked like a dead volcano. It was an empty place. We tried to make it full, but the desert is stubborn [. . .]. The desert does not give, it takes. (178)

Both the external narrator and the sculptor learn from Sarah's resilient "acceptance of contingency," refocusing their postmodern art on those details and "meanings that human subjectivity constructs," making their "value [. . .] increase with their tenuousness, with their hint of mortality" (Ermarth 109). Contemplating Sarah's stubborn desert garden, the sculptor concedes that her effort to "carve out these fields" (To Whom It May Concern 182) was worthier than his lonely wrestling with pieces of metal and stone. Sarah's desert ethos teaches the artist to appreciate a simple life freed from intellectual frills ("Here what is not a necessity is an encumbrance." [180]) The narrator also learns from Sarah to worry less about the enormity of his task and to delay his need to tell the story once and for all, allowing himself time to reimagine all the "crossroads," "layers," and historical "threads" (97) he disentangles.

4. IMAGINING THE STORY PROPERLY:
FREEDOM OF INVENTION VS. NARRATIVE CLOSURE

Imagining the story "as it should be imagined" means for the Federman narrators find-ing "the correct words" to bring it into the open (*To Whom It May Concern* 143), freeing it from the "shadow box" of history (*The Voice in the Closet* 19; my pagination). Their ex-istential and cultural survival depend on a liberating language and practice of writing. The authorial figure is called upon to create "a book of flights speak traps evasions question of patience determination" (16). Echoing J. M. G. LeClézio's *Book of Flights*, this motif of mythopoetic liberation is central to Federman's self-representation as a writer. As Dowling summarizes, "[T]hroughout his work Federman plays with the idea of himself as tripartite: mythical Daedalus, the hero of flights; the prosaic man with the feather who escapes conscription—the coward; and 'hombre de la pluma,' the artist who writes it all with his feather pen" (354). Federman's fiction, however, warns us to take the theme of "escape" as a self-problematized proposition. The old man in *The Twofold Vibration* pretends to agree with his friends that the goal of his writing has been "escape" and "transcendence." But the example of his life and work suggests oth-erwise: The writer can escape neither history nor language, having to submit both to a process of critical reformulation that frees some space for self-expression. Only through such reformulation can the "remade self" be reconciled with the "unself pre-sent" (*The Voice in the Closet* 17; my pagination). The writer undertakes his intense self-questioning work within a "primary closet" that is both historical and literary, hoping to turn his "wordshit" (18) into liberating expression.

Federman's entire aesthetics hinges on this tension between anality and creation, entrapment and liberating narration. By contrast to other postmodern writers, from Burroughs to Pynchon and Reed, who have used excremental imagery to denounce modern society's anality—its obsession with "shit, money, and the Word" (Pynchon, *Gravity's Rainbow* 28)—Federman turns excrement into a metaphor of survival and transformation, connected autobiographically to his closet experience. In this "pri-mal" scene, alluded to in every novel, young Federman escapes the Nazi "round up" hidden in a closet by his mother. During his confinement, the thirteen-year-old boy defecates on a newspaper that he places on the roof when he finally ventures out of the closet twenty-four hours later. The package of warm excrement becomes a metaphor for the artist's struggle with the survivor's guilt and with a narrative medium that distorts his experience. Federman's fiction illustrates the dynamic of blockage and release, "miserable excrement" and transformed evacuation described by Roberto Maria Dainotto. The prototypic Federman narrator is determined to "CRAP LIE OR DIE" (to quote the subtitle of chapter XXI in *Take It or Leave It*), using narration as an outlet. This "insignificant signifier, the guy without a name and without a respondent, the living dead in reprieve [. . .], without country or home" (*Take It or Leave It* 105; my pagination), turns his handicap into a strength, "leapfrog-ging" out of his closet through imaginative writing. At the same time his work con-tinues to remind us that, in the words of William H. Gass, "It is always from the point of view of the confined, the shut-in, that the work is performed; and the scenes of public life we see when we look through the pen appear only at the ink end where there seems to be a light" (*Habitations of the Word* 87).

Federman's ceaseless mapping of the interior space of writing (cubicles, boxes, closets, besieged rooms) suggests an obsession with closure and entrapment. In

"Imagination as Plagiarism," Federman ascribes a "primordial closet" complex to the writers who interest him most: Proust, Kafka, Céline, and Beckett. In our postwar "âge concentrationaire" (*Double or Nothing* 67), fiction has been inevitably forced to work in a "closed" field, both existentially and textually. Literature, as *Take It or Leave It* defines it aphoristically, is a matter of:

> four walls
> a table
> a chair
> paper
> pencils
> (or else a typewriter if you compose directly on the typewriter and)
> (tictac tictac tictac tictac tictac tictac tictac tictac tictac)
> and after that hours and hours days and nights weeks and
> months even years and years banging on it (on the damn machine)
> banging your head against the wall
> your ass on the chair
> alone—yes—alone [. . .]
> that's where—and how—literature begins. (180; my pagination)

Narration, we read further in *The Voice in the Closet*, starts in *sequestration*, with the reconstruction of an original "closet" situation (traumatic experience). By confining himself "within the limits of four walls," the first-person narrator hopes to regain the "interiority" of history, isolating the past from superfluous variables: "The purpose and function of the room are that of a container within whose constraints the past of the Third Person is to be (re)created" (Wielgosz 95).

Self-imposed confinement and imaginative escape are interdependent motifs in Federman's fiction. The concrete poem, "Reflection on the Walls" (*Take It or Leave It* 264; my pagination), evokes a Melvillean sense of universal immurement:

> [. . .] DIVIDING WALLS   ENCLOSURES
> DEFENSIVE WALLS
> CELL WALLS   TOWN WALLS   SURROUNDING WALLS
> GREAT WALLS   PAPER WALLS   CHINA WALLS
> INNER WALLS   OUTER WALLS
> ENOUGH TO DRIVE YOU MAD NUTS   CRAZY
> AND   YOU   FEEL CORNERED   LOCKED IN   ENCLOSED   STUCK IN
> THERE   PRISONER
> WALLED IN   ALIENATED
> AND YOU FEEL SICK SAD LONELY COMPRESSED MORBID CLAUSTROPHOBIC
> AND YOU'RE FED UP UP TO HERE ABOVE THE HEAD AND YOU
> WANT TO GET THE HELL OUT TAKE OFF GONE
>
> AND SO YOU JUMP OVER THE FIRST WALL
> OR ELSE YOU GO STRAIGHT THROUGH IT [. . .]
> BUT ON THE OTHER SIDE ANOTHER WALL AND ON THE OTHER
> SIDE OF THE OTHER SIDE ANOTHER ONE AND ANOTHER ONE
> AND ANOTHER ONE AND ANOTHER ONE
> AN INFINITY OF WALLS
> AN ETERNITY OF WALLS

Clearly, Federman is no naive "literaturist" (to use his own disparaging term) who rejoices in the autarchie of his text. His emphasis on improvisation is accompanied by the sobering thought that every narrative detail has a way of becoming inevitable. In its transition from an earlier typographic "acrobatics" to the uncompromising compactness of *The Voice in the Closet,* Federman's own fiction illustrates the growing pressure of narrativity. Against it the writer has deployed an array of disruptive procedures to increase the "unlikeliness that the narrator will be overtaken trapped engulfed by the tidal events of his narration" (*Take It or Leave It* 358; my pagination). But in the process, he has created another confining system: a narrative closet or "débarras" of linguistic and cultural references.

Federman's central philosophic concern has been with the relation between *chance* and *determination* in narration. A recent electronic "micro-fiction" depicts an ideal model of orderly disorder:

> [. . .] some people who had previous experience with other lines said that in spite of its disorganization this was a rather good line perhaps the best line they had ever joined because of its casualness and lack of regulations for indeed in spite of its disorder this line was remarkably smooth and easy going and as such acceptable to most [. . .] ("The Line," *Loose Shoes*)

The continuous (vertical in the on-line version) flow of this narrative sentence mimics the line's "slow endless process." Whatever its original purpose, this multicultural and all-ages line becomes a successful vehicle for narrative sharing: People exchange anecdotes, and the line changes shape "becoming thinner here or thicker there [. . .] because someone had stopped to tell a story or a joke and a crowd had gathered around or elsewhere someone had just finished telling a story or a joke and the people who had been listening were now moving on." Some would make speeches, but generally "people preferred the one-to-one conversation moving with the flow of the line two people would casually talk about [. . .] the usual banalities of life." This spontaneous, microstructural harmony is broken one day as a macrostructural order is imposed upon it. The line turns "into an ugly state of mutual suspicion simply because [. . .] there would be order now in the line oh yes alphabetical order unhappily [. . .]." The alphabetic order disrupts the local animation of the narrative line. And yet, if I understand this piece correctly, Federman does not simply oppose a formlessness that accommodates the contingencies of life to a metanarrative structure. As Charles Caramello has suggested, Federman's fiction is more ambiguously positioned between a desire to promote freeplay at a microstructural level and a commitment to the recovery/rewriting of personal history:

> Federman cannot write "about" the event of his family extermination any more than he can cease striving to do so. He can only [. . .] weave elaborate digressions [. . .], repetitions, interrogations, and interpretations [around it] in an attempt to discover and recover the event in the sedimentation. These constant circlings and layerings are, on the one hand, means of affirming the play of language, but they are also, on the other, strategies for arresting that play. (*Silverless Mirrors* 133)

Paradoxically, the effort to "arrest" the play of language and recover the historical event creates further distortions/drifting. This explains why Federman's fiction unfolds in a "conditional field of narration," to borrow Mas'ud Zavarzadeh's term (40–41), where a number of key autobiographical experiences are tentatively encircled and revised. As the writer told McCaffery, "My fiction emerges out of that unfinished, unsettled conditional statement, which means that the fiction cannot pass for remembered events but it is truly invented, even if some of the facts belong to the past—my past" (LeClair and McCaffery 129). But in the same interview, Federman described his use of an inaugural, "threshold sentence" in terms that emphasized his belief in fiction's capacity to find its own shape:

> Once I have that first sentence, I continually examine it, scrutinize it for its implications—not only of meaning, but of tone, tonality, structure, temporal twist, etc., for in fact the entire novel is already contained there. I begin to detect in that sentence some of the details of the plot [...] and even though I basically know what will happen in the novel, since it is usually loosely drawn from my own experiences, nonetheless what I am looking for in that sentence is the structure, the rhythm of the entire book. (129)

Federman demonstrates this process of narrative self-determination in *Smiles on Washington Square*. After a "rather problematic beginning," the narrator tries to establish his protagonists on a trajectory that would make a second and third encounter between them not only likely but also *necessary*. For, once outlined and accepted, a story "bears its complete organization in itself even before it has been fully formed. For this reason despair over the beginning of a story is unwarranted" (49). This principle of *inevitability*, presented to us as a quote from Kafka, is borne out by Federman's novel. What starts as a comic-speculative project gradually acquires validation. Thus, several episodes hypothesized by the narrator are later acknowledged as having come true and ascribed a position in the unfolding scenario of a "love story of sorts." This story remains appropriately open-ended, avoiding the "plain/normal/regular/readable/realistic/leftoright" form of narrative that *Double or Nothing* (85) mocked. The narrator prefers to build his story through such oblique techniques as *paralipsis* (passing reference) or the *palinode*. (See chapter 1.) He gets the narrative under way in brief hypothetical sallies, written in the present indefinite or in the future: Moinous will meet Sucette again, sharing with her "their first quiet cups of coffee together in a long series of such interludes until the final disappointment" (15); "Moinous comes to visit in her apartment, after they meet again and eventually fall in love" (15); "A few months later Sucette proposes a trip to Boston for the weekend to visit her family. [...] What an extraordinary weekend it will be" (19–20). Moinous finally "sees himself [...] walking away from Sucette's apartment, his black suitcase in his hand, their love story having reached its conditional disappointment" (68). Thus *Smiles on Washington Square* weaves a "love story of sorts" in which the real theme is the (im)probability of a second (mediated, meaningful) encounter with reality.

Finding an effective inaugural sentence that will set this process in motion is all that the fictional author in *Double or Nothing* can hope for. But while "inventing" various beginnings that he abandons a few paragraphs later, the fictional author is invaded by "real" stories that get partly told in spite of him. The extradiegetic narrator (the one closer to Federman, reporting on the trials of the fictional author) hints at the autobiographical source of these "stories": "I suppose I'll have him do muchthesamething I did. [...] **it's quite obvious that we may converge or merge**. And little by little we'll coincide. We'll overlap. *He & I*" (31). Against this recuperative narrative system, the internal narrator mobilizes strategies that "de-create" autobiography, reinventing it in the hope of making "life bearable" (137). Chief among these procedures is narrative *interruption*: interruption of any story that becomes overbearing, threatening the "truth" of life; interruption of the narrator's anxious preparations for writing; even *coitus interruptus*, for the sexual theme cannot be allowed to take over completely.

Invariably, the first-person narrator dreams of freedom in self-imposed confinement. In the closed-upness of his room and "the darkness of the mind [he] awaits the flashes of light that come with imagination" (92). He composes his text within self-imposed typographic and linguistic frames that stimulate his desire to overstep limits. The narrator's experimentation with word pictures and "noodled" columns of text recall the denaturalizing strategies employed by the *nouveax romanciers*. Like Sukenick, however, Federman has preferred a more flexible model of "denaturalization" derived from improvisational jazz, to the abstract formalism of Alain Robbe-Grillet or Georges Perec. Still, in jazz, too, improvisation and pattern are closely intertwined. Every note, even when "erased" or restated, leaves its mark on the compositional flow. Federman's emphasis on formal improvisation is counterbalanced by an awareness of the inner pressures of composition:

> [W]hen the writer begins his battle against the linearity of syntax with wild strokes, all strokes are recorded, a word carries another word, the writer can always add another, cross it out, repeat it, but the rule of the game forbids him to come back upon what has already been done, a return to zero is not possible, [. . .] the played stroke has to remain played. [. . .] Everything here leaves a mark, but not a sign of something, or of something else, but a mark of multiplicity of events which none can ever fall back into nonexistence! (Federman, "Playgiarism" 112)

This Derridean notion of narrative cancellation (a putting under "erasure" that leaves marks) becomes even more important in Federman's subsequent books. In *Take It or Leave It*, the narrator "dash[es] from one parenthesis to another" (291; my pagination), hoping to derail the linear march of narration. He sets up all kinds of obstacles (typographic, syntactic, rhetorical) to curb the "masturbatory recitation" of his "funny beautiful sad and useless stories" (290) that nevertheless keep pressing ahead. His is an art of "traps and survival," struggling "to find strength in what is destroyed / to set the final period / to end it all / to call the story finished / cancellation" (363). A good example of this art in action is provided by the sample love letters written by Frenchy for his army chums: These delectable pieces parodying the epistolary tradition digress freely within arbitrary formal constraints (sixty letters per line). The larger economy of the novel works in similar ways: Nothing is allegedly "planned decided drawn contemplated predicted sketched spoken discussed manipulated in advance" (48). And yet the narrator acknowledges the predictable progress of his "recitation," resorting continually to prolepsis. The "self-propelled" story that emerges almost in spite of the narrator maps the ground between "the madness of sketching all these possible words, the desire and the need to add more, the excitement of chance, but also [. . .] the cool restraint, the control, the necessary calculation, the extreme reserve and the cunning that this game without return presupposes! Irrational balance!" (9).

The peak of this "irrational balance" is reached in *The Voice in the Closet*. This circular fiction-essay, moving through "questions affirmation texture designs negation speculations" (18; my pagination), balances precariously "enunciation and denunciation" (11). Struggling to rationalize his escape from the original "closet," the autobiographical "scribbler [. . .] clumsily continues to fabricate his designs in circles" (7), hoping that "his words will eventually stumble" over the meaning of his survival (12). Though denounced by the protagonist of those experiences (the boy-in-the-closet) as "reducible to nonsense excrement," his narrative words manage to reach—after many digressions and displacements—a provisional understanding of history's implausible events.

The narrator's attempt to extricate a "real story" from the "other side," the "closet" of his past experiences (1; my pagination), is thwarted by the multiplying boxes that, like a visual counter, frame every page. Not only thematically but also textually this book spans a succession of closets. The adult scribbler, hammering away on the keys of his "selectricstud," is trapped in his original "shadowbox of guilt" (19): guilt for having escaped the Nazi extermination while the rest of his family perished in the gas chambers; guilt for "exterminating" that original event once more through his "fraudulent [. . .] edifice of words." The two biographical closets (of the boy's survival, and of the author's later self-sequestration to write his life story) merge in a third, *textual closet* that cancels distinctions between past and present, inside and outside, truth and fiction. The author's escape from his "shadowbox of guilt" is possible only through a successful narrative articulation that will account for his survival and bring some closure to his life story. Federman's work thus encapsulates the essential paradox of traumatic survival, as described by Cathy Caruth: For a Holocaust survivor, "Trauma consists not only in having confronted death but in having survived, precisely without knowing it" (*Unclaimed Experience* 64). Narrative reenactment is an attempt to "claim one's own survival" (64) and make sense of it.

In spite of the book's focus on encapsulation, there is textual flow from page-box to page-box. The dividing "walls" are infringed by narrative "slidings" from one temporal plane to another and from the English text to the French. The first-person monologue of the boy-hero registers this breakthrough, changing its own tone from suspicious resistance to acceptance of the game of narration:

> [. . .] he would like it to be my fault if his words fail to save me I resist curious reversal of roles whereby the rustle of his lies above my head leaves me storyless but through a crack in the wall of my closet I see his hand draw a tree and on a branch a bird a scared mockingbird the shape of a leaf I loved that bird so much that while my scribbler stared at the sun and was blinded I opened the door and hid my heart in a yellow feather to blank his doodling words mimicry of my condition [. . .] (12–13; my pagination)

Later, in *The Twofold Vibration,* the "old man" (a fictionalized Federman) clarifies the overall intention of his "closet" book, its celebration of freedom in face of stark limitations: "[I]t's precisely the fact of the physical text that promises a potential freedom, the closet exists only as a sequence of squares, of doors if you prefer [. . .]" (18). He also scolds his buddies for missing the vibrant life that pulsates in the language of his "closet" book:

> It's all there, you schmucks, inside the words, teller and told, survivors and victims unified into a single design, if you read the text carefully then you'll see appear before you on the shattered white space the people drawn by the black words, flattened and disseminated on the surface of the paper inside the black inkblood, that was the challenge, never to speak the reality of the event but to render it concrete into the blackness of the words. (118)

The Twofold Vibration explains effectively Federman's poetics of the vibrant *closet-text*; but as a novel, it no longer practices it with the energy of Federman's earlier books. The writer seems to have overcome here some of his misgivings about historical narration, confronting the Holocaust and its aftermath more directly. Through the "old man," a "notorious fiction maker" himself, he reconstructs a larger portion of his traumatic life story than in previous books. Still, this novel stops short of providing an explanation for historical disasters, leaving unsettled "the twofold vibration" that haunts it.

In spite of its cavalier acceptance of "love-storyism," not even *Smiles on Washington Square* can settle the conflict between freedom of invention and formal constraints. This book about an accidental exchange of smiles between two strangers cannot lead to a story "unless Moinous and Sucette meet again, by chance" (36): a highly improbable occurrence "[d]uring the Cold War. At the peak of the red scare" (71). The most logical expectation is that of a nonmeeting, that is to say, of a *nonstory*. As a "foreigner" in the "erratic, skittish, discontinuous, monstrously disappointing life of America" (61–62), Moinous's chances for meaningfully relating to Sucette are slim. Moinous is too absorbed in an exile's struggle with "narrow-mindedness and bigotry" (87) and Sucette too busy plotting "a story [that] is not going anywhere" to be easily "decoyed" into a love story. Against this logical conclusion, Federman employs authorial fiat to contrive "a love story of sorts": "Let's say that it is Tuesday the first time they meet. Almost meet. And let's say that it is raining. Cats and dogs, for the sake of the mood. [. . .] Moinous could carry an umbrella and thus avoid rainy depression" (3, 5). As we read on, we are drawn into the paradoxical story of a "destined" meeting by "chance." The grammatical constructions that prognosticate a second encounter are equally ambivalent: Moinous and Sucette exchange a "complicitous smile, *as if* [my italics] they know they are destined to meet again" (3); "that's where they *will probably* meet and speak. At the Librairie Française" (15); "Two weeks later, by pure chance, *they'll probably* come face to face again, but this time they speak to each other, and become acquainted" (69). What Federman highlights here is the conflict underlying any work of fiction, between an author's wished-for freedom of invention and the strong cultural and generic conventions that control his narration.

From this point of view Federman's improvised "love story of sorts" is no less constrained or "boxed" than his other narratives. Moinous's love story is locked in a predictable plot, moving unavoidably from chance encounter through "intensity of hope," postulated consummation, and "conditional disappointment" ("for as soon as a love story begins it has already begun to end" [67]). It is also constrained by genre stereotypes, with coffee interludes after "moments of sensual pursuit," intellectual conversations and confessions. The Moinous-Sucette plot is mirrored in Sucette's story-within-story about Susan meeting a young foreigner called Moinous. By the end of the book, Federman's narration is caught in a triple "bind": Moinous will not be able to consummate his love with Sucette until she brings her own story of Moinous and Susan to some resolution. And everything is enclosed in Federman's own "iffy" narration that explores the narrative and cultural conditions of "love-storyism."

To Whom It May Concern also begins hesitantly, looking for "the kind of opening that sets everything in place and makes the rest of the story happen by itself" (18). The story, told in anticipation with modal or future tense verbs, is as "iffy" as other Federman narratives. Still, a stronger sense of necessity governs this story about people "dragged into history in spite of themselves" (128). The subjunctive mood ("this should be a joyful occasion," "they should be joking [. . .], wondering what he will think of them") is neutralized periodically by a diegetic present or preterite—by a logic of "reality." Various events, from a terrorist attack in Tel Aviv to the "unforgivable enormity" (99) of the Holocaust, impinge on Federman's story, becoming his "inescapable facts." But these events are submitted to narrative re-creation that characteristically hopes to "stumble" over a larger truth of the writer's own making. There is, however, an important change in emphasis from previous Federman texts: What this novel demonstrates is not a process of fictionalization, as in *Smiles on Washington Square*, but rather one of narrative "realization": "We're not talking about a fairy tale here, we're talking about a story which in the process of being told might become the absolute truth" (19). We can understand Federman's distinction in terms of Lyotard's description of postmodern narrative realization as a process that combines the task of "perpetually flushing out" artifices of representation, with the task of reinventing the rules and categories of thought, seeking a new "realization" (*The Postmodern Condition* 79, 81). Federman's recent fiction tries to do both. The narrator's problem in *To Whom It May Concern* is no longer just "how to launch the story of the two cousins" (155), but also how to bring it to a "compelling end" without letting it fall into predictable plots. The post-Holocaust/post–Cold War writer cannot avoid "facing toward an unthinkable end" (166), just as he cannot prevent himself from "looking over [his] shoulder [. . .] toward the noise from which one came" (156), at the risk of being turned into a pillar of salt like Lot's wife. What he can do, however, is play one "vision" against the other, keeping fiction within the realm of endless revision and "uncalculated postponement" (155). By delaying the resolution of the story, the novelist hopes to avoid Lot's fate and also arrive at a provisional understanding of history.

5. "Exile[d] into this Recitation": Voices vs. Texts

An important part of Federman's effort to rethink the relationship between narration and history is his complicated mise en scène of persons and voices. The process of narration usually assumes a dialogic aspect in Federman's fiction, splitting/multiplying narrative functions (tellers-retellers-protagonists-narratees) and establishing competitive relationships between them. The effects of these divisions are often comic, but there is also a serious purpose behind Federman's theatricalization of narration. Though often regarded as a print-oriented novelist who uses new technologies to redefine the novel as an object, Federman has also been interested in recovering the oral immediacy of narration, which—to borrow Peter Brooks's description—can put us "in touch again with a lived situation of exchange between narrator and narratee, creator and public" ("The Tale vs. the Novel" 287). His effort to recuperate patterns of oral performance is motivated by an awareness of the epistemological limitations of writing. Federman's surrogate writers would probably agree with Walter J. Ong that writing is "a particularly pre-emptive and imperialistic activity that tends to assimilate other things to itself," "tyranically lock[ing words] into a visual field for ever" (12). Against what Ong calls the "diaeretic or divisive" semiotic of writing that separates knower from known (46), vision from hearing (72), surface from depth (129), Federman's "tellers" revert to interactive narration. By emphasizing "storytelling events" rather than simply the story and "participatory rather than objectively distanced" modes of expression (Fleischman 121), these tellers hope to narrow not only the discursive gap among teller, protagonist, and listener, but also the deeper epistemological gap between narration and life.

The effort to return narration to its "natural habitat" in a "real, existential present" (Ong 39, 101) may appear hopelessly nostalgic in our age that, as Walter Benjamin warned half a century ago, has lost "the ability to exchange experiences," bringing "the art of storytelling to its end" (Benjamin 91). Perhaps our faith in the "aggregative," community-building value of orality is somewhat unrealistic. Locked in a performative present, oral narration admits little revision or reflexivity. The sense of plenitude it creates is provisional and even fake: "After the speech was delivered, nothing of it remained to work over. [. . .] [T]here is nothing outside the thinker, no text, to enable him or her to produce the same line of thought again or even to verify whether he or she has done so or not" (Ong 10, 34).

Even though Federman's (re)tellers sound periodically a Platonic note, denouncing of the artificiality and disconnectedness of the written word, they are no naive defenders of phonocentricity. If Federman is at all successful in reclaiming the interactive dimension of storytelling, it is not because he subscribes to an idealistic notion of oral plenitude, but rather because he recovers—in Ong's words—the "agonistic dynamics of oral thought processes and expression," rejoining narration to the arena of human struggle (43–44, 45). Federman's fictions emphasize the confrontational, "platformatic" side of narration, involving competing agents at every level: "tellers," "retellers," "scribblers," and "rectificators," protagonists and their alternative selves. A *second-person pronoun* is often used to designate these divided positions. The second person can represent a "dialogic" splitting of the narratorial voice, emphasizing the conflicting selves of a character-narrator (*Double or Nothing, Take It or Leave It*). It can also refer to the fictional protagonist, creating empathy and/or uncertainty at the level of figural narration (*Double or Nothing, Take It or Leave It*). It can likewise dialogize a character's internal perspective, with the acting self identified as "I" addressing backward his later narrative self as "you" (*The Voice in the Closet*); or it can designate the communicative relation situated at various levels of narrating and listening: intradiegetically, as a "'relay point' between the narrator and the implied reader" (Genette, *Narrative Discourse Revisited* 131; for example, *Take It or Leave It, The Twofold Vibration, Smiles on Washington Square*), or extradiegetically, "as a relay point with the real reader" (*Take It or Leave It*). We can find further divisions at the level of the implied reader, with the second person designating an ideal respondent as in *To Whom It May Concern*. Together with other narrative positions (the first person of the reteller, the third person of the protagonist), the second person of the narrator/narratee is involved in a renegotiation of the narrative process, pluralizing and problematizing it:

> In place of authoritative narrative voiçes, [postmodern narratives] propose stammering, self-canceling ones; in the place of the Flaubertian ideal of impersonality and impassibility, they present idiosyncratic and engaged narratives. Above all, they appear—often in a mood bordering on desperation—to propose a dialogue with the reader, to ask for the reader's response in exchange for counsel given. What is difficult to judge is whether this attempt to engage a dialogue marks a renewal of the oral communication situation in the novel, or rather a last, desperate, and doomed action to react against decontextualization and consumerist reading. (Brooks, "The Tale vs. The Novel" 291)

At once self-assertive and self-controverting, Federman's "voices" dramatize the difficulty of articulation in a culture that has commodified narration. Unlike the

French *nouveaux romanciers,* who usually reduce narration to a textual dynamic, Federman has emphasized the tribulations of the enunciating voice: "[O]f course, there is always a matter of [. . .] degree of presentness of that authorial voice. But even the fact of pretending to write a piece of fiction which doesn't reveal the voice of the author is a way of pointing to that voice, or to the absence of that voice" (interview in LeClair and McCaffery 135). The method pursued by Federman against "a certain type of realistic novel [. . .] which functions on the basis of a closed form" has emphasized the presence of "all sorts of writers (storytellers) telling little stories (one could almost call these essays) which reflect on the main story" (135).

The process of narration in Federman's novels goes by the name of "recitation," a term borrowed from Beckett that designates a self-conscious interplay of oral performance and (re)writing. The question raised by Federman's fiction is whether this type of narrative recitation can recapture the rich circumstantiality of the world, or whether it is condemned to remain a mere simulation. Evidently, there is no clear answer to this question. The multiplication of voices in Federman's fiction indicates both a genuine dialogic pursuit and a distrust of voice. For example, in *The Voice in the Closet* experience gains meaning through a paradoxical interplay of speech and print. The "voice" of the boy-in-the-closet can be heard denouncing his author in the impersonal narrative stream "ejaculated" by the Select IBM typewriter. He can rant, "I am alive promising situation I am my beginning in this strange gestation" (9; my pagination) only as long as the "scribbler" is willing to hit the typewriter keys and play the role of "artificer of fledgling birth in retrospect for remade self caught in unself present" (17). Outside the narrative discourse of his "creator," the boy has no reality. Ironically, even within that discourse his identity remains uncertain, defined by the scribbler's "wordshit of fabulation" (14). The protagonist dreams of an alternative form of self-articulation:

> [. . .] there must be a better way to manifest myself to assert my presence in his exercise-book speak my first words on the margins of verbal authenticity I will step into the light emerge run to some other refuge survive work tell the truth I give you my word resist I will abolish his sustaining paradox expose the implausibility of his fiction with cunning expedients stratagems that will cure him of his madness even if the act of telling my own tale sends him to oblivion [. . .] I will step out of my reversed role speak in my own voice at last even if I must stretch myself to the unattainable [. . .] (14–15)

But his attempt at reversing diegetic roles and recovering a "truer" story ends circularly, "on the edge of the abyss stalled words in regress without destination an historic fiasco" (15). Federman's re-creative discourse remains suspended between a potential speaking subject who is never made fully present and an equally problematic, depersonalized authorial "voice."

The hesitation between different "voices" creates a continuous crisis but also dialogic animation in Federman's novels. In *Double or Nothing* the character-author (designated as "you") both "invents" and "suffers" the narrative events described. Being closest to the process of narrative production, this "second person" is confronted with the monumental task of reconciling living with writing, reinventing his "life" in the form of a narrative about a half-fictitious "third person." Consequently, the voice of this "paranoid and confused" second person has to be mediated by other voices according to an interactional pattern explained in Federman's next novel:

> [T]his first novel [. . .] juggles four "voices": first, a rather stubborn and determined middle-aged man who decides to record word for word the story of another (second) man, rather paranoid and confused, who decides to lock himself up for a year (365 days, more or less), subsisting entirely on noodles (that's right), in order to write the story of yet another (third) young man, shy and naive, who comes from Europe (perhaps from France) to America and who (if the second voice can pull itself together sufficiently to write and be recorded by the first voice) will experience various adventures and so on but who must for the time being wait until he is charactered—all of which implies a fourth voice managing the glorious, sacred, gimmicky confusion craftily jumbled. (*Take It or Leave It* 361; my pagination)

In spite of this division of narrative tasks, neither the involved second-person character-writer nor the extradiegetic first-person "scribbler," nor for that matter the "overseeing" fourth person whose main contributions are the "Summary of the Discourse" at the end of the book and the mock introduction called "This Is Not the Beginning," can get their novel about the third-person hero under way. The voiceless "third person" vainly awaits adequate description while the "second person" remains concerned with the shape of his projected novel and the "first person" engages the "second person" in speculations about how to survive as a transplanted writer in the fifties America:

> Have you ever considered the possibility of borrowing money? Just enough for the room the noodles and all the rest **365 days**. Which means you've got to have an exact figure **down to the last penny**. Of course, it doesn't work unless you borrow with the intention of never returning the money. [. . .] Good twist! [. . .] But assuming I borrow the money. And I don't mean borrowing from a private party. That's out of the question. Nobody would trust me. I mean borrowing from one of those loan places. [. . .] Takes a long time to find out about these places when you are new in America. [. . .] Takes a good five years even more before you learn all the ropes. Neither did I in the beginning. That's why the first five years are so tough. For him. For me. For anybody in our situation. (*Double or Nothing* 104–5)

As this passage suggests, the multiplication/hesitation of voices and persons is caused by the pressures of exilic life. Federman delegates the narrative function to a dialogized author-character, a "you" who holds the promise of sharing the burden of rearticulating self and world. This approach complicates the type of "auto-bio-heterodiegetic" narrative described by Genette (*Narrative Discourse Revisited* 133) on the model of Philippe Lejeune's "second-person autobiography" (36), in which "je est un autre" ("I is an other"). Federman's narrators may want to write a "MONOBIOGRAPHY," but they have to acknowledge that "all fiction is digression" and "a biography is something one invents afterwards" (*Take It or Leave It* 97; my pagination). This is particularly true for a survivor of the Holocaust and self-exiled writer like Federman, who has had to retranslate his life story into new narrative propositions and languages. His voice can be only "a voice within a voice," which "plays hide and seek with its shadow" (*Critifiction* 76). Ideally, the reporting "I," the participating "you," and the figural "he" meet—as in the above quote from

Double or Nothing—in the shared space of "us." Narrative sharing has in the mean-time become Federman's chief reason for employing the second person. In his first novel the interaction of voices still marked a space of hesitation wherein author, narrator, and character exchanged identities, becoming equally absorbed in the uncertain project. The second person is employed in *Double or Nothing* either to in-dicate a comic crisis at the level of authorship:

> You've got to invent something more credible more reliable. [. . .] [W]hen you get the guy convinced but you start mentioning the toilet paper for sure the guy will throw you out of his place. [. . .] You can't take a chance and run out. That would really be a hell of a note. [. . .] If one runs out of toothpaste one can do without. Sugar also even if the coffee tastes like piss. Noodles that you can always stretch out a bit even though it's at the core of the system. Essential in fact. That's survival. *But toilet paper?* [. . .] How es-sential is it? (105–6)

or to recapitulate—in a questioning mode—the fragmented life story of the charac-ter-author:

> you   ?   crazy   or
> irresponsible Rooms and suitcases that
> ? fine but a tenor saxophone that ? too
> much Rooms and suitcases it ? living
> in one place and traveling sometimes
> A   whole   life   ?   contained   between
> rooms and suitcases The room ? you
> ? in one place The suitcases that you ?
> from one place to another That ? the
> way   to   ?   movement   Movement   in
> time The time element ? important
> too Space and time in other words [. . .] (198)

The use of the second person as a self-problematizing technique should not sur-prise us. As Brian Richardson notes, "[o]ne would assume that the inherent artifi-ciality and instability of second-person narrative is ideally suited for the genre of autobiography, a mode that is both theoretically falsifiable and necessarily fabri-cated" ("The Poetics and Politics of Second-Person Narrative" 324). As a result of this use, the biographical "author," as a traditional authority figure, is problematized and absorbed into his fictional text, becoming—in Barthes's words—"a paper being, [. . .] matter for *connection*, and not *filiation*" (*S/Z* 217). Federman comments on this process on an unnumbered double page inserted between pages 146 and 147 of *Dou-ble or Nothing*: "Through all the detours that one wishes, the subject who writes will never seize himself in the novel: he will only seize the novel which, by definition, ex-cludes him [. . .]." Nor will the author "seize" his character, allowing him to develop a consistent life story. The *second-person* pronoun covers that space of indeterminacy and hesitation wherein narrator and character merge, exchanging identities, becom-ing equally absorbed in the uncertain project of the text.

In Federman's subsequent novel, second-person narration is more clearly con-nected with a number of projective scenes that the Federman persona, designated as "you," and the implied reader are encouraged to participate in as part of their initiation into the mysteriously inviting image of America, a "big fat broad" (*Take It or Leave It* 114; my pagination) that "you" must seize boldly, passionately. This use comes closer to the "typical" form of second-person narration defined by Brian Richardson as "a story told, usually in the present tense, about a single protagonist who is referred to in the second person," and where the "'you' also designates the narrator and the narratee, though [. . .] there is frequently some slippage in this usual triumvirate" ("The Poetics and Politics of Second-Person Narrative" 311). Federman is well aware of the ambiguity this mode creates through its violation of narrative boundaries. As Monika Fludernik explains, the extradiegetic narrator "addresses a character (an entirely non-realistic, deliberately anti-verisimilar procedure, violating the boundary between discourse and story)" while at the same time "instantiat[ing] an existential bond with his or her former (discourse) self" ("Second-Person Fiction" 222). Federman adds an interesting cultural twist to this paradoxical narrative epistemology: Retold in different versions through his novels, the scene of the narrator's furtive exchange of glances with a subway girl is amplified and fictionalized until it becomes a metaphor for the foreigner's encounter with a darker, richer side of the New Continent. Dialogic narration marks simultaneously a "discourse of opportunity" and a chance missed by the protagonist:

> [N]o idea what America was about (who does?), geographically speaking [. . .], no idea particularly of the size. [. . .] Unbelievable the colors and the spaces. [. . .] You've got to see that to believe it. [. . .] ENORMOUS distances—between places! [. . .] Between people too! Between words also! And all these people (all these words) all of them Americans who look (what a way to start!) at you, who scare you shitless [. . .].
> (19; my pagination)

> [. . .]But New York [. . .] it's like a spider web a huge stadium a huge arena with all kinds of athletic events the kinds that force you to excel to surpass yourself or else! [. . .] Crawling on your stomach doesn't matter QUICK got to make it back! [. . .]
> And as soon as you arrive
> and as soon as you're over the bridge out of the tunnel QUICK (never fails) a telephone booth!
> The little dime in the slot! QUICK! Nervous fingers, and the familiar voice quickly connected far away at the end of one of the threads QUICK the spider web! HELLO! (196)

The re-creative effort of the novel passes periodically through the discursive space of the second person, but it does not rest there, moving subsequently either into a first-person "teller" mode or a third-person figural mode. Neither mode is able to sustain itself for more than a few pages, dropping back into an interpersonal narrative space that overlaps voices or, as in the above New York vignette, into a "personless" prose of nominal notations that give the impression that "this story will now tell itself alone without the support of the person (pronominal or otherwise)," moving "without efforts by a simple horizontal (but vertical too) accumulation of signs and facts" (295). Federman's novels, like those of Joyce or Beckett, circumscribe "un espace poliphonique/cacophonique, un espace plein des ruptures," in which voices and textual silences compete continuously (Durand, "Le continu de la fiction" 126–27). But their conflict is not allowed to reach the extremes of cacophonous noise or solipsistic silence. While "so many of Beckett's voices are alone," Federman's "main voice is always surrounded by noisy listeners," mediated by secondhand tellers who raise the told "to epic proportions"; "Beckett's voices, trapped in a language that is not theirs, agonize over the need to speak, Federman revels in this situation" (Pearce, *The Novel in Motion* 119–20).

Take It or Leave It makes this confrontation "platformatic," pitting the "narrator's voice (varied and disguised, to be sure)" against "those of various unnamed but easily identifiable others (the TEL QUEL boys, some odd strangers, plus everyone else Hombre has ever known or imagined)" (361; my pagination). Described by Ronald Sukenick as "the most successful attempt I know of to recognize and resolve the tension between the communicative and visual aspects of fiction" (*In Form* 45), this novel distributes the tasks of narration to competing agents, adding dialogic texture to an otherwise flat and problematic story. As Federman explains, "it is by a system of double-talk that the story rises from its banality to what can be called a level of surfiction" (*Take It or Leave It* 179; my pagination). The traditional triumvirate narrator/hero/reader is redefined here as main teller/character-teller/listener, to suit the economy of a "recitation"; it is then complicated through further splitting/doubling. For example, the semiautobiographical protagonist "Frenchy" is given an American alter ego, Moinous, to contain his "foreignness." Federman is represented by a "teller" (with punned names such as Corporal Hombre della Pluma and "feathermerchant") who is quite finicky about authenticity, promising "real life" instead of a "crummy reproduction." He nonetheless manipulates Frenchy's story through reinvention and "exaggeration." At a certain point he exits the book to visit his buddy Ron Sukenick, so that he is replaced temporarily by a "Re-Teller" who insists on getting the story right but who eventually admits his failure to rectify a story mistold from the outset. Both tellers (finally collapsed) promise more than they can deliver. Their authenticating effort is circular, self-controverting:

> Everything I am telling you of course is true. Naturally, it is somewhat distorted from reality. But in general it follows the broad lines of life. Evidently it is possible that there are errors, and exaggerations in this tale. False reflections. Chronological deformations. Confusions! Paddings! In other words, all kinds of things which, normally, ought not to be found in such a tale, and yet, inevitably, cannot, must not be left out! Because (as I firmly believe) all fiction is digression. (99; my pagination)

This contradictory art of "recitation" is challenged by "a bunch of disgusting rectificators inquisitors interrupters [...] auditors" (109) addressed as "you." Though they are never quoted directly, we glean from the teller's angry pleadings that these narratees harangue him with questions and protestations: "Dammit! If you guys keep talking all the time / and at the same time / we'll never get it straight! Do you think it's easy to tell a story? [...] Particularly when it's not YOUR story—a second hand story!" (17). The narratees, we are told, are faddish "literaturists" and "pseudostructuralisators" (100) with no stomach for "simple" narration. (In the French version of the novel their identity is appropriately adjusted to include the "nouveaux romanciers" and adepts of "bouquinage.") They allegedly abandon their narrative level, landing in the middle of the story as participative "critispies."

Such metaleptic transfers from one narrative level to another undermine the tellers' jealously guarded distinctions between writing and reading, reality and fiction. The repeated interference of narratees makes it increasingly difficult for either teller or reteller to regain control of their "progressive" story. There are a few rare exceptions where a more cooperative relationship is suggested, one that encourages the emotional participation of the narratees in the experiences described by the tellers:

> Can you feel abstractly the electricity that circulated through the air. Can you feel how Benny suddenly looked mean and angry [. . .]? (239; my pagination)

> [A]nd above all if you are Jewish yourself, and on top of it, as in my case, you're one of the lucky ones who escaped the holocaust, and you've just arrived from the old country [. . .] then it's even better, [. . .] the New York Jews they really feel sorry for you and want to partake in your misery in retrospect. (265)

On the whole, however, Take It or Leave It emphasizes the confrontational rather than the supportive side of the second person, the latter aspect entering only gradually Federman's fiction as part of a difficult process of reconciliation with history and its narratives. The confrontation with real or imagined narratees forces the tellers to a process of reflection that helps refine their narrative poetics. Against both dogmatic realism and the excesses of experimentalism, Federman's narrators propose a form of "laughterature" (176) that can chuckle at "all the shitmerde of life and death" (177) but also at itself. By way of their "delirious writing laughing up and down the pages [in] words that move and crack and giggle" (186–87), Federman's character-authors manage to insert "through the cracks in a written text that which was originally excluded for aesthetic or ethical reasons" (342).

If in Take It or Leave It Federman's "plural voices" seemed to thrive within their "intramural set-up" (334, 344) as they contested a set of stories, in The Voice in the Closet this type of narrative confrontation heightens anxiety. There is a good thematic explanation for this increased tension: While the two preceding novels dealt with Federman's American initiation (his work in "shit city Detroit," his training in a bigoted paratrooper unit, his departure to fight in an "illegitimate [Korean] war" [158, 220]), this compact text takes on the much-deferred theme of Federman's survival of the Holocaust. The level of anxiety is also enhanced by the confrontation of "voices" and texts that dramatize for us the problems of historical writing. This remarkable experiment in polyphony pits the protagonist (Federman as a boy, survivor of the Nazi roundup) against the "scribbler" (Federman as an adult writer). The logical sequence of past self and present self, narrator and protagonist is thus reversed, with the protagonist (as the "first person") talking back to his creator (as the "second person"). The "counter-voice" of the hero "reflects backward-wise on the hesitant progress of the central voice-text, displacement and denial being the technique here" (Take It or Leave It 361; my pagination). The confrontation between protagonist and scribbler turns narration into an endless "enunciation and denunciation" (11):

> [. . .] federman achieve the vocation of your name [. . .] cut me now from your voice not that I be what I was [. . .] but what I will be [. . .] the self must be made remade caught from some retroactive present apprehended reinstated [. . .]
> (3; my pagination)

> [. . .] but where were you tell me dancing when it all started where were you when the door closed on me shouting I ask you when I needed you the most letting me be erased in the dark at random in his words scattered nakedly telling me where to go how many

times [. . .] must he foist his old voice on me his detours cancellations [. . .] repetitions
what really happened ways to cancel my life digressively each space relating to nothing
other than itself [. . .]
(5; my pagination)

As these passages suggest, there is a continuous shift in voice from first person to
second and third and even to impersonal constructions that reproduce the mechanical flow of the text from the scribbler's Selectric IBM typewriter. The voice
of the protagonist, addressing his creator in the second person, pleads with him
to find the "right aggregate" (5) of words that will illuminate his story. But just as
often the protagonist lapses back to a third-person address, as if doubting the efficacy of his interpellation of his creator. The protagonist remains a "divided I
who speaks both the truth and the lie of [his] condition" (11) and who in turn depends on ambiguous, half-cooperative second- and third-person positions. Even
before he can start denouncing his creator, the boy-in-the-closet has to earn his
forbearance, pleading with him for a voice and liberating plot: "create the true me
invent you federman with your noodles gambling my / life away double or nothing in your verbal delirium don't let anyone / interfere with our project" (4). Protagonist and scribbler remain interdependent, coming alive at the intersection of
first-, second-, and third-person positions that conflict, converse, and begin to
cooperate.

The Voice in the Closet marks a turning point in Federman's re-creative approach
to historical narration. "Leapfrogging" over conflicting voices and narrative styles,
this text tries to open history's plots up to "infinite possibilities" through "digression" and repeated "displacements" (6). The Twofold Vibration pursues a similar narrative dialectic, but manages to enhance its cooperative side. The
teller-reporter-scribe, later identified as "Federman," is helped in his effort to
piece together the history of a Holocaust survivor and "Jewish space hero" ("the
old man") by two detective-narrators addressed as "you" (Moinous and Namredef) and by the protagonist who is a novelist himself, claiming the same works as
Raymond Federman. The first-person reporter-scribe seems more in control of
his story than previous Federman (re)tellers, more engaged with the theme of the
Holocaust and exile to the New Continent. It is also true that his expectations
have diminished: All he asks is "a little elegance, a touch of decorum, style [. . .] on
the outskirts of darkness, that endlessness of survival on the edge of the precipice,
leaning against the wind" (66). He shares more willingly his reconstructive task
with Namredef and Moinous who, as the protagonist's buddies, can report "on the
anxiety, the self-doubt, the fear of our old man, but the joy too, [. . .] the mad radical laughter of his existence" (66).

Even when approached cooperatively, the task of making sense of history remains forbidding. How can the Holocaust be revisited except in the guilty, melo-
dramatic "merdier littéraire" (114) that Federman himself briefly illustrates when
recounting the old man's visit to Dachau? How can a sequence of historical fiascoes (the Holocaust, the Cold War confrontations, the projected deportation of
undesirables to the space colonies at the end of the millennium) be told except in
"words abandoned to deliberate chaos and yet boxed in an inescapable form"
(116), like those of The Voice in the Closet? In a polemical exchange with his buddies,
the old man defends his right to produce disarticulated narratives, spinning count-

less versions of the same story. As a fictional exploration of recent history, *The Twofold Vibration* inevitably runs into "tautological disjunctions," becoming a "text of rupture" in Sollers's sense of the term (6–7). Expanding on a motif from Edmond Jabès, the scribe "Federman" suggests that recent history can be summed up as a plot of existential and cultural displacement:

> [W]e are all displaced persons surviving in a strange land, in life as well as in fiction, and so why not ask, even if it is in vain, as it is asked in The Book of Questions, Old Man tell us the story of your country, and speaking for us all he would answer, as Yukel does, I have no country, I am an old man, and my life is the story. (150)

Displacement is the ultimate subject and condition of fiction itself. A historical narrative "is always something one invents afterwards, after the facts, [. . .] usually from beyond the grave, outre-tombe [. . .], for the truth of this world is death, one must choose, to die or lie" (70).

But *The Twofold Vibration* also envisions a possible reconciliation with history in the "borrowed land" of fiction. The old man, his two buddies and critics, and the teller-scribe Federman are all exiles, seeking an uncertain "home" in their own narratives that allow them to "live a deferred life," in "transition from lessness to endlessness" (147). Their voices, kept distinct for a while, "overlap within the twofold vibration of history" (the pronominal marker becomes appropriately "we"), cooperating toward a common goal: rearticulating the old man's dramatic past so as to give him a future. Echoing the Freudian mapping of the psyche, each member of the "triple-headed" narrator (122) contributes an important aspect to this rearticulation of history: a speaking subject through "Federman," the teller; rationalization through Moinous, the "philosopher"; and comic relief through the impish Namredef.

There is, thus, a significant evolution in Federman's use of multiple narrative persons that further reflects on the writer's narrative philosophy. The early novels employed different narrative persons for dialogic reasons, to enhance the confrontational tension among tellers, protagonists, and narratees, but also for cultural reasons to generate tensions between history and narration, fact and fiction. By contrast, Federman's more recent novels have employed the second person to enhance the cooperative, integrative aspect of narration. A character-author addresses his story to a sympathetic narratee (the old man's buddies in *The Twofold Vibration*, Sucette in *Washington Square*, an unidentified fellow writer in *To Whom It May Concern*) who is invited to receive, but also to contribute to the (re)writing of the story. The narratees are all students or practitioners of writing. Their contributions become an essential part of the theme and structure of each novel. This shift is most noticeable in *Smiles on Washington Square* and *To Whom It May Concern*. If *The Twofold Vibration* expresses the darker, troubled side of historical rewriting, *Smiles on Washington Square* emphasizes the comic, conciliatory side of the process. Thematically, this book still brackets an uncertain existential space. The accidental exchange of glances between Moinous and Sucette on March 15, 1954, is actually a missed chance for further interaction: "[T]hat day they do not speak. No. They smile at each other, nothing more" (3). And yet the promise of that initial encounter is enough to generate "a love story of sorts," filling 145 pages. The "love story" becomes possible because of the cooperation between several tentative plots: Federman's frame narrative, but also the stories wished into existence by his two characters, Moinous and especially

Sucette who, as an aspiring writer herself, is at the time she spots Moinous engrossed in writing a romantic piece about imaginary "Moinous" and Susan. Moinous and Sucette perform important narrative functions, naming themselves and inventing their own amorous encounter:

> Sucette already loves the Moinous she has created in her story. And certainly, for Moinous, all the delights of flesh and spirit will be contained in the name of Sucette, which he will repeat to himself time after time when he is alone, trying to kiss it with his lips whenever is passes through his mouth. And Sucette too will continue to murmur affectionately as she brings her story to its conclusion. (132)

The pronominal distribution in *Smiles on Washington Square* further emphasizes the theme of cooperation. Moinous's name suggests a partly successful synthesis of voices, a viable compromise between the subject position "moi" (me) and what Jean Ricardou (in a discussion of Sollers's fiction) calls the "nous/we of the text" (118–19). This synthesis is obtained via "a sense of togetherness with you" (89), where "you" stands both for a replenishing other (Sucette), who gives Moinous spiritual support and a vocabulary to understand the bigoted discourse of Cold War America (Sucette reveals to him in mockery America's lexicon of invectives against its political and ethnic others), and for an earlier alienated self of Moinous whose experiences as an exile in America and soldier in Korea are recalled in second-person narration:

> You meet people. Some of them very friendly. Mostly housewives because the husbands are at work during delivery hours. Sometimes they invite you in for a cup of coffee or something. Tell you their stories. Complain about how life is boring. Show you the kids' pictures who are in school. Ask about your life. Notice your French accent, and immediately tell you how charming and sexy it is. In fact, that's how Moinous got into trouble again. (63–64)

The true "encounter" between Moinous and Sucette takes place on paper, in a fictional/textual convergence that defines several relationships: between a character (he/you) and an embracing text (she/you; as her name suggests, Sucette is a "pacifier" for Moinous); between an author-within-the-text (Sucette) and a narratee (Moinous) who critiques her story in progress; or between two characters with their parallel narratives meeting only on the imaginary horizon of intertextuality. This dialogic structure of characters and pronominal positions works almost by itself: Sucette is created by Moinous's need as a foreigner with "[n]o past. No future " (132), and in turn creates him and his need to indulge in "the vain dream of love, the inexhaustible torrent of fair forms, the sterile and exquisite torture of understanding and loving" (130). Moinous (Federman) seems more eager here to sustain the flow of words that will "hold together" reality and fiction in a love story of sorts. Federman's Ishmael has learned to smile upon his "exile into this recitation" (*Take It or Leave It* 365; my pagination).

Federman's most recent English novel, *To Whom It May Concern,* completes this exploration of the conflictual and cooperative uses of narration. This novel begins in a familiar dialogic mode, with an apostrophe to a narratee who is probably a friend and fellow writer: "Listen [. . .] suppose the story were to begin with Sarah's cousin delayed for a few hours in the middle of his journey [. . .] stranded in the city where he and Sarah were born" (9). The second person designates a cooperative reader-respondent who is conjured up to share the questions of the authorial narrator as he launches the "book of Sarah & her Cousin" (37) that would reinterpret not only their personal stories as survivors of the "absolute erasure," but also the "lamentable [collective] history" of the last fifty years that impacted it. But this narratee appears at times to be little more than a figment of the narrator's imagination, so that the second person may actually indicate a form of self-interrogation or self-address. This may have been the case also in some of Federman's earlier fiction, whose "tellers" and "scribes" often imagined themselves to be in competition with alternative narrators and harassing narratees. In Federman's latest novel, however, the need for a second-person position has a stronger motivation: The monumental nature of the task at hand makes it impossible for a single narrator to handle it. The burden of history must be shared, the interpretive duties distributed to a number of complementary agents and pronominal positions: an emotional first-person voice, speaking like Dostoevsky's "underground man" from the "basement of [his] own despair" (30); a responding interlocutor ("you"); and the two protagonists of the projected book ("he" and "she"), themselves participants in the reimagining of their stories. From this perspective, Federman's second person plays more than an "apostrophic function (in which the speech act of address is an exclusively rhetorical device and the addressee cannot be envisaged as present on the same communication level with the addressor)" (Fludernik, "Second-Person Fiction" 218). Its role is to signal a shift from a subjective to an intersubjective model of historical re-creation.

Narrative cooperation undergirds the entire novel. There is first of all cooperation between the authorial narrator and his protagonists whose post-Holocaust story of separation and reunion the narrator tries to re-create. The characters take on part of the burden of understanding their own life stories. The scenes are recounted from the distinct perspectives of the two main characters as they anticipate their reunion, one at home in Israel, the other "stranded at the airport of the city [Paris] where it all started" (14). In their individual reminiscing, both protagonists "repeat the same words, ask the same questions," whisper "fragments of their story," circling around its mysteries much like the extradiegetic narrator does with their combined narratives. After a while, it is no longer clear whether the story of Sarah's survival from the war is recounted from her memory or from the combined perspectives of her cousin and the authorial narrator, both "mixing [their] own survival, [their] own story [. . .], [their] own words with hers" (49). The narrator and his characters make repeated efforts to complete a "story [that] refused to be spoken" (110), saving it from the "incomprehension" into which it periodically slips.

The effort of narrative rearticulation is as heroic here as it was in *The Twofold Vibration*. What the cousins are seeking "is the meaning of [their] separation—the meaning of their absence from each other" (40). Their reunion can take place only in a narrative form that offers them a cathartic encounter with the collapsed structures of history and the "meaning of an absence. [. . .] The suffering of Sarah and her cousin was never adequate, it dissipated into the incomprehension of suffering. That is why the void of their lives can only find its fulfillment in the circumstances of that void" (108). The narrator struggles to fill the void created by history, speaking his characters back into meaningful existence. His ambition is to create "a stereophonic effect" in the linear discourse of history: "If only one could inscribe simultaneously in the same sentence different moments of the story. [. : .] That's how it feels right now inside my skull. Voices within voices entangled within their own fleeting garrulousness" (76–77). His stereophonic narrative invokes an ideal narratee, a fellow writer to whom letters containing ideas, queries, fragments of the projected narrative are addressed periodically. The reconstruction of history becomes a joint enterprise, a process of shared speech and responsive listening.

"Leaning against the winds over a precipice" (6), the teller of *Take It or Leave It* struggled hard to realign his innovative syntax with the "constraints" of history and the limitations of story form. *To Whom It May Concern* gives new meaning to Federman's theme of "leaning": The historical rewriter no longer leans "against the winds," but rather on a community of fellow (co)writers whose responses support the work of "carving, the molding, the scratching away, the erasing" (157) that a narrative performs on history. Both in his candid reliance on collaboration and in his unpretentious concept of narrative rearticulation, Federman's most recent "teller" recaptures something of the traditional ethos of storytelling. By asking simple questions and concerning himself with both the momentous and the "trivial details of the cousins' adventure," he manages to complete and validate his story (157), wresting some significance from traumatic historical events.

CHAPTER 6

Translating a History of "Unspeakable" Otherness into a Discourse of Empowered "Choices"

Toni Morrison's Novels of Radical Rememory

It is a peculiar sensation, this double-consciousness, this sense of always looking at one's self through the eyes of others, of measuring one's soul by the tape of a world that looks on in amused contempt and pity. One ever feels his twoness,—an American, a Negro; two souls, two thoughts, two unreconciled strivings; two warring ideals in one dark body, whose dogged strength alone keeps it from being torn asunder.

—W. E. B. Du Bois. *The Souls of Black Folk* (3)

Now that Afro-American artistic presence has been "discovered" actually to exist, now that serious scholarship has moved from silencing the witnesses and erasing their meaningful place in and contribution to American culture, it is no longer acceptable to imagine us and imagine for us. We have always been imagining ourselves. [. . .] We are the subjects of our own narrative, witnesses to and participants in our own experience and, in no way coincidentally, in the experience of those with whom we have come into contact. We are not, in fact, "other." We are choices.

—Toni Morrison. "Unspeakable Things Unspoken" (208)

I. "RIP[PING] THE VEIL" OF COLOR AND GENDER: BLACK WOMEN'S SEARCH FOR SUBJECTHOOD IN THE BLUEST EYE AND SULA

In the most radical act performed in Toni Morrison's fiction, the fugitive slave Sethe "collected every bit of life she had made, all the parts of her that were precious and fine and beautiful, and carried, dragged them through the veil, out away, over there where no one could hurt them" (*Beloved* 163). Her infanticide takes one unnamed daughter outside the reach of slavery and almost succeeds in carrying the other three children as well. Reconstructing her scandalous mercy killing, Morrison's own novel "rips that veil drawn over 'proceedings too terrible to relate'" in traditional slave narratives (Morrison, "The Site of Memory" 113). Morrison's double act of "unveiling" has been read in connection to W. E. B. Du Bois's penetrating diagnosis of the American Negro's psychocultural condition (see the first epigraph). Born with the "Veil of Color" instituted by a segregationist ideology that "ended a civil war by beginning a race feud" (Du Bois 29), the American Negro is caught between the desire to achieve "true self-consciousness" and the necessity to "see himself through the revelation of the other world." In order to regain visibility, Negroes must rip the "veil," striving for self-definition.

Among Morrison's characters, the fugitive slave mother Sethe performs the most extreme form of unveiling, fulfilling Du Bois's prophecy that "there shall yet dawn some mighty morning to lift the Veil and set the prisoner free" (213). Not all escapes through the "Veil" are as radical or violent. Toni Morrison's first two novels, The Bluest Eye (1970) and Sula (1973), illustrate the baleful effects of the "veil of color" imposed by the dominant culture on African American women, but also the latter's effort to use what Du Bois called the gift of "second-sight" (3) to escape through the "veil." Song of Solomon (1977) and Tar Baby (1981) confront subtler forms of "veiling" that occur when blacks "play the game" of middle-class integration (Ellison, Shadow and Act 64), trapping themselves in a "double life, with double thoughts, double duties, and double social classes, [that] must give rise to double words and double ideals, and tempt the mind to pretense or to revolt, to hypocrisy or to radicalism" (Du Bois 144). This deleterious duality demands a comprehensive "ripping of the veil," engaging characters in a cultural "archeology" (Morrison, "The Site of Memory" 92) that unearths the roots of their "double consciousness." Tar Baby and Beloved (1987) begin that process of unearthing, revising Du Bois's metaphor of the veil to include not only "a division between races" but also "a division within the race" (Mae Henderson 63): a gender division in Tar Baby, "a psychic and expressive boundary separating the speakable from the unspeakable and the unspoken" (63) in Beloved. Jazz (1992) and Paradise (1998) take the process further, using narrative rewriting to disrupt a vision of the past as an "abused record with no chance but to repeat itself" (Jazz 220). Though these more recent novels have characters who rip the "veil" in self-destructive ways (the dispossessed mother Rose Dear flings herself into the well, Dorcas dies secretly from a gunshot wound, and the Convent women deliberately draw the wrath of the Ruby males with their defiance), their unconventional narration re-creates both the ghostly story of the victims, humanizing them, and the story of the survivors, giving them a future.

Morrison's view of history resulting from these acts of unveiling/reimagining has avoided reductionism ("In American literature we have been so totalized [. . .]. We are not one indistinguishable block of people who always behave the same way" [Morrison, "The Art of Fiction" 117]), offering alternative narrativizations of black experience from slavery to the civil rights movement. Haunted by the specter of violence and war (the Civil War in Beloved, World War I in Sula and Jazz, World War II in The Bluest Eye and Sula, the Vietnam War in Paradise) Morrison's fiction has sought nonconflictive ways of negotiating racial identity and the relationship between black and white America. Morrison's fiction has also engaged racialized gender categories, challenging the stereotype of the black woman as overbearing "mammy" and of the black man as a figure of "rawness and savagery" (Playing in the Dark 44). While gender and race remain important categories for Morrison, her more recent fiction has called into question the Cold War narratives of "othering" that have polarized American society, regarding identity categories as dynamic, mutually constitutive categories. This dialectic approach reflects also at the level of Morrison's poetics that renegotiates traditional poles, moving beyond "the social organicism of the black arts movement and the formalist organicism of the 'reconstructionists,'" into what Henry Louis Gates Jr. calls a "new black aesthetic" ("African American Criticism" 309). Like Clarence Major, Morrison is interested in re-creating rather than merely representing black experience, foregrounding those aspects left out by a traditional aesthetic and politics of black art.

Morrison's novels attempt not only to bear witness (a weighty task in itself since, as Morrison stated in an interview, "I have this creepy sensation [. . . that] something is about to be lost and will never be retrieved. Because if *we* [. . .] black women, if we Third-World Women don't know [what our past is], then, it is not known by anybody at all" [Shange 52]), but also to reimagine the history of blacks so as to promote them from American culture's "others" to its empowered "choices." The latter task is made difficult by a white culture that has for so long based its concept of identity on the "exorcism and reification and mirroring" of an "Africanist persona" (Morrison, *Playing in the Dark* 39). Confronted with this endless production of an "unspeakable" otherness, black fiction must resort to an "unpoliced, seditious, confrontational, manipulative, inventive, disruptive, masked and unmasking language" (Morrison, "Unspeakable Things Unspoken" 211). Each novel uses this disruptive/revelatory language both to speak the "unspeakable" and expose "a posture of vulnerability" (220) in traditional representations of (African American) history.

In being able to rearticulate both black and white culture, "orchestrating [a] sense of connectedness between cultures rather than attempting to dissolve the differences, Morrison's successful career appears to have transcended the 'permanent condition' of double consciousness that afflicts her fictional characters" (Heinze 10). But this success has not been purchased at the cost of sacrificing the experiential and expressive needs of black culture. The early novels suggest that the task of constructing an effective identity is considerably more difficult for African American women who learn early enough that they are "neither white nor male, and that all freedom and triumph [are] forbidden to them" (*Sula* 44). As "doubly other" (Davis 13), black women occupy a subordinate position inside both white and black culture: "Everybody in the world was in a position to give them orders. White women said, 'Do this.' White children said, 'Give me that.' White men said, 'Come here.' Black men said, 'Lay down.' The only people they need not take orders from were black children and each other" (*The Bluest Eye* 109). Coming of age for any woman in patriarchal society involves some diminution as she gradually submits to the prison bars of "femininity" enforced by law and custom (Beauvoir 306–47). For black women, entry into adulthood is even more traumatic: Their education in *The Bluest Eye* includes incest, spousal violence, and an endless litany of "hard times, bad times, and somebody-done-gone-and-left-me-times" (24). To be sure, there are variations in this bluesy Bildungsroman: "good Christian colored" women (48) and free-spirited prostitutes fare somewhat better than "very black and ugly" (61) women who—like Pauline Williams-Breedlove—"never felt at home anywhere, [n]or [. . .] belonged anyplace" (88). Still, "the line between colored and nigger" (71) constantly breaks down, allowing the tales of Morrison's women to meet in a common "recitative of pain" (109).

The education of Morrison's black men is no less dramatic, whether it follows a traditional path, as with Cholly Fuller who, abandoned in a junk heap by his mother, cast aside by his father for a craps game, and forced to have sex under the emasculating flashlights of white hunters, struggles his entire life to bear the "cloak of ugliness" and rejection (34); or an "emancipated" path, as in the case of Elihue Micah Whitcomb, who "passes" for white believing with the "father of racism," Joseph Arthur de Gobineau, that his infusion of white blood prevails over his "weaker" African and West Indian heritage. Elihue adopts the superficial characteristics of the master culture but finds no better use for "his exposure to the best minds of the Western world" (134) than the detestable life of a misanthrope, quack-healer, and pedophile. And yet both Cholly and Elihue enjoy a freedom unavailable to the black women in Morrison's first novel: Elihue is free to enjoy vice, filth, and disorder (136) under the protection of his white education; Cholly is free to take his hatred of his white persecutors out on the black woman "who bore witness to his failure, his impotence" (119). Cholly's dangerous freedom to "be tender or violent" (125) to his daughter Pecola contributes to her ruin. As the narrator comments "[T]he love of a free man is never safe. There is no gift for the beloved. [. . .] The loved one is shorn, neutralized, frozen in the glare of the lover's inward eye" (159–60).

Morrison's women are victimized also by subtler cultural pressures, such as "romantic love" and the obsession with "physical beauty" (97). The effects of these two "most destructive ideas in the history of human thought" (97) are illustrated in the lives of Pauline Williams-Breedlove and her daughter Pecola. Bored with her lusterless life, Pauline takes her hopes to the movie house, but her experience there predictably "bind[s] her mind," giving her "self-contempt by the heap" while teaching her to regard "love as possessive mating, and romance as the goal of the spirit" (The Bluest Eye 97). This mix of self-loathing and possessiveness dooms Pauline's relationship with Cholly, "curtailing freedom in every way" (97). She gives up her husband and children, seeking "power, praise, and luxury" in the Fishers' white household (101). The story of Pecola is even more poignant: A victim of the two "most destructive ideas" of romance and racialized beauty (97), Pecola tries to reconstruct herself on the model of the blue-eyed Shirley Temple and her screen self Mary Jane promoted by the racially polarized culture of the 1940s, but in the eyes of that culture she remains irredeemably different. Allowing herself to be defined by the aesthetic norms of the white culture and the prejudicial "colorism" practiced by the lighter-skinned members of her own community, Pecola can only play the role of an inassimilable Other, "trying to discover the secret of [her] ugliness" (39). Instead of claiming a positive identity, Pecola tries to make her own "static and dread" blackness disappear. Ironically, Pecola's attempt at obliterating her body stops short of her eyes. Hoping that by making "those eyes [. . .] different, that is to say, beautiful, she herself would be different" (40), Pecola turns to the conjurer Elihue who tricks her into believing he has given her blue eyes, precipitating her lapse into madness.

As Wilfred D. Samuels and Clenora Hudson-Weems have argued, the problem with Pecola, Pauline, and Cholly Breedlove is "their failure to transcend the imposing definition of 'the Other's' look. Reduced to a state of 'objectness' [. . .], each remains frozen in a world of being-for-the-other and consequently lives a life of shame, alienation, self-hatred, and inevitable destruction" (10–11). Against this failure of vision, the novel's narrator (Claudia MacTeer) develops a confident first-person position that challenges Pecola's submission to the culture's objectifying gaze. Her effort is impressive especially if considered against the many hurdles that black female writers face:

Given the kind of racist, sexist iconography in our culture that always presumes that black women should serve the interests of others, whether it's black children or black men or the larger society, it's very hard for black women to claim that space that is the precursor to writing, the space where you can think through ideas. (conversation with bell hooks, Olson and Hirsh 112).

But Claudia seems better prepared to claim that space. By contrast to Pecola who stares at things with "great uncomprehending eyes [. . .] that questioned nothing and asked everything" (*The Bluest Eye* 75), Claudia trains her eyes and understanding on the stories of Lorain, Ohio (Morrison's own birthplace), retelling them in ways that foreground their deeper implications.

The Bluest Eye begins the important operation of clearing some space for the creative subjectivity of black women with one of the culture's most basic texts, a Dick and Jane reading primer. Parodying Jane's naive description of her suburban paradise, the novel suggests that even a white middle-class girl has difficulties growing up in the 1940s, in spite of the fact that "Mother is very nice," Father "is big and strong," and the kitten and dog quite spirited (7). Bits of Jane's text are used subsequently as contrapuntal titles for the sections that recount Cholly Breedlove and Pauline Williams's short-lived domestic harmony in agrarian Alabama and their transplantation to a decentered life "shredded with quarrels" (94) in the urban North; also for other sections that represent the counterfeit comfort of the "colored" women, or Pecola's dialogue with her imaginary playmate in the mirror. Each component in the cultural vision of suburban felicity is deconstructed ironically by the African American experiences related in the book.

The second short text with which the novel opens is the narrator's rumination on the marigold seeds planted by her and her sister Frieda in the summer of 1941. Meant as an offering to God to spare the child conceived by Pecola with her father, these seeds shrivel and die, like Pecola's premature baby. The narrator realizes retrospectively that the fault may lie in the culture and "the earth itself [that] might have been unyielding," shriveling the "seeds" of innocence (9). Therefore, though when Pecola's dramatic story is done "[t]here is really nothing more to say—except why" (9), the narrator undertakes the challenge of understanding the "why" and of looking for better ways to define black identity than self-effacement or destructive rebellion. Her approach is dialogic in at least two ways: by involving a "countertextual dynamic" that pits an incipient "oppositional discourse" (Gibson 20) against the texts of the dominant culture (the Dick and Jane primer, the Shirley Temple iconography, de Gobineau's racist theories); and by engaging the communal perspective on the fate of individual characters. The novel alternates between third-person narration, several first-person voices (Claudia, Pauline Williams, Jane in the opening paragraph), and a choral "we" that skips around "like a gently wicked dance" (*The Bluest Eye* 15), converting sickness, loss, and fear into "a productive and fructifying pain" (14). This patchwork technique is explained early in the book as follows:

Each member of the family in his own cell of consciousness, each making his own patchwork quilt of reality—collecting fragments of experience, pieces of information there. From the tiny impressions gleaned from one another, they created a sense of belonging and tried to make do with the way they found each other. (31)

While the whole town of Lorain—"a melting pot on the lip of America" (90)—would certainly profit from such patchwork, this dialogic approach is indispensable for black girls like Pecola who shrivel under the "othering" gaze of white culture. Their "secret, terrible, awful story" requires the solidarity of many voices to be brought out. Though *The Bluest Eye* stops short of the radical break with patriarchal culture that Sula Peace experiences in the next novel, "floatin' around without no man—nor children" (*Sula* 80), it defines a form of solidarity among women that builds on their individual victories and failures, while still allowing for racial and class distinctions.

In *The Bluest Eye*, the closest we come to rebellious women is with China, Poland, and Miss Marie (Maginot Line). By contrast to "those generations of prostitutes created in novels, with great and generous hearts, dedicated [. . .] to meliorating the luckless, barren life of men" (47), the "three merry harridans" of Lorain abuse their male visitors and scoff subservient women. These impish middle-age women provide an early example of an unconventional female community, but their resistance proves finally as ineffective as the World War II Maginot line after which one of them is named. *Sula* (1973) challenges more thoroughly the roles available to African American women, emphasizing the title character's refusal to settle for the "painful and unattractive" career of "mother and laborer" (Lester 49), living for herself rather than "for others." But while making possible the defiant character of Sula, Morrison's novel also foregrounds the unresolved tensions between Sula's desire to become an autonomous individual and the traditional coming-of-age plots available to black women.

With its unconventional character and plot, *Sula* challenges the traditions of African-American fiction, focusing "less on conventionally defined 'protest' than on a depiction of the black experience—but an experience that is at once 'rebellious' and *anti*traditional [. . .] disputing the communalistic, socio-centric claims and 'verities' of much African-American literature" (Robert Grant 92). The "feminist slant" of this novel also disrupts the reassuring linearity of realism. The novel's preamble refuses to offer a "seductive safe harbor" (Morrison, "Unspeakable Things Unspoken" 221), describing instead the destruction of the black neighborhood of Bottom, "uprooted" to make room for a white golf course in the town of Medallion, Ohio. After less than a page "Nothing is left of the Bottom" (3), except its taunting name that commemorates the master culture's deception on the former slave culture. Started as a "nigger joke" (4), with a ex-slave receiving for his efforts not God's "bottom of heaven-best land there is," as promised by a white farmer, but a plot of hilly land where "planting was backbreaking, [and] where the soil slid down and washed away the seeds" (5), the community of Bottom still struggles to enter history fifty years after emancipation. Since this community has hardly any evolutionary narrative, Morrison looks for alternative ways to tell its story. Abandoning gradually the pretense of chronological reconstruction (the chapters jump from 1919, 1920, 1921, 1922, 1923, to 1927, 1937, 1939, 1940, 1941, and 1965), Morrison reorients her attention toward a mostly female "history" that valorizes matrilineal and interfemale relationships. The novel makes room for female creativity in various forms, from Eva Peace's ever-expanding boardinghouse, to Reba's Grill, Irene's Palace of Cosmetology, or the "dark woman in a flowered dress doing a bit of cakewalk, a bit of black bottom, a bit of 'messin roun' to the lively notes of a mouth organ" (*Sula* 4). Sula herself plays the role of a disrupter/innovator, "extracting choice from choicelessness" (Morrison, "Unspeakable Things Unspoken" 223). She challenges the culture that destroyed Pecola in *The Bluest Eye* with her unorthodox attitudes and friendship with Nel Wright, which is conspicuously free of pa-triarchal pressures (they "never quarreled [. . .] the way some girlfriends did over boys or competed against each other for them" [72]).

The androcentric orientation of a traditional chronicle novel is successfully dis-rupted to create room for an "[i]mprovisational[,] [d]aring, disruptive, [. . .] mod-ern, out-of-the-house, outlawed, [. . .] and dangerously female" imagination and relationships (223). Several male characters suffer accidents (Little Chicken drowns, Mr. Finley chokes on a bone), are sacrificed (heroine-addicted Plum is put out of his misery by his mother when he wants to "crawl back into [Eva's] womb" [62]), or be-come diminished in body and spirit (Shadrack goes crazy over the memory of the "headless soldier" running "gracefully" [8] across a World War I battlefield; the three forty-eight-inch-tall "Deweys" are reduced to one mind). Even Sula's lover "Ajax" carries false mythic promises (as Sula finds out after his disappearance, his real name is Albert Jacks). With "no men in the house, no men to run it" (35), the center stage is occupied by strong, sexually liberated women like Eva Peace, daugh-ter Hannah, and granddaughter Sula organized in a "three-woman household" that recalls for Barbara Christian "an older mythic family structure" ("Layered Rhythms" 24). Connected through a resilient matrilineal string, these women gain increas-ing—if somewhat dubious—freedom through "amputation": giving up a leg, a son, marriage, sentimentality, traditional domesticity, and replacing them with an uncon-ventional enjoyment of their femaleness. Eva Peace, who returns after eighteen months spent away from her three children "with two crutches, a new black pocket-book, and one leg" (34), never disguises "the empty place on her left side. Her dresses were midcalf so that her one glamorous leg was always in view" (31). Having sacrificed a leg to ensure the survival of her family abandoned by her husband Boy-Boy, Eva is the first incarnation of Morrison's gurulike woman who is "both the law and its transgression" (Morrison, "Nobel Lecture" 267). Her daughter Hannah re-fuses to remarry after her husband dies, welcoming "a steady sequence of lovers, mostly the husbands of her friends and neighbors" (Sula 36); and she forfeits the role of good mother, confessing that she "just [didn't] like" Sula. Sula amputates her fin-ger with a paring knife when four Irish boys accost her and Nel: "If I can do that to myself, what do you suppose I'll do to you?" (54–55). She later learns to use "the cut-ting edge" of sexuality (122–23) to her advantage. Sula's whole life becomes "exper-imental" by dint of its freedom from the normative:

> She had no center, no speck around which to grow. [. . .] She was completely free of ambition, with no affection for money, property or things, no greed, no desire to com-mand attention or compliments—no ego. For that reason she felt no compulsion to verify herself—be consistent with herself. (118–19)

According to Morrison's own description, Sula is a figure of contestation, "perfectly willing to think the unthinkable" and challenge the "law" (Stepto, "Intimate Things in Place" 216–17).

A similar restructuring goes on at the narrative level: Sula evokes images that sug-gest known symbolic orders (historical, ethno-social, mythic), but these appear "de-capitated," emptied of their conventional content. At the same time they are refilled with the fluid reality of "femaleness." For example, the motif of initiation is recon-figured around the experiences that the two young women, Sula Peace and Nel

Wright, share: "In the safe harbor of each other's company they could afford to abandon the ways of other people and concentrate on their own perceptions of things" (55). Their intimacy disrupts the patriarchal narrative that imagines Nel "lying on a flowered bed, [. . .] waiting for some fiery prince" (51). In a succession of scenes that "dust off [. . .] clichés, dust off the language" (Morrison interview in LeClair and McCaffery 254), the twelve-year old girls are shown performing an ironic phallic ritual: "undressing" twigs, digging with them two separate holes in the ground until "they were one and the same," and finally burying their broken phallic tools together with other "small defiling things" in a common symbolic "grave" (58–59). This is followed by a dramatic scene in which Sula lets the body of a little boy she is swinging around "slip from her hands and sail away out over the water" (60–61), watching with Nel the swift river water close "peacefully over the turbulence of Chicken Little's" frail body (61, 170). This incident cements the girls' relationship around a complicity of guilt, but also the exercise of what has previously been the prerogative of the male spectator: an *interested gaze*.

Thus not only the coming-of-age plot but also the traditional specular economy of fiction privileging the male "look" that "objectifies and masters" (Irigaray, *Speculum of the Other Woman* 50), are disturbed. After an early scene in which Helene Wright adopts—to the embarrassment of her daughter Nel and two black soldiers who witness the scene on the New Orleans train—the traditional "to-be-looked-at" posture of women (Mulvay 10), smiling "coquettish[ly] at the salmon-colored face of the [white] conductor" who confronts her (*Sula* 21), the novel repositions women in more active roles of watchers/focalizers. One-legged Eva enjoys power not only as a "creator and sovereign" (26) of her household but also as "overseer" of the community from her top-floor rocking chair; Sula watches unperturbed her mother die in flames or herself making love; and Nel promises not to allow a man's gaze turn her into "jelly" like her mother (22). In Nel's case, the subversion of patriarchy remains superficial. Twice in the novel, Nel renounces her friendship with Sula because of a man: first, when she marries Jude Greene at the end of part I; and ten years later, when she blames Sula of seducing Jude and breaking up her marriage. By contrast, Sula shows no "sense of possessiveness or conventional identity" (Byerman 67), refusing to be defined by traditional expectations.

But *Sula* should not be read merely as a deconstructive novel. As Elizabeth J. Ordóñez has argued, the use of "a heretofore buried or subversively oral matrilineal tradition, either through inversion or compensation" opens a space for "alternate mythical and even historical accounts of women" (17). *The Bluest Eye* featured a few quasi-mythic women like the ancient midwife and healer M'Dear, or the chorus of wise matrons "who were through with lust and lactation, beyond terrors and tears," and who "alone could walk the roads of Mississippi, the lanes of Georgia, the fields of Alabama unmolested" (110). In *Sula*, Eva Peace "recalls her Biblical foremother, then shifts our perspective [. . .] toward matrilineal autonomy and bonding" (Ordóñez 19). Sula herself is a sort of a temptress-witch, completing and challenging Nel. In her presence, Nel feels increasingly estranged from her mother's "oppressive neatness" (29) and embarrassed denial of her Creole descent, bonding with Sula's household, where "a pot of something was always cooking on the stove; where the mother, Hannah, never scolded or gave directions [. . .] and a one-legged grandmother named Eva handed down goobers from deep inside her pockets or read you a dream" (29). This three-generational household gives Nel a new appreciation for a vital female reality that proliferates eccentric structures (Eva keeps adding rooms to her ramshackle house according to the needs of her racially diverse boarders) and redefines the world as a "throbbing disorder constantly awry with things, people, voices, and slamming of doors" (52).

The fact that Eva cannot keep her domain from falling apart eventually suggests that such antipatriarchal formations are difficult to sustain. Nevertheless, Morrison's fiction has stayed committed to the retrieval of ignored mythic and historical possibilities from both the African and the Western traditions. These alternative modes of organization and perception replace the dualistic thinking of the "patriarchal/capitalist" order with a "profusion of personas, beliefs, and values that defy categorization" (Heinze 57, 150). In *Sula,* the title character illustrates the disruptive impact of folk myth and dark supernaturalism upon the conventional order of things. As a young girl Sula bears a mysterious mark above her right eye, which is interpreted successively as a "stemmed rose" (52), "tadpole" (156), "copperhead," or "rattlesnake" (103, 104). Her return to town as an adult coincides with an ominous "plague of robins" (112). The black community receives these hints of evil with an ambivalent attitude, fearing but also welcoming her "ungodly" (118) behavior because it redeems their own failures.

Just how important Sula was for the stability and cohesion of this community becomes clear after her death in 1940. The citizens of Bottom, who, while Sula was alive, "cherish[ed] their husbands and wives, protect[ed] their children, repair[ed] their homes and in general band[ed] together against the devil in their midst" (117–18), lose their focus after her demise. Even Shadrack is shaken out of his madness long enough to wonder whether he should continue to observe his National Suicide Day (January 3) with a "solitary parade" (13), as he has done every year since 1920 in the hope that by devoting a day to random death "everybody could get it out of the way and the rest of the year would be safe and free" (14). While Shadrack has second thoughts, the citizens of Bottom decide to join in the commemoration. Prodded by "their need to kill it all," they march against a local construction site they were prevented from contributing to, going "too deep, too far" (161) into the collapsing tunnel. Bottom is subsequently torn from its "roots" (3) to make room for a new golf course. But the novel does not end in death and destruction. Nel and Sula are reunited spiritually in the 1965 coda, in response to Nel's search for some meaning in the history of Bottom and in her own life. From the evolving perspective of the 1960s, Nel finds the town of Medallion more open racially ("You could go downtown and see colored people working in the dime store behind the counters, even handling money with cash-register keys around their necks" [163]), but misses the old solidarity she felt in Bottom. Her search for the root of her attachment to Bottom takes her predictably to the colored section of the Beechnut cemetery, where she reclaims Sula as her lost sister: "We was girls together [. . .]. O Lord, Sula [. . .] girl, girl, girlgirlgirl" (149). Nel's invocation breaks the linearity of the plot, providing no real closure as it expands in a "fine cry" without bottom or top, "just circles and circles of sorrow" (174).

The entire novel proceeds similarly, expanding the space of femaleness in circles, but leaving this process of restructuring unfinished. The divergent careers of the two female protagonists cannot be easily evaluated: Sula defies patriarchal expectations, refusing to marry and raise a family ("I don't want to make somebody else. I want to make myself" [92]), but she dies alone in her grandmother's sealed-up bedroom.

Sula bears out bell hook's observation that marginality can be a site of opportunity as well as of oppression (Olson and Hirsh 113). However appealing, the idea of an exclusively "female society [. . .] without prohibitions, free and fulfilling" remains an "a-topia, a place outside the law, utopia's floodgate" (Kristeva, "Women's Time" 202). By contrast, Nel never really leaves the patriarchal establishment, marrying Jude because "greater than her friendship [with Sula] was this new feeling of being needed by someone who saw her singly" (84). The intimate friendship between Sula and Nel enables both girls to "see old things with new eyes" (95), but their gyno-centric semiotic does not displace the patriarchal symbolic order. Sula is a provoca-tive postmodernist text that makes good use of a "force of discontinuity" (Robert Grant 94) to create a new, albeit "unfinished" and creative-destructive concept of womanhood. The whole novel works like a "deliberate hypothesis" that molds, through an ironic rewriting of traditional roles, a "character of possibility," a hypo-thetically "free woman" (Hortense Spillers 184). If typologically this project exceeds the work carried out in The Bluest Eye, narratologically it is constrained by Sula's inca-pacity to find adequate forms of self-expression that would leave a more enduring mark on her culture. As she puts it ruefully, "There aren't any more new songs and I have sung all the ones there are. I have sung them all" (137).

2. INTERROGATING OPPOSITIONAL PARADIGMS: HYBRID IDENTITIES AND ETHNIC HISTORIES IN SONG OF SOLOMON AND TAR BABY

Read against the context of Morrison's subsequent work, Sula does not settle the conflict between the impulse to adapt established narrative forms to black experi-ence and the search for alternative modes of articulation that rewrite more radically the dominant tradition. This pendulation between narrative paradigms (realistic and fantastic, modern and postmodern, male and female) continues in Song of Solomon (1977; the National Book Critics Circle Award) and in Tar Baby (1981). Both novels illustrate a more traditional narrative economy, but they also break away from con-ventional representations of African American culture, complicating definitions of ethnic identity and recovering forgotten histories—from the mythic African roots to the cultural readjustments of the 1970s, in the aftermath of the Cold War and the civil rights movement.

Dedicated to "Daddy" and the memory of Morrison's maternal grandfather, Song of Solomon recovers a narrative tradition subverted in Sula: the male coming-of-age and quest plot. The book's epigraph, "The fathers may soar/And the children may know their name," connects the name of the son to the success of father's story. The world of fathers provides thus the framework within which the novel's protag-onist, Macon Dead III ("Milkman"), undergoes his education. Part I of the novel follows Milkman's ambiguous education in the "urban Eden" created by his entre-preneurial father in Southside Mercy. Part II traces Milkman's quest for identity, driven by personal versions of the "grail"—materialist at first (a sackful of gold), spiritual in the end (his ancestral heritage)—that reflect the conflicting values in-herited by Milkman from the paternal and maternal tradition.

With a few notable exceptions, women hold subaltern positions in Song of Solomon. Milkman's mother, Ruth Foster Dead, plays the role of a self-effaced child-mother "pressed into a small package" by the "great big house" in which she lives (124), first as "Dr. Foster's daughter," then as Macon Dead II's wife since the age of sixteen. Milkman's sisters Magdalena and First Corinthians are brought up with the illusion of an upward-moving life, but the only outlets they find for their creative talents are making artificial roses. While First Corinthians manages to break out of her "doll baby" (196) life, shacking up with a commoner, Milkman's cousin-lover Hagar is crushed by her feelings of insecurity. This Pecola-like character, who needs "a chorus of mamas, grandmamas, aunts, cousins, sisters, neighbors, Sunday school teachers, best girl friends, [. . .] to give her the strength life demanded her" (307), undergoes a disastrous sentimental education that leaves her with "no self [. . .], no fears, no wants, no intelligence that was her own" (137). Hagar dies echoing Pecola's self-deprecation: "[Milkman] don't love [her hair] at all. [. . .] He loves silky hair. [. . .] Penny-colored hair. [. . .] And lemon colored-skin. [. . .] And gray-blue eyes. [. . .] And thin nose" (315–16).

This "foolish" world of mothers and daughters is "stunned into stillness" (11) by a contemptuous materialist father, Macon Dead II, who fondles the ownership keys he has been accumulating since the age of twenty-five, enjoying the phallic power they give him over his working-class tenants. But his plutocratic domain, built on an imitation of white capitalism, is challenged by a few eccentric women like Pilate, "natural healer" (150) and conjure woman who conspires with Ruth to make possible Milkman's birth against his father's will; and Circe, the ancient midwife who delivers Milkman's father and later cures his son of his materialist delusions. Pilate, who on one occasion sticks a knife into her daughter's abusive friend to remind him that a mother has the right to get angry when "a grown man starts beating up one of us" (94), is a version of Eva Peace. Upsetting conventional expectations both with her appearance ("short hair cut regularly like a man's" [138]) and occupation (bootlegging that "allowed her more freedom hour by hour" than "any other work of a woman" [150]), Pilate reconstructs herself as a strong female archetype, a priestess of nature "whose deeds and stories bring healing and knowledge to her community" (Mobley 2). Circe herself has mythic implications: Her decrepit body hosts the "strong, mellifluent voice of a twenty-year-old girl" and her house exudes a mixture of "a hairy animal smell, ripe, rife, suffocating" and a "sweet spicy perfume" (*Song of Solomon* 240, 238). Circe symbolizes the victory of imagination over materialism, both white and black: She makes sure she survives her white masters long enough to see their estate go up in ruins. She also challenges the acquisitiveness of a "colored man of property" like Macon II whose philosophy is reduced to "Own things. And let things own other things. Then you'll own yourself and other people too" (55).

Another alternative to Macon's materialism is offered by Milkman's childhood friend, Guitar Bains. The two share a friendship in many ways as unconventional as that between Nel and Sula. Guitar introduces teenage Milkman to the "forbidden": first by bringing him to Pilate's house where Milkman is seduced by his aunt's "unbelievable but entirely possible stories" (36); then by taking him to Railroad Tommy's barber shop, where an alternative black politics takes shape, hostile to both white bourgeois culture and black imitators of it. There Milkman hears the first account of racially motivated murder (the stomping to death of Till Emmett in the early 1950s) and participates in a male community of debate and storytelling. A decade later, as he becomes aware of his lack of "coherence, a coming together of the features into a total self" (69), Milkman turns again to Guitar for a clarification of his identity. Guitar's misogynistic theory ("Everybody wants the life of the black

man. [. . .] White women, [. . .] want us, you know, 'universal,' human, no 'race consciousness.' Tame, except in bed. [. . .] And black women, they want your whole self.
Love, they call it, and understanding" [222]) resonates with Milkman's fear of domestic commitment. Milkman resists, on the other hand, Guitar's polarizing vision
according to which all whites are predisposed biologically to violence against blacks
and his terrorist politics as a member of the Seven Days vigilante group that kills
whites randomly to keep the "Ratio. Balance" (158) of victims proportionate. But
while he rejects the solutions offered by Guitar or the "red-headed Negro named X"
(160), Milkman cannot come up with a more satisfactory remedy to his "slave status" than a "dream-bitten" vision of escape.

At the age of thirty-one, Milkman abandons Hagar after a twelve-year relationship
and gets involved in a larky plan to steal Pilate's sack of gold nuggets taken from a
Pennsylvania cave. When Pilate's "treasure" turns out to be a bag of family bones,
Milkman is forced to expand his quest, tracing his family roots back to Danville,
Pennsylvania (his father's home), and Virginia (home of his legendary ancestor
Solomon). In Danville, Milkman finds that his father is revered as a hero who "out-
ran, outplowed, outshot, outpicked, outrode" (234), and more recently outwitted
everybody in business. However, Milkman's encounter with the midwife Circe reorients his attention from the feats of his capitalist father to those of his farmer grandfather. Dispatched by Circe to the Hunter's Cave in search of grandfather Jake's
remains, Milkman undergoes a "threshold" experience (according to Victor Turner,
caverns are ideal catalysts for such experiences, conjoining "birth and death, womb
and tomb" [On the Edge of the Bush 295]), shedding his bourgeois trappings as well as his
pride in his father's capitalism. Milkman's subsequent journey to Virginia completes
the shift from his father's capitalist values to those of a premodern rural community.
This stage involves a "journey of immersion into the 'deeper recesses' of the Black
Belt," in search of a true "communitas and genius loci" (Stepto, From Behind the Veil 67).
His journey begins in the small town of Shalimar where, outfitted in World War II
fatigues, the "city boy" accompanies a party of hunters into the woods and learns to
rely on his "[e]yes, ears, nose, taste, touch—and [. . .] the one [sense] that life itself
might depend on" (277). These newly discovered skills help Milkman escape the pursuit of his demonic double, Guitar, and the constraints of a competitive individualism in favor of the cooperative rituals of a traditional black community.

But it is at the level of storytelling that Milkman's transformation is most significant. In the course of his journey, Milkman evolves from a reluctant listener to his
family's "stranger" tales (74) to a seeker of "gossip, stories, legends, and speculations," who, "bit by bit, with what [people] said, what he knew, and what he guessed,
put it all together" (323). Like an anthropologist, he constructs a revelatory intertext
by linking Pilate's favorite blues tune, "O Sugarman don't leave me here," to the
"Solomon" ring game he hears children play in Shalimar (300) and decoding his
family history from their garbled clues. This history centers on the legendary cotton
slave Solomon who one day flew back to Africa, leaving his wife Ryna and twenty
children behind. He took along only his youngest son Jake (Macon Dead I) but accidentally dropped him on the way. Rescued by a free Indian, Jake grows up alongside her daughter Singing Bird, with whom he later flees to Pennsylvania to start a
new life as a farmer. Since many folk in Shalimar (or "Shalleemone" [261]) bear the
ancestral name of Solomon, Milkman's family history coincides with the sociogene-
sis of this black community. The path followed by Milkman's forerunners, from their

slave days in rural Virginia to farming in Pennsylvania and middle-class assimilation in urban Michigan, recapitulates black history in the nineteenth and twentieth century. Milkman's own quest, begun in a desire to "beat a path away from his parents' past" (180), returns him to that communal narrative, enriched by personal exploration and understanding.

Though *Song of Solomon* ends with Milkman's sacrificial leap off the same ridge whence Solomon took flight, the coming-of-age plot and the broader narrative that reconstructs the history of a prototypic black community manage to break out of conventional frames. The recurrent imagery of flying (see Hovet and Lounsberry), the patterns of questing that draw on both Hellenistic and African myth, and the various Freudian reveries that Milkman has in the novel introduce a modernistic preoccupation with a mythopoetic layer that transcends superficial realism. There are also techniques and emphases in the latter part of the novel that suggest a postmodern agenda: retrieving a "pre-industrial social space, anterior to modernity as the resolution to modernity's alienation and fragmentation" (Hogue x); or resorting to a version of "magic realism" that establishes new relations "between the spirit world and the material world, between the living and the dead" (Marshall 180). The retrieval of a premodern ethos is not left unqualified. The premodern, the modern, and the postmodern mix in Morrison's novel, undercutting distinctions among history, oral narrative, and myth. Historical representation is complicated with "word-of-mouth news" (*Song of Solomon* 3) and an enhanced African American cosmology in which "ghosts" trigger important events and people obey mysterious cosmic and bodily rhythms. Combining the role of cultural archivist with that of imaginative rewriter, Morrison rescues black history from representations that have reduced it to a narrative of disasters, "extracting [new] choice[s] from choicelessness" (Morrison, "Unspeakable Things Unspoken" 223).

Significant in this respect are Morrison's revisions of modernist representations of American experience. The hunting episode in Shalimar echoes/rewrites Faulkner both at the level of theme and that of style. Like a black Ike McCaslin, Milkman realizes in the woods that "all a man had was what he was born with, or what he learned to use" (278). Accordingly, he sheds his white-inspired middle-class trappings and relearns to experience nature's "Braille" through his "finger tips" (279). *Song of Solomon* can also be read as a response to Ellison's *Invisible Man*, revising the latter's "existential journey from invisibility to the alienated self with a journey from the alienated self to the affirmation of self and community. Acts of naming and being supersede running and escaping" (Mobley103 n. 24). Storytelling itself changes emphasis in Morrison's rewriting, from the self-exploratory and often monologic accent it had in Faulkner and Ellison to the dialogic and communal orientation of *Song of Solomon*.

Morrison's novel bears out Barbara Christian's observation that "[P]eople of color have always theorized—but in forms quite different from the Western form of abstract logic. [. . . O]ur theorizing [. . .] is often in narrative forms, in the stories we create, in riddles and proverbs, in the play with language, since dynamic rather than fixed ideas seem more to our liking" ("The Race for Theory" 52). In *Song of Solomon*, this type of narrative rethinking involves several areas: the search for a liberating imagination embodied in the imagery of flying and shedding; the emphasis on dialogic storytelling as a way of reclaiming personal and collective history; and the regrounding of black culture through an active interpretation of names. Expanding the "search for redemptive flight first articulated in slave songs and narratives and then imagined in texts by Wright, Ellison, and LeRoi Jones" (Dixon 165), Morrison involves her protagonist in a reevaluation/reenactment of this effort. Milkman's birth coincides with a black insurance agent's failed attempt to fly, which Milkman interprets as a collapse of imagination: "To have to live without that single gift saddened him and left his imagination so bereft that he appeared dull even to the women who did not hate his mother" (9). Pondering the meaning of this initial crash, Milkman realizes that his own inability to fly is due to the cultural baggage of division and false pride that weighs him down. Like the white peacock stuck on the roof of the Nelson Buick building, he has "too much tail" and "vanity" (178). Milkman can attempt to fly only after he sheds his bourgeois pride, takes some distance from Guitar's racial hatred, and considers the mixed blessings of the narrative of flying. As his family history attests, flying always leaves behind "a body" (333), a sacrifice that has to be acknowledged. Solomon's flight back to Africa causes the suicide of his forlorn wife; the flight of his youngest son Jake from Virginia to Pennsylvania ends up in a cave, with the legendary farmer destroyed by a jealous white culture; Macon and Pilate's flight from rural Pennsylvania to urban Michigan leaves behind Jake's bones, which Pilate is later summoned by her father's ghost to retrieve: "You just can't fly on off and leave a body" (332). Milkman's own attempt to fly off Solomon's Leap is both a "triumph and [a] risk," leading to the "abandonment of other people" (Morrison qtd. in Shange 48). While endorsing her characters' flight to freedom, Morrison supplements the "male mode of heroism" (Davis 23) with an alternative female model of flying.

This model is embodied in the wise woman who first awakens Milkman's appetite for familial narratives, Pilate. "[W]ithout ever leaving the ground she could fly" (*Song of Solomon* 336) in her storytelling. While the stories told by Macon, Ruth, and Guitar reinforce divisions, Pilate's stories provide opportunities for reconciliation and bonding. Pilate's liminal existence, on the margin of both white culture and the "elaborately socialized world of black people" (149), allows her to play nonconventional roles of boundary-crosser, healer, and "griot." Her storytelling triggers a web of narrative responses, giving Milkman's quest a focus and encouraging the silenced women who have nurtured and protected him (Ruth, First Corinthians, Lena) to speak their own stories. Even Macon Dead II is seduced by Pilate's ethos, recalling a more innocent time spent on his father's farm, "Lincoln's Heaven," where animals named after Union and Confederate war heroes lived peacefully together.

Morrison's characters can regain this utopian space of reconciliation through dialogic narration, trading "oral, meandering" stories (Morrison, "Rootedness" 341) that allow a redemptive revisitation of the past. For example, the events that occurred in the Pennsylvanian cave where Pilate and Macon took refuge after the murder of their father are retold from different angles until Macon is absolved of the crime he thought he committed (killing a white man in self-defense), and the bones Pilate removed from the cave are identified as belonging to their father rather than to a white man. Morrison's characters learn from this process that the truth of history is in its retelling and that you need to carry the "bones" of your past "right there with you wherever you go. That way, it frees up your mind" (208).

The names and stories of black people are the "bones" that need to be reclaimed periodically. The fact that they are often mistaken for the white man's "bones" or are regarded as "improper" is not surprising. Most African American names are the result of "an onomastic *impropriety*," given "by the *proprietor* (the slaveholder) without any concern for their identities" (Moraru 190). But Morrison's novel suggests ways in which names can be "reappropriated" through genealogic and narrative reconstruction. One reappropriative strategy is illustrated in the anecdote of "Not Doctor Street": Known to the Southside community as "Doctor Street," after the only colored doctor (Ruth's father) who moved there in 1896, the town's Main Avenue is renamed in the oral lore "Not Doctor Street" when its original name is prohibited by the white city legislators. Through this witty resignifying that obscures the "apparent meaning" of the white term while operating a "formal revision and intertextual relation with it" (Gates, *The Signifying Monkey* 51, 55), the black commemorative name is recovered in the very act of its denial. An alternative strategy used by African Americans (see Morrison's interview in LeClair and McCaffery 259) is the recourse to biblical names picked blindly in the hope that they will offer protection to their bearers. Yet, with the possible exception of Solomon, other Morrison characters (Rebekkah, Hagar, Magdalene, First Corinthians) garner no benefits from their biblical names.

Pilate, who also bears an improper name given to her by an illiterate father who copied a pleasant sequence of letters from the Bible, illustrates the most successful approach to names in this novel, "recreating the meaning of proper names [. . .] that cannot or should not be changed" (Moraru 193). Instead of indulging in fruitless reflections on "foolish misnaming" (18), like her brother Macon Dead who inherits a name erroneously given to his father by a drunken office clerk, Pilate makes good on her name. Her straight, proud body authenticates the original impression left by the printed word on her father (that of a row of tall, protective trees); and her uncanny powers, including that of flying, make alternative readings of her name such as "Pilot" possible. Milkman himself adopts this authenticating approach as he retrieves the stories behind familial names, turning naming "errors" into motivated signs. Names for him are "legacies" to be deciphered but also narrative shells to be fleshed out/reimagined by their bearers. The latter can do this only if they are allowed to become subjects and narrators of their lives, contributing to the "vast genealogic poem that attempts to restore continuity to the ruptures or discontinuities imposed by the history of black presence in America" (Benston 152).

Addressing this question from the post–civil rights perspective of the late 1970s, *Tar Baby* offers two examples of strong black agency: one male (Son) the other female (Jadine). Neither is entirely satisfactory, reflecting the contradictory legacy of the civil rights movement that pursued simultaneously the aspiration of "the middle-class person of color [. . .] to be accepted, to 'pass' as middle-class instead of being excluded as racial other" and his struggle to "re-proclaim his racial identity in order to advocate for his Enlightenment-derived rights" (Hogue 22–23). This conflicting agenda has spawned division: While some people of color have attempted "to reject their race completely," others have found that their "only defense against racism and inequalities is to return to a racial collective, to an 'outdated' racial or cultural narrative, which represses their non-racial or other class identities" (23). Morrison's novel exemplifies this with Son and Jadine. Even though he enters the novel as a renegade, Son (Willie) Green reaffirms a "racial collective" that is more militant and traditional than Milkman's. The cosmopolitan Jadine Childs, on the other hand, rejects a traditional definition of black womanhood, feeling alternatively emancipated and alienated from her race. The issue of black agency is further complicated by being placed within a multicultural Caribbean environment that interplays white and black, Anglophone and Francophone, colonial and postcolonial mentalities. Son and Jadine have to negotiate their identity not only in relation to each other but also to the cultures they traverse.

The novel's "multidimensional matrix" (Ryan 64), conjoining personal questing (male and female) with the parallel exploration of options within Euro-American, African American, and Afro-Caribbean cultures, puts a strain on the narrative discourse. The most obvious break comes after the prologue that introduces the yet-unnamed Son, showing him jumping ship and crawling aboard a fifty-six-foot yacht navigated by two women "with polished finger nails" (7)—Margaret and Jadine. After this suspenseful interlude, the fugitive black man disappears, literally submerged in the narrative. The center stage is taken by the retired millionaire Valerian Street because white men "run the world" (Morrison interview with Ruas 225). Son is discovered three chapters later in the closet of Valerian's wife, Margaret, and has to be tackled and defined by every member of Valerian's household before he can play a significant role in the novel. Another major narrative rift follows the Streets' Christmas dinner. Instead of confirming Valerian's household as a big interracial "family" (49), the dinner enhances divisions, sending characters on diverging paths through the last section of the novel: The candy maker Valerian focuses on his dying, Margaret on coming to terms with her past, Jadine uproots herself one more time, and Son shifts attention from Jadine to his quest for mythic ancestors.

The hothouse atmosphere of the Isle de Chevaliers, where everybody is a seasonal laborer, exile or tourist, provides the illusion of an edenic retreat but also a challenge to individual identities. Himself an exile from Philadelphia, the white master Valerian brings laborers from Haiti to build the "most handsomely articulated and blessedly unrhetorical house in the Caribbean" (11). But the fragile order of this (post)colonial house, suggestively called "l'Arbre de la Croix" (Tree of the Cross), is threatened continually by maiden ants, emperor butterflies, or human intruders such as the black man discovered in Margaret's closet. The "old-fashioned family Christmas" (32) organized by Valerian Street and his second wife Margaret Lenore has to welcome the "uninvited" (butterflies, ants, Son) because those invited (the Streets' son Michael, neighbors) never show up. This triggers a series of tensions and dislocations, forcing each character to reexamine his or her position in the (post)colonial house.

Valerian's Christmas reunion fails because it is superficial and incomplete, excluding the indigenous blacks. (Thérèse and Gideon are fired just before Christmas because they craved a few of Valerian's apples.) The significance of their exclusion does not escape Son:

> Son's mouth went dry as he watched Valerian chewing a piece of ham, [. . .] approving even of the flavor in his mouth although he had been able to dismiss with a flutter of the fingers the people whose sugar and cocoa had allowed him to grow old in regal comfort [. . .] [and] move near, but not in the midst of, the jungle where the sugar came from and built a palace with more of their labor and then hire them to do more of the work he was not capable of and pay them again according to some scale that would outrage Satan himself, and when those people wanted a little of what he wanted, some apples for *their* Christmas, [. . .] he dismissed them with a flutter of the fingers, because they were thieves, and nobody knew thieves and thievery better than he did. (202–3)

Son's denunciation of Valerian as a neocolonial master for whom "waste was the order of the day and the ordering principle of the universe" (203) dispels the myth of postcolonial reconciliation. Valerian prides himself with being an enlightened master who, at sixty-five, trades white corporate culture for an unspoiled Caribbean island; but like Prospero, he overlooks the fact that his continued comfort and "love of [higher] things" (51) is obtained through systematic exploitation of a world of Calibans. In his role as patronizing husband for Margaret, rich white protector for Jadine, and benevolent master for servants Sydney and Ondine, he wards off serious challenges, until the fateful Christmas dinner that gives unexpected voice to the marginals, allowing them to break through the "mask [that] sometimes exists when black people talk to white people" (Morrison interview in Ruas 218). Their remarks, from Son's charge that "Two people are going to starve so your wife could play American mama and fool around in the kitchen" (205), to Ondine's attack on Margaret as an intruder and "baby killer" (209), break the composure of the white master and mistress. Margaret begins a fitful retelling of her "unspeakable" past of child abuse (235) for which, like *Beloved*'s Sethe, she "did not have the vocabulary to describe what she had come to know, remember" (236); and Valerian reacts to these disclosures with "knees [. . .] trembling," unable to shut the past out and not "strong enough to hear it" (232).

The failure of Valerian's attempt at interracial fraternization reinforces divisions, bringing the "guest" Jadine and the intruder Son together to plot their escape to New York, making Sydney and Ondine walk "on glass shards, afraid, angry, sullen" (235), and returning Gideon and Thérèse to Dominique to utter incantations against the Americans. L'Arbre de la Croix becomes a "house of shadows," with couples "locked into each other or away from each other" (235). But as the novel's Corinthians 1:11 epigraph reminds us, the "contentions" that bring the "house of Chloe" down are as much intracultural (internal to each family or group) as they are intercultural. Class, gender, and generation divide not only the white culture but also the black, suggesting how difficult it is to maintain an "African diaspora [in] the western hemisphere" (Gilroy 15). As "one of those industrious Philadelphia Negroes—the proudest people in the race" (*Tar Baby* 61), Sydney treats condescendingly not only Son but also the uneducated "outdoor" hands Gideon and Thérèse. Ondine keeps "Yardman" (Gideon) out of her kitchen and refuses to call him by his real name. In turn, the "natives of Dominique [do] not hide the contempt they felt in their hearts for everybody but themselves" (110). Gideon is especially put off by American "yallas" like Jadine who "don't come to being black natural-like"(155). But he is also disdainful of his Dominique-born wife, mocking her semiblindness and wild imagination. A more vicious form of sexism informs Valerian's relationship with Margaret. By contrast to their black servants Sydney and Ondine, who sleep "back to back" a "tranquil, earned" sleep (43), the seventy-year-old Valerian and fifty-year-old Margaret face their nightmares and disappointments in separate rooms.

The island "wilderness" exaggerates these internecine conflicts ("Too much light. Too much shadow. [. . .]" [69]), but also provides characters with a delimited space wherein they can concentrate on their "ticky-tacky" thoughts and scraps of narrative remembrance (55). Just like in *Song of Solomon*, self-narration offers Morrison's characters a chance to redeem themselves. Margaret welcomes the opportunity to reveal, in "bite-sized pieces" and "fleeting sentences" (235, 236), the story of her sadistic abuse of infant Michael, relating it to the anger she felt as a neglected wife. After she

forces old secrets out, she feels less fragmented as a person. Valerian is also transformed by these revelations, gaining a critical awareness of his own "crime of innocence" (242), which prevented him from knowing what "was inconvenient and
frightening" (242). Even relatively static characters like Sydney and Ondine undergo
a subtle revision of roles. Ondine admits her own inadequacy as a nurturer for Jadine. By sending her niece to European schools and objecting to her relationship
with Son, she has not only cut Jadine off from her cultural roots but also prevented
her from learning "how do be a daughter" (281). Ondine is also more forgiving of
Margaret's deficiencies, allowing the reconciliation of white and black mothers to
begin from a new understanding of the failures of women in patriarchy.

Jadine also undergoes a useful reexamination of her options. By challenging her
aunt's idea that "there ain't but one kind of woman" (281) and refusing to be a dutiful "daughter" for her, Jadine commits "matrophobia—rebellion against the imposed
female image" (Rich, *Of Woman Born* 235), but her attitude has certain empowering
features that not surprisingly connect her to the novel's insubordinate "Son." Both
are disruptive liminal figures. Jadine's racial and cultural hybridity allows her to play
unconventional roles in relation to white and black culture, celebrating mobility and
change. Viewed in this light, Jadine suggests an end point in the development of the
biracial woman figure from the "tragic mulatta" (the alienated Zoe type) introduced
by white abolitionist writers like Lydia Maria Child and Elizabeth Livermore to ac-
tivistic heroines empowered by their "in-betweenness," as in the novels of Frances
Harper, Pauline Hopkins (see Washington 76), Paule Marshall, and Michelle Cliff.

Son is even more of a challenge to the cultural order set up like a traditional
"plantation" (220) around a white master with a Roman emperor's name. A discharged Vietnam soldier, refugee, and vagrant worker since 1971, Son is a liminal
character in the deepest sense: "A man without human rites: unbaptized, uncircum-
cized, minus puberty rites or the formal rites of manhood. [. . .] He had attended no
funeral, married in no church, raised no child. Propertyless, homeless, sought for but
not after" (165–66). Reluctant to follow conventions, he joins the "great underclass
of undocumented men,"

> the international legion of day laborers, and musclemen, gamblers, sidewalk merchants,
> migrants, unlicensed crewmen on ships with volatile cargo, part-time mercenaries, full-
> time gigolos, or curbside musicians. [. . .] Some were Huck Finns; some Nigger Jims.
> Others were Calibans, Staggerlees and John Henrys. Anarchic, wondering, they read
> about their hometowns in the pages of out-of-town newspapers. (166)

Because he is both a liminal-anarchic character and an advocate for ethnic values,
Son poses a challenge not only to the white culture that wants to erase his "funky"
identity (Willis 156), but also to members of his generation with whom he feels a
certain affinity: like Michael, a "poet and a socialist" (91) with multicultural sensi-
tivity who champions a precapitalistic culture of crafts and barter; or Jadine Childs,
who like Son is a "child" of the postwar emancipation of African Americans. The
reader looks forward to a meeting between Son and Michael, but the fact that this
never takes place problematizes the promise of a cross-cultural "fraternity" of liminal positions. A "cultural orphan who sought other cultures he could love without
risk or pain" (145), Michael participates inadvertently in the "multicultural" obfus-
cation of real ethnocultural differences.

The limits of a transculturalism based on a denial of roots are illustrated also with Jadine. At first sight, this diligent woman who by the age of twenty-five completes a degree in art history and makes "those white girls disappear" (40) with her successful international modeling can be considered a cross-cultural success. Jadine shuttles with relative ease between two different sets of "adoptive" parents (Sydney and Ondine, Valerian and Margaret) and two different cultures (the upstairs white culture where she sleeps and socializes, and the downstairs black culture where she seeks familial interactions). Similar to Gloria Alzaldúa's "mestiza," Jadine seeks a "pathway 'beyond' difference," "migrating through many forms and types of cultural, racial, gender and aesthetic hybridity as a performative resistance to fixity" (Susan Friedman, *Mappings* 94). But Jadine's liminality is as much a weakness as it is a strength. Her mixed identity translates into a "double consciousness," with some of the negative implications attributed by Du Bois to that state. Jadine's sleep in Valerian's house is troubled by the vision of an African "woman's woman—[. . .] mother/sister/she" (66) glimpsed in a Parisian supermarket. The "unphotographable beauty" of this African queen, comfortable in her tar-black skin (65), mocks Jadine's commodified beauty; her graceful walk with three eggs in her hand questions Jadine's infertility. A more direct challenge to Jadine's cultural assimilation comes from Michael and Son, who admonish Jadine for "abandoning [her] history" (72), and from the cohort of black mothers "with pumping breasts" (288) who haunt her barren room at night. Neither challenge is unproblematic: Michael's exhortation that Jadine abandon capitalism, embracing the traditional art of the "mask-makers" (74) makes sense as long as the artist understands (the way Morrison does) that masks need to be ripped and renewed periodically. Likewise, the haunting "mammies" want to return Jadine to a traditional motherly role for which she has little appetite. By dedicating her book to mothers, grandmothers, "and each of their sisters, all of whom knew their true and ancient properties," Morrison seems to endorse their viewpoint. But while decrying Jadine's ignorance of "ancient properties," Morrison does not curtail her effort to find alternatives to them. As the writer commented in her interview with Rosemarie Lester, women's aspirations to motherhood and a profession are not "mutually exclusive [. . .]. Black women are both ship and safe harbor" (49).

The problem is that Jadine's choices are still not empowering enough for a black woman coming of age in the 1970s, caught between the awareness of her ethnic identity and a desire to assert herself in the larger culture. Jadine's best option in the novel would have been an artistic career, but she lacks the skill to practice the "difference between fine and mediocre" (182). Without this self-expressive outlet, she is returned to an Isabel Archer type of question, pondering which of her "three raucous men" to choose (48): "clear-headed—independent" Michael (72), who fails to show up in the novel; untamed Son with his essentialist views on "blackness"; or the "white but European [Ryk] which was not as bad as white and American" (48) who wants to marry "a black girl [. . .] any black girl" (74). Feeling at twenty-five "too old for teenaged dreaming, too young for settling down" (159), Jadine defers decision, fleeing first to the Caribbean islands to "sort out things before going ahead" (49), then back to Europe at the end of the novel. Son disturbs her fragile self-control most profoundly. With his "skin as dark as a river-bed, his eyes as steady and clear as a thief's," and his "[u]ncivilized, reform-school hair" (113), Son represents for Jadine the threatening masculinized figure of her race. Jadine's fear of Son is influenced by "colorism" but also by displeasure with the traditional male role that he plays initially. On their first encounter, Son threatens Jadine with rape and he enjoys seeing her scared because "it made him feel protective and violent at the same time" (177). Jadine rejects Son's advances several times, but finally commits to his hand "large enough, maybe, to put your whole self into" (212) after her fantasy of interracial harmony is shattered by the failed Christmas dinner.

Like in Song of Solomon, tensions specific to modern African American culture undermine Jadine's relationship with Son, pitting Son's Southern small-town experience against Jadine's metropolitan aspirations. Except for an early passionate interlude in New York, when the couple managed to build on their complementary experiences, their relationship is fraught with controversy. According to Morrison herself, the problems Jadine and Son confront "are caused not so much by conflicting gender roles as by the other 'differences' the culture offers," such as "what work to do, where and when to do it, [. . .] what they felt about who they were, and what their responsibilities were in being black" (interview in McKay 421–22). Still, gender views clearly play a role, inflecting the characters' cultural discords. Son begrudges Jadine her economic power and seeks her area of "bird-like defenselessness" that was "his to protect" (220). He thinks of himself as a redeemer, rescuing Jadine from Valerian's imperial "plantation" and a corrupting New York that drives old people into "kennels," "childhood [. . .] underground," and "beautiful males who [. . .] found the whole business of being black and men at the same time too difficult" (216) to destruction. But he wants to return Jadine to the small-town culture of his native Eloe, Florida, that Jadine finds oppressive because of its "paleolithic" gender attitudes (257). Jadine embraces New York instead, with "an orphan's delight" that knows that "if ever there was a black woman's town," encouraging participation in all professions, "New York was it" (222).

In spite of certain concessions both of them make, Son and Jadine remain trapped in a battle of stereotypes: "mama-spoiled black man" vs. "culture-bearing black woman" (269), or "cultural throwback" steeped in "white-folks-black-folks" oppositions (275) vs. "tar baby" (269) who betrays her own culture (270). These gender stereotypes undermine the characters' efforts to find a common ground, but this failure should not be blamed on Jadine alone. After all, she is the one character in the novel who—carrying on the challenge that the mulatta has always posed to unified concepts of ethnicity (Chow 70)—tries to envision a postcolonial community without socioethnic polarizations. If she fails, it is both because she is "irrevocably separated from her cultural heritage and has no means of negotiating the conflict between that heritage and her adoptive one" (Mobley 143), and because Son refuses any racial intermingling, convinced that "people don't mix races; they abandon them or pick them" (Tar Baby 270). While Son deplores Jadine's assimilation to white culture, he cannot offer her a better way of mediating gender and racial oppositions. According to Morrison herself, both positions "are inadequate [. . .] today [. . .]. She hasn't enough of what he has and he doesn't have enough of what she has" (Samuels and Hudson-Weems 87).

After they separate, Son and Jadine return independently to the Caribbean to complete their quest. Jadine's options are noticeably more limited at this point than Son's: While she savors her new independence, she has few opportunities to exercise it creatively. As she prepares to embark for Paris, she realizes that she cannot run away from the "diaspora mothers with pumping breasts" who try to "impugn her character" (288). She plans to "let loose the dogs" in Paris, "tangle with the woman in yellow [. . .] and all the night women who had *looked* at her," giving up her "dreams of safety" (290) as a step toward maturing. Her final vision, however, is of a regimented world of soldier ants that barely remember their magic coupling with a flying suitor as they work diligently for their survival. Son's return to Dominique is more enlightening, providing him with two ideal guides (Gideon and Thérèse) for his final cultural quest. Gideon is not only a man-of-all-trades but also a cross-cultural traveler who once immigrated to Quebec, married an American Negro nurse, and returned to Dominique bringing gifts from other cultures for Thérèse. His wife is equally versatile. Attuned to the folk culture of the Caribbeans, she weaves intricate stories about the blind African horsemen, descendants of runaway slaves, who roam the hills challenging the ghosts of the French imperialist chevaliers.

Like the legend of the flying tribe in *Song of Solomon,* the tale of the "blind race" (152) is a mythopoetic metaphor for an alternative ethnic history and for the workings of narrative imagination. The slaves brought to Dominique by the French colonialists went blind on seeing the beautiful island, causing their ship to sink. The half-blind were recaptured by the French; the others hid in the mountains, mating at night with the "swamp women" (153), themselves mythic representations of slave mothers. As Judylyn S. Ryan has argued, the blindness of the Africans "must be seen as a willed and self-conscious act of survival" 69), freeing them from the trauma of slavery that would have looked hellish in a subtropical paradise. By closing their "short-sighted natural eyes in order to see, as Gideon says, 'with the eye of the mind'" (Ryan 70), the former slaves develop an alternative, runaway imagination, a redemptive "second sight" (Du Bois 3).

Paralleling Jadine's involuntary encounter with the swamp women, who bait her back into the pungent substance of their heritage but also teach her how to perform a life-saving dance with a tree, Son's final trip to the blind race is both empowering and threatening. Its placement at the end of the novel makes this episode look like a response to Jadine's failed immersion in her heritage. The swamp women claim Jadine as a "runaway child [that] had been restored to them" (183), but Jadine struggles to free herself from their "permanent embrace," "to be something other than they were" (183). She walks away with a tarlike substance on her dress that turns her into a "tar-baby"—a figure that for Morrison represents "the black woman who can hold things together" but which in Joel Chandlers Harris's retelling of the "The Story of the Rabbit and the Fox" plays the troubling role of a trap "used by a white man to catch a rabbit" (Morrison interview in LeClair and McCaffery 27). While Jadine is not yet ready to work out the implications of her ambivalent role (instead of assuming the positive definition of "tar-baby," she tries to rub the tar off her legs and clothes), Son accepts his role as the pilfering rabbit in the "Uncle Remus" version of the tar-baby fable he tells Jadine. Taken by Thérèse to the rocky side of the island, Son feels his way up the cliffs and starts running "lickety-split" (*Tar Baby* 306) toward his tryst with the "blind race." By contrast to Jadine's quest that leads to more wandering, Son seems to have found his destination.

Even so, the fact that Son remains "revolted by the possibility of being freed" also from Jadine (305) and that his reunion with the mythic figures of his race takes place in an uncertain ontological realm suggests that Morrison did not wish to end the novel "with easy answers to complex questions" (Morrison interview in McKay 420). Jadine's own story is left unfinished (after all, she is only twenty-five), and her return to Paris is haunted by the need to play more varied roles than those of "mammies" or "tar babies" and to integrate herself in the larger post–civil rights/post-Vietnam culture without sacrificing her own. At the end of the novel Jadine still faces what Morrison has called the "problem of the rescued" who have "internalized the master's tongue [. . .]. Unlike the problems of survivors who may be lucky, fated, etc., the rescued have the problem of debt" to the dominant culture ("Friday on the Potomac" xxv). By contrast to the "rescued" Jadine, survivors and runaways like Son or the ex-slaves who roam the hills seem to have a "choice" (306). And yet, as Morrison's following novel Beloved makes clear, this choice is not necessarily more liberating. The "problems of survivors" can be as paralyzing as those of the "rescued."

3. "UNSPEAKABLE THOUGHTS" SPOKEN: RECLAIMING THE OTHER'S HISTORY IN Beloved

Narrative, Morrison told Tom LeClair, "remains the best way to learn anything permanently, whether history or theology" (Leclair and McCaffery 255). If Morrison's earlier narrators "bear witness" to the history of the black "civilization that existed underneath the white civilization," attesting "who the outlaws were, who survived under what circumstances and why, what was legal in the community as opposed to what was legal outside of it" (254), her more recent storyteller-witnesses address the traumas of slavery, dislocation, and assimilation that have threatened the very survival of that civilization. Confronted with a "painful," "lost," or "unspeakable" past (Beloved 58), Morrison's storytellers engage in laborious process of "remembering, repeating and working through" (Brooks, Reading for the Plot 111) that finds no clear resolution. This is not unusual for narratives that tackle traumatic historical events that, as Slavoj Žižek has shown, "resist symbolization, totalization, symbolic integration" (6).

Based on the historical account of a fugitive slave from Kentucky who, when caught in Ohio in 1856, killed her three-year-old daughter with a butcher's knife and attempted to slay her other three children to prevent their reenslavement, Beloved (1987; winner of the Pulitzer Prize for fiction and National Book Award finalist) had an arduous task ahead of it. As Morrison told Marsha Darling, "I did not do much research on Margaret Garner other than the obvious stuff, because I wanted to invent her life, which is a way of saying I wanted to be accessible to anything the characters had to say about it" (5). Giving voice to an infanticidal mother proved particularly hard since participants in such "unspeakable" dramas must overcome not only the rules of decorum that discourage slave narrators "from dwelling too long or too carefully on the more sordid details of their experience" (Morrison, "The Site of Memory" 113), but also their own traumatic "disremembering." Survivors of great ordeals waver "between the compulsion to complete the process of knowing and the inability or fear of doing so" (Laub and Auerhahn 288). For them the "crisis of death" ("the unbearable nature of an event") is inextricably tied to "the crisis of life" ("the unbearable nature of its survival"), so that any narration becomes "an impossible and necessary double telling" (Caruth, Unclaimed Experience 7, 8).

The "double telling" of trauma witnesses translates into a nonlinear, anxiety-ridden narrative that alternates moments of intense involuntary recalling with evasions and silences. Trauma narration is by necessity dialogic, interplaying many partial perspectives; it is also incomplete, haunted by a "spectral" infrastructure that disrupts the surface plot with unresolved, ghostly messages. Beloved illustrates both aspects. Morrison's novel "rip[s] that veil drawn over 'proceedings too terrible to re-late,'" recovering suppressed events and filling in the gaps "on the basis of some in-formation and a little bit of guess-work" (Morrison, "The Site of Memory" 90, 92). Yet even this "literary archeology" has its "specter"—Sethe's third-born child, Beloved, who haunts the novel as a representation of the victim's need for narrative but also as a demonstration that "only when the events of the past can be imagined not only to have consequences for the present but also to live on in the present [. . .] they can become part of our experience and can testify to who we are" (Michaels 7). In this sense Beloved represents historicity itself, the repressed past we need to re-cover in order to have "some kind of tomorrow" (*Beloved* 273).

As a manifestation of the "[d]isremembered and not accounted for," Beloved is an aporetic figure who "has claim, [but] is not claimed" (*Beloved* 270) and who can only be named posthumously. ("Beloved" is not a proper name but a truncated version of the priest's address at the baby's burial liturgy, carved on her headstone.) The "bot-tomless longing" (58) in Beloved's eyes reflects a yearning for completion through the stories of survivors who are encouraged to step back into the victim's footsteps, be-cause "they will fit" (275). But as James Phelan has correctly argued, Beloved also functions as a figure of "recalcitrance," simultaneously inciting narrative and "pre-vent[ing] it from being completely successful" ("Toward a Rhetorical Reader-Response Criticism" 232). She takes on conflicting identities in the novel, from ghost of Sethe's murdered daughter to "embodied spirit" of those lost in the Middle Passage who return to be memorialized (Christian, "Fixing Methodologies" 13), and from in-carnation of Sethe's guilt to a figure of female energy. The presence of an authorial narrator who fills in the gaps with clarifying information does not help settle Beloved's definition. This authorial voice creates "tension between the fictional past and the moment of narration" and between the "pre-modern" (oral) and "postmod-ern" (experimental) narrative discourses "it assumes and transforms" (Pérez-Torres 104, 108, 109). The novel's perplexing opening ("124 was spiteful. Full of a baby's venom" [3]) challenges readers to accept that this narrative is driven both by the trau-matized memory of the former slaves and by the ghost of Beloved who haunts Sethe's home at 124 Bluestone Road as a reminder of the radical otherness of the past.

The novel's characters have to undergo a similarly strenuous process of adjust-ment to the otherness of their historical experience. As the montage of "unspeakable thought, unspoken" in part 2 suggests, each of the three focalizers (Sethe, Beloved, and Denver) must struggle alone with the contradictions and silences of her story. Illustrating features of trauma narratives, with a "not fully conscious address" and "departure[s] from sense and understanding" (Caruth, *Unclaimed Experience* 24, 56), the silent monologues of Sethe, Denver, and Beloved confront issues specific to each speaker: how to "look at things again" (201) from the perspective of an unwilling survivor, in Sethe's case; how to make a story out of a truncated, unrealized life, in Beloved's case; how to diffuse the threat of infanticide, in Denver's case. But these monologues also confront a common problem that makes them intelligible to one another: the question of how to speak the "unspeakable," how to negotiate a collec-tive trauma. The central episode of the infanticide is circumscribed by several nar-rators, including Sethe, who "knew she could not close the circle, pin it down for anybody who had to ask" (163). The former slaves take all kinds of precautions to suspend memory and ensure the "stillness of [their] soul" (5). When Paul D wan-ders back into Sethe's life, eighteen years after her escape from Sweet Home Farm, she is an expert at the "serious work of beating back the past" (73). As the last male survivor of the slave farm, Paul D himself has also "shut down a generous portion of his head, operating on the part that helped him walk, eat, sleep, sing" (41).

But the peace gained through "escape and avoidance [. . .] as coping strategies" (Creamer 60) is illusory. Sethe's house is "palsied by the [. . .] fury" (5) of the baby daughter sacrificed by her, but also by the clamor of thousands of other "black and angry dead" (198). Fleeing 124 Bluestone Road is not an option because, as Sethe's mother-in-law wisely states, "Not a house in the country ain't packed to its rafters with some dead Negro's grief" (5). Therefore, Sethe decides to "call forth the ghost" and end the persecution with a "conversation [. . .] an exchange of views or some-thing" (4). Later, when Beloved enters the novel in flesh, Sethe experiences what trauma psychology describes as a "literal return of the event against the will of the one it inhabits" (Caruth, *Unclaimed Experience* 59). She responds to this assault of "remem-ory," as the novel calls it, with a compulsive-repetitive effort to account for past events: "[E]ven when Beloved was quiet, dreamy, minding her own business, Sethe got her going again. Whispering, muttering some justification, some bit of clarifying information to Beloved to explain what it had been like, and why, and how come" (225). The unresolved traumas of the past demand close interaction between mother and daughters in order to move the reconstructive process from mere "recollection" to "an interpretive elaboration or working-through whose role is to weave around a rememorated element and entire network of meaningful relations that integrate it to the subject's explicit apprehension of itself" (Laplanche and Leclaire 128).

Paul D's own process of rememory—triggered by Beloved who evokes "something [. . .] I'm supposed to remember" (234) but also by Sethe who pries open the "close portion of his head [. . .] like a greased lock" (41)—focuses on the fate of the five male slaves on the Sweet Home farm: "One crazy [Halle, who goes mad after witnessing Sethe's rape], one sold [Paul F], one missing [Paul A], one burnt [Sixo], and me lick-ing iron with my hands crossed behind me" (72). Even without the burden of infan-ticide, Paul D's memories are as traumatic as Sethe's. His recollections circle around his last hours on the farm, bridled and bitted like a mule after a failed escape attempt; his chain gang days in Alfred, Georgia, after trying to murder his new owner; the con-fusion of the war years, when he is sold to the Confederate Army, and of his seven years of wanderings that take him finally to Cincinnati. But these traumatic experi-ences also teach him elemental survival skills, such as "talking" through the chains while making his escape from the mud prison in Georgia or finding his way to the "Magical North" (112) by following the progression of blooming trees.

In foregrounding the dramatic history of male slaves, Paul D completes the process of "rememory" started by the former slave mother, Sethe, but also seeks an empathetic realignment between male and female experiences. His effort is shared by Sethe, who finds that "[h]er story was bearable because it was his as well—to tell, to refine and tell again" (99). Other characters are drawn into this process of coop-erative reconstruction: Encouraged by Beloved, Denver recalls the story of her own birthing, turning it into a narrative of personal survival and interfemale cooperation. In turn, Denver asks Beloved if she "disremembers everything" (119), and is treated to a cryptic account of crouching among dead bodies, being snatched away from her "woman," and waiting perpetually on a bridge, which sums up the story of a Middle Passage child. In the end of the novel the entire community joins in, releasing their own repressed memories. As Laurie Vickroy argues, the dialogic cooperation between narrators "is vitally important to helping individuals emerge from traumatic stasis and repression. [. . .] Dialogism has thus three purposes: to enable testimony, to juxtapose different perspectives, and to confront the viewer or reader with the complexities of traumatic experience and its interpretation" (132).

What concerns Morrison in *Beloved* "is not what history has recorded in the slave narratives" but what it has "disremembered" or repressed (Samuels and Hudson-Weems 96). To begin with, the Middle Passage is vaguely present in Sethe's memory, in a language whose "code she no longer understood" (*Beloved* 62), and in Beloved's equally obscure recollections. By filling in the gaps, the novel allows us to discern a few patterns in the story of the Middle Passage and slavery on the American continent: traumatic severance from parents, premature exposure to life's hardships, and the reduction of black men and women to chattel ("Anybody [. . .] who hadn't run off or been hanged, got rented out, loaned out, bought up, brought back, stored up, mortgaged, won, stolen, or seized. So Baby [Suggs]'s eight children had six fathers" [23]). Rape is also a common thread in the slave experience of both sexes: "[R]ape is the trauma that forces Paul D to lock his many painful memories in a 'tobacco tin' heart (113), that Sethe remembers more vividly than the beating that leaves a tree of scars on her back, that destroys Halle's mind, and against which Ella measures all evil" (Barnett 418).

By comparison to the Middle Passage generation represented by Baby Suggs, who lost all her children ("four taken, four chased" [5]), or Sethe's mother, who lost her life, Sethe's generation fares initially better on the Sweet Home farm in Kentucky. The Garners run "a special kind of slavery" (140), treating their slaves as "men" who could "[b]uy a mother, choose a horse or a wife, handle guns, even learn to read if they wanted to" (125). But when the widowed Mrs. Garner hands over the administration of the farm to her brother-in-law ("Schoolteacher") and his two nephews, slavery is restored to a vicious technocratic form. In the name of capitalist progress, Schoolteacher replaces the "patrimonial and sentimental version of racial domination" practiced by his precursor with a "rational and scientific racism" (Gilroy 220). As a result of Schoolteacher's "scientific" management, the Garner slaves are scattered and destroyed: Paul F is sold to pay off the farm's debts, the fugitive Sixo is burned alive, Paul D is handed over to a new slave master who drives him to murderous revolt, Halle loses his mind over Sethe's rape, and Sethe runs away to be later recaptured and forced to commit infanticide.

In retrospect, the difference between Garner's "special kind of slavery" and Schoolteacher's new managerial system proves as tenuous as that between soft-spoken and loud slave masters. As Halle warns Sethe, "What they say is the same" (*Beloved* 195). The dehumanizing racist message conveyed by masters like Schoolteacher who regard their slaves as only three-fifths human does not change radically after emancipation. In 1874 when Baby Suggs dies, worn out by the tragedies witnessed around her, "whitefolks were still on the loose. Whole towns wiped clean of Negroes; eighty-seven lynchings in one year alone in Kentucky; four colored schools burned to the ground; grown men whipped like children; children whipped like adults; black women raped by the crew; property taken, necks broken" (180). The perpetuation of racism is ensured not only by fringe groups in the South but also by the larger culture that constructs a "jungle" (199) around black identity, "othering" it.

In reconstructing the traumatic history of slavery, Morrison's narrator-witnesses hope to achieve not only personal catharsis but also a redefinition of their roles, renaming themselves as subjects. Since "definitions belong to the definers—not the defined" (190), Morrison's narrators must begin by challenging official history that has sanctioned the oppressive objectification of blacks. Rewriting/renaming becomes a choice strategy for both Morrison and her characters. An important area of redefinition in Beloved concerns the role of women during and after slavery. Against the traditional slave narrative in which "more often than not the [slave woman's story] had been told by the black male narrator whose focus was primarily upon his own journey to wholeness" (Samuels and Hudson-Weems 97), Morrison gives voice both to Beloved as haunting victim and to Sethe and Denver as survivors who undergo a rebirth in the novel. Sethe's name is a feminine inflection of the Egyptian god who broke taboos and dismembered Osiris and of Eve's third child born "in place of" the dead Abel, whose descendants were Noah and his unjustly cursed son Ham. Like the falcon-headed Seth, Sethe can "fly" when in danger and proves destructive when threatened. Cooperation between women is also crucial to the survival of Sethe and Denver, from the time Denver is born assisted by the white fugitive Amy, through the period when Sethe and Denver are sheltered in Baby Suggs's house, to the end when Denver finds her way back into the community helped by female neighbors. Morrison's novel pits heterosexual family units (Halle, Sethe, and their children during slavery; Paul D, Sethe, and Denver after slavery) against female trios (grandma Baby Suggs, mother Sethe, and daughter Denver, before the novel starts; mother Sethe, reincarnated daughter Beloved, and daughter Denver, in the narrative present). The latter prevail whenever Beloved puts pressure on the traditional heterosexual relationship, chasing the male figures away (first brothers Howard and Buglar, then surrogate father/lover Paul D). Since Beloved's haunting is responsible also for Baby Suggs's withdrawal and death, Beloved can be credited with further converting a three-generation female household into a symbiotic relationship mother-to-daughter, sister-to-sister, and woman-to-woman that upon occasion (e.g., when Sethe is being "licked, tasted, eaten by Beloved's eyes" [57], or touched in sleep with a "touch no heavier than a feather but loaded [. . .] with desire" [58]) suggests "one of the most violent forms taken by the rejection of the [patriarchal] symbolic," lesbianism (Kristeva, "Women's Time" 204).

Paul D manages briefly to break "up the place" charged with female emotions, "shifting it, moving it over to someplace else, then standing in the place he had made" (39). He wants Sethe to make room in her life for him, allowing his story to interact with hers. On the day he takes Sethe and Denver out to the carnival, Paul D feels hopeful that "We can make a life, girl" (46). Upon their return home, however, they find a "soft and new" Beloved (52) waiting for them, ready to displace Paul D "imperceptibly, downright reasonably" (114). What Beloved starts, Sethe's confession of infanticide finishes, forcing Paul D to take refuge in the basement of the Church of the Holy Redeemer.

In chasing Paul D temporarily away, Morrison's novel does not simply illustrate a "sentimental feminist ideology" (Crouch 205), as a few critics have complained. Beloved refuses to treat female slaves as mere victims. The four women with whom Paul D competes for attention—Baby Suggs, Sethe, Beloved, and Denver—play unconventional roles that challenge the quietly enduring or "crazy" female slave in traditional slave narratives. Baby Suggs has an unusual career for a woman who bore the brunt of slavery. Bought out of slavery by her last surviving son, Halle, Baby Suggs is allowed by the abolitionist Bodwins to settle in their childhood house, "in return for laundry, some seamstress work, a little canning" (145). The former slave turns 124 Bluestone Road into a "cheerful, buzzing house where Baby Suggs, holy, loved, cautioned, fed, chastised and soothed" (86–87). She plays the role of "uncalled, unrobed, and unanointed preacher" (87) for the black community, teaching the men to dance, the women to cry for "the living and the dead" (88), and all of them to love their bodies because "Yonder they do not love" them (88). The ceremonial space she creates in a forest clearing outside of Cincinnati encourages egalitarian bonding and unconventional imagination. Her holy period culminates in an epic feast during which, like a bountiful mother, Baby Suggs multiplies the blackberry pies to feed ninety neighbors. Soon after, Baby Suggs's utopian vision of a world with "not one touch of death in the definite green of the leaves" (138) is rudely disrupted by the arrival of the white slave hunters in pursuit of Sethe and by Sethe's infanticide. With her heart strings broken by the "whitefolks" (89) but also by the "community of other free Negroes" (177) whose jealousy of her bountifulness kept them from warning the family of the approaching danger, Baby Suggs takes to bed and silence. She speaks only once to impart the skeptical lesson learned from sixty years as a slave and ten as a free woman: "[T]here was no bad luck in the world but white people" (104).

Sethe's life is marked by as much drama as Baby Suggs's and even more obstinate self-reliance. Paul D remembers her as a young woman "with iron eyes and a backbone to match" (7). Sethe's capacity for survival is proven during her molestation by Schoolteacher's nephews, her birthing in the wilderness, and her later infanticide. As the nephews brutally assault her and suck her life-sustaining milk, Sethe tries not to bite her tongue for fear she might "eat [herself] up" (70), finishing the job started by the boys with "mossy-teeth" (70). She survives this episode of vampiric rape that bell hooks urges us to see "as part of overall patterns of genocidal assault" (Olson and Hirsh 133) and makes her escape across the Ohio River. During her strenuous journey to freedom, she survives a premature birthing helped by the fugitive Amy Denver, daughter of a white indentured servant. In what reads like a feminist revision of *Huckleberry Finn* and of the biblical tales of Nativity and the good Samaritan, Amy (whose name is derived etymologically from the French "aimé," "beloved") and Sethe experience a moment of profound solidarity between "two throw-away people, two lawless outlaws—a slave and a barefoot white woman with unpinned hair—wrapping a ten-minute-old baby in the rags they wore" (*Beloved* 85). Though divided by race and cultural opportunity (unlike the "nigger woman," "Miss" Amy Denver can dream of a future comfort embodied for her in the carmine velvet she hopes to purchase in Boston), the two women manage to interact "in other terms than those of heroism and helplessness" provided by the "racialized, gendered, and class-marked discourse of American romance" (Moreland 160, 176). Amy is capable of naming not only Sethe's hurt (she describes her scar as a chokecherry tree), but also her own suffering at the hand of a white master. She plays the role of midwife and healer, restoring Sethe's "loaves of flesh" (her feet [30]) to life and helping Sethe deliver her first free child. In doing this, she ushers in a world of hope that gives new meaning to her dream of "velvet": "Velvet is like the world was just born. Clean and new and so smooth" (33). Encouraged by Amy's dream of velvety newness, Sethe herself imagines a better future for her infant baby to whom she gives Amy's last name. In a further disruption/rewriting of the American romance, Morrison's liminal females find their sense of self inside rather than against a nature whose every image (bed of leaves, water, antelope) suggests a rejuvenating motherly world.

While the relationship with Amy still suggests cultural inequality (the two women's dialogue seems to slip "effortlessly into yard chat [. . .] except one lay on the ground" [33]), Sethe benefits from full acceptance in the free black community across the Ohio river. Human solidarity saves Sethe and her newborn baby, ferrying them across the river on Stamp Paid's flatbed; welcomes her fugitive story in Ella's house, who fills empathetically in the "holes" the things unsaid and the people left behind (92); and finally deposits Sethe safely in Baby Suggs's house, who nurses her body back to life. Sethe's enjoyment of "unslaved life" (95) in an environment that, as Morrison told Claudia Tate, "offers an escape from stereotyped black settings. It is neither plantation nor ghetto" (119)—lasts for twenty-eight days. The allusion to the lunar cycle suggests a "period of regeneration and renewal" for Sethe (Samuels and Hudson-Weems 118) but also predicts the "flow of blood" in response to the arrival of Schoolteacher's posse to reclaim Sethe and her "pickaninnies." Like the historical Margaret Garner, Sethe kills her two-year old daughter with a handsaw and wounds her sons Howard and Burglar in order to save them from "what Ella knew, what Stamp saw and what made Paul D tremble" (Beloved 251): the dehumanizing experiences of slavery. Only Denver is saved by Stamp Paid, the Underground Railroad worker, who snatches the baby away before her head is dashed against the wall.

Though historically rare by comparison to other methods of resistance against slavery and condemned by the black community of Beloved as "prideful, misdirected" (256), Sethe's infanticide dramatizes the "damage African American families suffered both from racist institutions and sometimes from themselves, as their members acted under pressures from these institutions" (Berger 418 n. 1). For James Berger the novel's apocalyptic climax is the outcome of a duplicitous culture that perpetuated forms of racism even after emancipation, maintaining an original "historical trauma [. . .] at the center of American race relations" (414). In an ironic circularity, Sethe is saved from hanging by the same cultural system that forced her to infanticide in the first place: A petition drawn by the abolitionist Mr. Bodwin and a group of white women releases her from prison and returns her back "home" to be haunted by her baby ghost and ostracized by the black community. Sethe's postslavery reeducation instills in her the idea that her "twenty-eight days of having women friends, a mother-in-law, and all her children together; of being part of a neighborhood; of, in fact, having neighbors at all to call her own" was a "short-lived" and undeserved "glory" (Beloved 173). Paul D's flight at the end of part 1 reinforces her skepticism about the possibility of having a life of her own and strengthens her commitment to a female nucleus that now includes reincarnated Beloved.

The much shorter part 2 of the novel represents Sethe's effort to come to terms with her embattled past and resume her role as mother to Beloved. Sethe's graceful skating on the frozen creek holding hands with Beloved and Denver, followed by their sharing of hot sweetened milk, confirms Sethe in the role of "primal mother" (Moglen 29). Convinced that her entire "world is in this room" (Beloved 183), she severs her remaining connections with the outside world. This period of mother-daughter absorption is signaled textually through a montage of interior monologues exchanged silently by Sethe, Beloved, and Denver, and "overheard" only by Stamp Paid, who cannot make out their language since they probe intense female experiences inaccessible to a linear male logic. The "women speak in tongues from a space 'in-between each other,'" performing a "fugue-like ceremony of claiming and naming through intersecting and interstitial subjectivities" (Bhabha, *The Location of Culture* 17). But their effort at building an interpersonal space with "no crack or crevice available" (*Beloved* 188) is undermined by emerging tensions. Each monologue begins by asserting possessively the connection to Beloved ("Beloved, she my daughter. She mine"; "Beloved is my sister"; "I am Beloved and she is mine" [200, 205, 210]), creating competition between speakers. An important rift emerges between Denver and Sethe, the younger daughter feeling threatened by the possibility of her mother repeating her infanticidal act if present circumstances warranted it. Beloved's own unpunctuated monologue is estranged from the others in its idiosyncratic and self-absorbed "babble." After finding some common imagistic and emotional ground, the three monologues end indecisively, affirming and questioning the integration: "You are my face; I am you. Why did you leave me who am you?" (216).

The fact that Stamp Paid is denied access inside Sethe's house and language suggests to Marianne Dekoven the limitations of a female utopia that denies cross-gender solidarity and locates hope "in post-utopian times" rather than "within history" (119). Part 3 of the novel dramatizes these limitations. Going against her better instinct that made her resist Paul D's earlier request to bear him a child, Sethe succumbs entirely to Beloved's demand for motherly attention. She switches places with her presumptive daughter, Beloved looking more and more like a pregnant mother and Sethe like a powerless child. Sethe's fate bears out Kristeva's observation that "without the maternal 'diversion' toward a Third party," a "third realm supplementing the autoeroticism of the mother-child dyad," the "bodily exchange" between them is "abjection or devouring" (*Tales of Love* 22, 34). As a further irony, in spending her entire energy on Beloved and neglecting Denver and herself, Sethe forfeits her earlier promise to be a bountiful mother with "milk enough for all" (100).

The last part of the novel focuses on Sethe's tribulations as an embattled mother and her subsequent rediscovery of that replenishing "third realm" (community, outside world, Paul D). When word of Sethe's desperate condition gets out, the community welcomes the news as a fair retribution for Sethe's "outrageous claims" of "self-sufficiency" (*Beloved* 171). Still the neighbors intercede with the embodied ghost of Beloved, disturbed by the prospect of "past errors taking possession of the present" (256). With the paradoxical wisdom of a Pynchon character, Ella wants to keep ghosts within reasonable boundaries: "As long as the ghost showed out from its ghostly place—shaking stuff, crying, smashing and such—Ella respected it. But if it took flesh and came in her world, well, the shoe was on the other foot. She didn't mind a little communication between the two worlds, but this was an invasion" (257). Therefore, the exorcism that Ella and thirty other women perform in front of Sethe's house targets not only Beloved, as the most visible incarnation of a traumatic past, but also their own lurking ghosts. "Building voice upon voice until they found it" (261), their chant chases Beloved away, reinstating Denver as the rightful daughter. It also facilitates Sethe's cathartic reenactment of her encounter with the white master, allowing her to refocus her anger on Denver's employer Bodwin (whom she mistakes for Schoolteacher) rather than on her daughter. This reenactment purges Sethe's infanticidal impulses, the guilt of the neighboring women for having forsaken the troubled Sethe, and even the remorse of the former owner of 124, Edward Bodwin, for not bringing to fruition his father's democratic views on race. The hold the past has on these characters is finally broken because the process of "rememory" has managed to move from an individual to a collective focus, making possible a new "postcolonial" knowledge (Mohanty 61). However, the fact that this entire sequence of scenes, including Paul D's return to 124 to rescue Sethe from her deathlike retreat in Baby Suggs's room, is narrated in the present tense suggests that the collective recovery of a "genuinely noncolonial moral and cultural identity" (Mohanty 67) has just started. One question left undecided to the very end is whether Sethe will find a role beyond self-denying motherhood. When Paul D rejects her notion that Beloved was her "best thing" in life ("You your best thing, Sethe"), Sethe replies incredulously: "Me? Me?" (273).

Problematic is also the fact that Sethe's recovery and reintegration depend on the expulsion of Beloved as a figure of radical female otherness. Beloved's puzzling representations of her origin suggest a pre-oedipal female imaginary (Wyatt 481) in which the baby daughter yearns for the motherly "face" and "hot thing" and resents the interference of males who threaten to crush her body. The water imagery associated with Beloved (born from sea-depths, she drinks buckets of water, moves and speaks with a fluid cadence, and keeps her eyes glued like "rainwater" to her mother's face) also hints to an otherworldly water spirit, or "undine." This female archetype is snatched violently from the site of a maternal (pre-oedipal) semiotic, remaining—like Sethe herself—without the "language her ma'am spoke" (*Beloved* 62). In being murdered at the age of crawling, she is also denied access to the symbolic, missing the adaptive skills and self-narration that come with maturity. While her return, illustrating Freud's revenant logic of trauma, excites "shaped and decorated" tales from the people who saw her, hers "is not a story to pass on" (270) because of its "unassimilable" nature. As far as Beloved is concerned, Morrison's novel ends indecisively with a eulogy of the storyless victim whose fate is forgotten except for occasional "rustles" and "shifts" (275). But as Adorno warned, a culture that fails to "come to terms" with its history of injustice, rendering the "causes of what happened then [. . .] no longer active," is condemned to stay under "the spell of the past" (129). In *Beloved* neither the exorcism of the incarnated ghost nor the community's intercession with Sethe's assault on Bodwin resolves the larger issues of American interracial relations. Sethe is wrestled down when she attacks Bodwin, but her "confusion" of persons may be due to a subconscious connection she makes between the white slave owner (Schoolteacher) and the abolitionist Bodwin. The latter keeps on his shelf a small figurine representing a kneeling black boy, "his head thrown back farther than a head would go," his eyes "bulging like moons [. . .] above the gaping red mouth" that "held the coins needed to pay for a delivery or some other small service" (*Beloved* 97). This hideous caricature called "At Yo Service" perpetuates racist stereotypes beyond emancipation. Through Bodwin the novel comments ironically on the ambiguous legacy not only of nineteenth-century abolitionism but also of the "1960s liberalism" (Berger 415) that sought to deconstruct the racial hierarchies promoted by the Cold War. Reconsidered twenty years later, both movements appear "only a minor triumph in a larger story of defeat [. . . that] remains blind to the interests and culture of African Americans" (Berger 415–16).

The presence of the white abolitionist at the scene of final reckoning has another implication. "Although the fact barely enters his consciousness, Bodwin is intimately and irrevocably connected to the black community" (Berger 417). He shares with it a common "house" (124, his birthplace), some hybridity (with his white hair and black mustache he is called a "bleached nigger" by his political enemies [*Beloved*

260]), and a biography that converges on the same historical sites. But while Bodwin and Sethe remain oblivious to these commonalities, Denver acknowledges some of that intertwined history as she "steps off the edge" of her insulated world (243) at the end of the novel and reaches out to her neighbors, black and white.

The return of Paul D to 124 coupled with the expulsion of Beloved undermines but does not erase the claims of Sethe's female world. Paul D's return is literally and symbolically "the reverse of his going" (270), tuning him to a world of female needs and affects. He approaches the house from the back, where he can admire Sethe's flowers. He enters the "bleak and minus nothing" (270) of the house and negotiates each room as a "place [where] he is not." He seeks Sethe both as a rescuer and the one to be rescued, moving from self-pity to cooperative hope: "Sethe, me and you, we got more yesterday than anybody. We need some kind of tomorrow" (273). By proposing to put not only his slave story but also his story as a "freed" man close to Sethe's, Paul D suggests that he finally understands the depth of their need for sharing. The vanished ghost of history makes sure that he does: In spite of the novel's reassurance that Beloved has dissipated, leaving only "wind" in her wake (275), she may be "waiting for another chance" (263) to rekindle the characters' "rememories."

4. THE PAST IS NOT A REPETITIVE RECORD: THE REMAKING OF AFRICAN AMERICAN HISTORY FROM THE GREAT MIGRATIONS TO THE CIVIL RIGHTS ERA (JAZZ AND PARADISE)

In a critical self-appraisal toward the end of *Jazz* (1992), the narrator concedes that her view of the past as "an abused record with no choice but to repeat itself" was contradicted by her characters who "were busy being original, complicated, changeable—human, I guess you'd say, while I was the predictable one, confused in my solitude into arrogance, thinking my space, my view was the only one that was or that mattered" (220). Making amends, the narrator allows her characters to tackle their life stories and ponder the role that imagination can play in reinventing the world: "What's the world for if you can't make it up the way you want it?" (208). The narrator's own contribution to this re-envisioning is "to imagine what it must have been like," taking risks in figuring out the logic of events and her characters' "state of mind" (137). In the spirit of the title, both the narrator and her protagonists play variations on the theme of loss but also celebrate the potential for individual recovery and change.

Inspired by James Van Der Zee's 1926 photograph of a young black woman who died concealing from her friends the fact that she had been shot "by her sweetheart at a party with a noiseless gun" (Van Der Zee, Dodson, and Billops 84), *Jazz* foregrounds like *Beloved* the drama of a woman who "loved something other than herself so much. She had placed all of the value of her life in something outside herself" (Naylor and Morrison 568). Love and violence are again intertwined, this time in a triangular plot that ends with Joe Trace shooting his eighteen-year-old mistress, Dorcas, and with his fifty-year-old wife, Violet, trying to slash her rival's face at the latter's funeral. After Dorcas's death, the two surviving characters replay events in their imagination, thinking they "could solve the mystery of love that way" (5). Especially Violet behaves like an empathetic historian, trying to borrow not only Dorcas's photograph but also her life in order to understand its history. Violet applies a similar technique to Joe's and her own life story: Through a "recursive" form of narration typical of Morrison's novels since *Song of Solomon*, the present is revealed to be the product of unsettled past traumas.

Morrison's own historical approach is nonconventional, focused on the reverberations of unforeseen events in the lives of lowly characters. Though its plot is placed in the heyday of the Harlem Renaissance, *Jazz* omits any direct reference to the black artistic and political accomplishments of the 1920s. Morrison is not interested in re-capitulating the events of "History with a capital *H*" (Peterson "'Say Make Me, Re-make Me'" 205), but in re-creating (much like Pynchon) the small-case history of daily human strivings ignored by official historians. Like Morrison's latest novel *Paradise* (1998), which prompts new attention to the stories of "ordinary folk" whose contribution to the civil "rights movement" has been overlooked (212), *Jazz* reconstructs the feel of another important period in the history of African American culture by telling "a simple story about people who do not know that they are living in the jazz age" (Morrison, "Art of Fiction" 117). The big picture is not absent (among the events alluded to are the 1882 hangings in Rocky Mount, the July 1917 riots in East St. Louis, the race conflicts caused by the job shortage after World War I, or the emergence of clubs, societies, and print media in Harlem), but what the novel suggests is that the broader story is a composite of local narrative "traces": of Joe Trace's aspiration to be a "new Negro" in the manner suggested by Alain Locke's *New Negro* anthology (1925); of Violet Trace's effort to apply survival skills learned in the cotton fields of Virginia to her new metropolitan environment; of Dorcas's hunger for a redeeming love; and of all three protagonists' need to fill the central gap of their existence with bold ventures.

In spite of the narrator's self-advertised abilities ("curious, inventive and well-informed" [137]), the novel maintains an aura of mystery around human motives. The narrator's own identity remains uncertain, though passages such as the following suggest a female voice:

> I lived a long time, maybe too much, in my own mind. People say I should come out more. Mix. I agree that I close off in places, but if you have been left standing, as I have, while your partner overstays at another appointment, or promises to give you exclusive attention after supper, but is falling asleep just as you have begun to speak—well, it can make you inhospitable if you aren't careful [. . .]. (9)

More tellingly, the novel begins with "Sth, I know that woman" (3), using the ono-matopoeic syllable that *Beloved* describes as one of "the interior sounds a woman makes when she believes she is alone and unobserved at her work: a *sth* when she misses the needle's eye" (172). Based on the novel's epigraph from *The Neg Hammadi* fables and on the first paragraph of the final section of *Jazz*, where "the phrase 'I the eye of the storm,' and the statement 'I break lives to prove I can mend them back again' are heard," Eusebio L. Rodrigues has argued that "it is the thunder goddess who narrates the [entire] story" (261). But this interpretation ignores the narrator's forceful self-criticisms. It is more plausible to construe the narrator as a "choral" voice speaking for "the community or the reader at large," a procedure that Morrison has drawn from oral African American literature ("Rootedness" 341).

Whether confident or self-deprecating, this voice is always *engaged*, "in a reckless hurry to tell everything at once without stopping" (Rodrigues 246). Just like jazz,

Morrison's novel "keeps you on the edge. There is no final chord" (Morrison interview, McKay 429), forcing the reader to experience the raw energy of desire. In *Jazz* the mechanism of desire is attended by physical and psychological violence. The three protagonists are caught up in "spooky loves" that make them resort to violence (Joe and Violet) or self-violence (Dorcas and Violet) to "keep the feeling going" (*Jazz* 3). Though women are more often victims than perpetrators of violence, they learn not to be "defenseless ducks" any more (74). Alice Manfred (Dorcas's guardian) is uncomfortable with women using weapons and is especially put off by Violet, whose name suggests to her—much as it did to Freud—a "violent" streak. But while Freud saw in the dream of a girl who arranges the table with "expensive flowers; [. . .] lilies of the valley, violets and pinks or carnations," a confirmation of the masochistic nature of female sexuality caught between "virginal femininity" and a desire for "violent defloration" (5: 374–77), Alice comes to realize that Violet is a product of her culture. She also concedes that, unless rich, "attached to [an] armed [man]," or protected by "leagues, clubs, societies, sisterhoods," any "unarmed black woman in 1926 was silent or crazy or dead" (78). Morrison's own stand on the issue is more complex: Foregrounding a theme that goes back to her first novel, the author suggests that violence erupts when the dominant male desire denies women "the consciousness needed to act as a subject" and "sexual desire becomes the only desire operative when the fulfillment of other desires is denied" (Cannon 240, 235). Reduced to a state of "wanting," Violet resorts to violence against her competitor and to self-violence, trying to regain her husband's attention.

Though both forms of violence enhance rather than resolve her self-alienation, some of Violet's random acts challenge the patriarchal system itself: She sits down in the middle of the street, disrupting traffic and her own expected routines; she frees her caged birds, and she learns to empathize with Dorcas. *Jazz* ends with a more constructive form of self-violence that purges Violet of false icons: She rips out the "mole" (208)—the dream of a golden mulatto boy planted into her head since childhood—that has messed up her sense of identity. She also "kills" symbolically that "other Violet" for whom the world had a "weaponry glint" (89), seeking a more confident and less competitive "me" (209). This self-redefinition takes place in the company of other women (Alice and Felice), in an exchange of narratives that helps them regain a sense of subjecthood. Joe goes through a similar process of narrative reexamination, though in his case without the support of a cooperative narratee. These acts of retrospection allow characters to move from a Western concept of love—"full of possession, distortion and corruption [. . .] slaughter without blood" (interview with Tate 124)—to more fulfilling forms of desire that allow them to become "more like the people they always believed they were" (*Jazz* 35).

In its most rugged form, the discourse of desire is embodied in the "City" (New York and Harlem) during the "Jazz Age." The narrator and her characters are seduced by its energy that—like jazz—arises from a creative "friction" between different cultural or "harmonic systems" (Bastien and Hostager 150). They also embrace the utopian promise embodied in the 1920s New York, "when all the wars are over and there will never be another one. [. . .] Here comes the new. Look out. There goes the sad stuff. The bad stuff. The things-nobody-could-help stuff. [. . .] History is over, you all, and everything's ahead at last" (*Jazz* 7). But this celebratory imagery that echoes the visionary rhetoric of the Harlem Renaissance ("In Harlem, Negro life is seizing upon its first chances for group expression and self-determination.

[. . .] Harlem has the same role to play for the New Negro as Dublin has had for the New Ireland or Prague for the New Czechoslovakia" [Locke 7]) cannot hide a darker side of the City. For the narrator, Harlem is a place of contrasts: "Daylight slants like a razor cutting the buildings in half. In the top half I see looking faces and it's not easy to tell which are people, which the work of stonemasons. Below is shadow where any blasé thing takes place: clarinets and lovemaking, fists and the voices of sorrowful women" (7).

The Traces experience the "double consciousness" of the City with bittersweet intensity. They make their journey North in 1906, lured like so many other blacks by "a new vision of opportunity, of social and economic freedom, of a spirit to seize [. . .] a chance for the improvement of conditions" (Locke 6). As soon as their northbound train leaves Delaware, the "green-as-poison curtain separating the colored people eating from the rest of the diners" (31) is pulled back. Closer to New York, their bodies begin to resonate with the dancing train:

> They weren't even there yet and already the City was speaking to them. They were dancing. And like a million others, chests pounding, tracks controlling their feet, they stared out of the windows for the first sight of the City that danced with them, proving already how much it loved them. Like a million more they could hardly wait to get there and love it back. (32)

After a shaky beginning, both gain decent employment: Joe as a salesman of Cleopatra cosmetic products, Violet as an unlicensed hairdresser. But their hard-won middle-class status proves hollow. The lovers, who danced their way into the City, barely speak to each other twenty years later when the main events take place. Joe and Violet's lives are disturbed not only by the contradictions of the City, which "back and frame you no matter what you do" (8–9), but also by "re-memories" arising from their past. Neither of them can reconnect their lives in the City, haunted by aging and a crooked "love appetite" (67) to their beginnings in Vesper County, Virginia.

The contrast between country and city love is familiar to Morrison's fiction. Still, what *Jazz* proposes is not a return to beginnings that were not any more idyllic (the novel suggests that Joe and Violet were thrown together by a common need as "orphaned," disenfranchised laborers) but rather a reconciliation between the rural and the urban halves of the African American experience, as a first step toward reestablishing the coherence of the black family. Characters in *Jazz* are all literally and symbolically orphaned as a result of the "destructive influence of racism and oppression on the black family" (Heinze 97) and of further displacements incurred during the migration north. The few families in the novel are childless, transferring their parental feelings on surrogates: Malvonne takes care of her nephew Sweetness, Alice raises her niece Dorcas, and Violet tries to abduct a baby to "bring light into places dark as the bottom of a well" (22). The "bottom well" that haunts her is literal: Violet's mother, Rose Dear, threw herself in it after being dispossessed in 1888 of her house and stock. Neither the occasional returns of her estranged father to dispense "gifts and stories that kept them so rapt they forgot for the while a bone-clean cupboard and exhausted soil" (100) nor her grandmother's arrival to take charge of the orphaned household can erase Violet's memory of her mother's wretched death.

True Belle, the able grandmother who left Virginia a slave and returned to it a free woman to save her daughter's family, reminds us of Baby Suggs. But as a mother figure and example of the "benevolent, protective, wise Black ancestor" (Morrison, "City Limits, Village Values" 39), she has clear limitations. True Belle spends her adult life away in Baltimore, caring for her white mistress, Vera Louise Gray, and her bastard mulatto son, a "beautiful [fair-haired] young man whose name, for obvious reasons, was Golden Gray" (139). After returning to Vesper County, she fills Violet's head with tales about big-city life and about Golden Gray who set out "to find, then kill, if he was lucky, his father," a "black-skinned nigger" (143). These stories confuse Violet's sense of identity, suggesting to her that cultural assimilation to Western standards of beauty is desirable.

The fact that both Violet and Joe contribute to the perpetuation of those standards (Violet by straightening her black clients' hair and Joe by selling them cosmetics to make them "whiter") and that Joe commits adultery with "a younger [Violet] with high-yellow skin instead of black" (97) aggravates Violet's insecurities. She comes to realize that, in her relationship with Joe, she has been a substitute for the motherly recognition he never received, and he a substitute for the "golden boy I never saw either" (97). The fact that both are caught in a game of substitutions prevents Violet from using colorism as an alibi and blaming the "high yaller" Dorcas for her misfortunes. This complicates the issue of biracial identity beyond some of Morrison's earlier novels that opposed refined but conformist light-skinned women (Maureen Peal, Geraldine, Helene Wright, Ruth Foster, Jadine Childs), to powerful and transgressive black women (Eva Peace, Pilate, the African princess who haunts Jadine's dreams, or Baby Suggs—there were also significant exceptions, such as Pecola and Hagar), looking forward to the deconstruction of color hierarchies in *Paradise*.

Joe's own present is troubled by questions of identity that, as in Violet's case, go deeper, involving a genealogical gap left by his unknown father and "unrecognizing" mother. Joe's memory is haunted by his failed quest for his presumptive mother, Wild, a savage creature roaming the woods. She responds to his efforts to communicate with an "indecent speechless lurking insanity" (179), abandoning him to an "inside nothing" (38). Joe's affair with Dorcas is an attempt to fill that gap with somebody who "knew better than people his own age what that inside nothing was like [. . .] because she had it too" (37–38). The City's springtime streets encourage even straightlaced Joe to have "loose" thoughts (119), pulling him "like a needle through the groove of a Bluebird record" (120).

Dorcas responds to Joe's attentions out of an equally profound sense of emptiness that in her case combines a Pecola-like lack of selfhood with the traumatic memory of her parents' innocent deaths in the 1917 St. Louis riots: her father pulled off a streetcar and stomped under the feet by an angry white mob, her mother burned alive in her own house. As Morrison's novel suggests metaphorically, Dorcas internalizes the fire she witnesses as a child from across the street: a burning chip of wood ends up in her throat, slipping deeper into her body until it lodges somewhere below her navel (61), as a symbol of her inextinguishable desire. The City of New York fans the burning ember in her groin. Her aunt Alice tries to rein in this destructive desire, teaching Dorcas a self-denying "deafness and blindness." What Alice fears most are the City's unruly "[s]ongs that used to start in the head and fill the heart [but] had dropped on down [. . .] to places below the sash and the buckled belts" (56). Alice tries to offset the decadent jazz with the brash drums she heard in July 1917, when black men and women marched down Fifth Avenue to protest the two hundred dead in the East St. Louis riots. But she concedes that between the "gathering rope" of the marching drums and the visceral, "get-on-down" (58) jazz there is a subtle connection, jazz disguising a "complicated anger [...] as flourish and roaring seduction" (59). As a progeny of the violence that created that passionate musical response, Dorcas also tries to balance rebellion with seductive abandonment. In spite of the supervision exercised by Alice, Dorcas's mutinous "glow" manages finally to shine through. After her niece's death, Alice reflects on the ironies of city life that allow a "nice, neighborly, everybody-knows-him man" to kill Dorcas with impunity and his scatter-brained wife to desecrate Dorcas's funeral.

When Violet shows up at her door in February 1926, Alice is afraid that her own private world will be invaded by the ferocious passions of the City. But her subsequent conversations with Violet enable Alice to work out her anxieties (including her color bias against the "black as soot" man and woman who destroyed her "cream-skinned" niece [66, 76]), while also prompting Violet to release her emotions, "tighten[ed] up" by the City (81). Their relationship becomes cooperative, with Alice the "seamstress" mending Violet's torn coat to "normalize" her appearance and supporting Violet's disjointed life narration. Alice's obstinate stitching is paralleled by Violet's patchwork narrative that tries to reconnect her two selves (the third and first person Violet uses to talk about herself finally overlap), as well as her past in rural Virginia and her present in New York. As "the things [she] couldn't say were coming out of [her] mouth anyway" (97), Violet begins to realize that her life has been thwarted by patriarchal fears "seeded in childhood, watered every day since" (85), such as the fear of losing her man to a lighter-skinned woman. Instead of freeing her from this bondage, the City draws her into an "Electra" conflict in which the woman who steals her man is a symbolic emanation of the miscarried "daughter who fled [Violet's] womb" (109). The pressures of competing with her own "daughter" make Violet "scatty" and divided, watching "that other Violet" (90) act out her violent impulses.

In Alice's kitchen, Violet experiences an alternative, noncompetitive relationship. Alice helps Violet rediscover "serious laughter" (113) and teaches her not to fight life, but to "make it" through her imagination (113). In turn, Alice uses Violet's self-reflexive narration to probe her resentment against "women with knives" (85) and memories of a time when she herself fantasized revenge against her husband's mistresses. The two women experience a moment of deep recognition of their common origin and fate, as "the black and smoking ship" of Alice's pressing iron "burned clear through the yoke" (264). The allusion to a slave ship functions, paradoxically, as a liberating memory that burns both through the garment and their patriarchal "yoke."

By contrast to Violet, Joe must do his work of narrative retrospection alone. Having cut himself from his past, he has no adequate audience to support his process of "rememory." In spite of this, Joe manages to use narrative reconstruction to move beyond his shallow salesman's self, recovering those moments in his life when he successfully renewed himself: the first time, by naming himself "Trace," converting a biographical lack (his parents vanished without a "trace") into a self-defining attribute; the second time, by undergoing an initiation into the lore of the woods, guided by the wisest black man in Vesper County; the third time, by relocating from Vienna—"encouraged by guns and hemp" (173)—to Palestine, Virginia, where he married Violet in 1893; the fourth time, by buying his own piece of land in 1901, becoming an independent farmer; the fifth time, by boarding in 1906 the "Southern Sky for a northern one" (127), after the whites had run him off his land; the sixth time, by moving from a rat-infested lower Manhattan apartment into a spacious up-town home, in defiance of the Jim Crow law and the opposition of light-skinned renters; the seventh time, by landing a well-tipped job at a hotel and adding a "little side-line selling Cleopatra products in the neighborhood" (128–29).

Joe's self-renewals emulate the aspiration of the Harlem Renaissance to define progressive roles for African Americans. But Joe's imitation of the 1920s call to make it new, promoting himself "not only from countryside to city, but from me-dieval American to modern" (Locke 6), alienates him from his roots. It confirms thus Farrah J. Griffin's observation that the "New Negro" metaphor involved both "self-creation" and loss (197). Morrison's novel also reminds us that the rhetoric of the "New Negro Man" left little room for the affirmation of female selfhood. As a result, women (unless they were themselves artists like Zora Neale Hurston, Nella Larsen, or Jessie Fauset) were often left behind by their metamorphic husbands. When Violet starts sleeping with a doll, Joe suspects that he changed one time too many. With the desperation of somebody who experiences his last renewal, Joe plunges into a relationship with seventeen-year-old Dorcas and shoots her when she leaves him for a younger lover. In a section of his narrative monologue addressed to dead Dorcas, Joe blames his violence on the "sooty music" heard in blues parlors and on the unfair competition of the city's "young roosters" (132). Yet this violent de-nouement reflects more directly his own failures: the failure of a male subject to move beyond his notion of womanhood as gratifying quarry; the failure of a "New Negro" to stay "new and [. . .] the same every day the sun rose and every night it dropped" (135). Joe is a product and victim of the same patriarchal economy of de-sire that destroys Dorcas and thwarts the life of his wife.

Loosely related to this exploration of the patriarchal roots of violence is the story of Vera's mulatto boy, Golden Gray. The role of this story, featured obliquely in Vi-olet's recollections, is to connect more clearly the issue of gender to those of race and class and to allow the narrator to undergo a process of self-examination that calls into question her preconceived ideas on these matters. As she follows her naive bira-cial character on his quest for his biologic father, the narrator imagines him stopping his elegant phaeton to pick up the unconscious naked body of a black woman (Joe's savage mother, as we learn later). In a scene that repeats with a reversed cast the first encounter between cream-skinned Jadine and Son with "skin as dark as a river-bed" (Tar Baby 113), Golden Gray is nauseated by the black woman's odorous body that conspires to "become something more than his own dark self" (Jazz 146). The nar-rator mocks the young man's performance as reluctant rescuer, but as soon as we be-come uncomfortable with the narrator's heavy editorializing, she interrupts her caricature and admits that her character is "young and [. . .] hurting" (155). The nar-rator decides to waive judgment and allow her character to shed his bourgeois "armor," "dar[ing] to open wide, to let the layers of [his] petals go flat, show the clus-ter of stamens dead center for all to see" (160).

Racist or at least racially confused, Golden Gray is a difficult test case not only for the narrator but also for the author who has similarly wrestled with the question "of how free I can be as an African-American woman writer in my genderized, sexual-ized, wholly racialized world" (Playing in the Dark 4). Both author and narrator try to work around their cultural biases, empathizing with this hybrid other. As the narra-tor makes clear, "Not hating him is not enough; liking, loving, him, is not useful. I have to alter things," becoming the "language that wishes him well, speaks his name, wakes him when his eyes need to be open" (*Jazz* 161). Her narrative foregrounds startling connections between Golden Gray's story and the stories of the novel's protagonists: The naked woman that Golden Gray rescues on his trip is Joe's presumptive mother, Wild, a legendary haunter of the woods; Golden Gray's father turns out to be the equally legendary Henry Lestory a.k.a. Hunters Hunter, a Faulknerian character who helps the wild woman deliver her baby and later teaches the boy Joe the lore of the woods. The conversation between Golden Gray and Hunters Hunter polarizes the young man's options—"choos[ing] black" (173), as his biological father urges, or remaining a racially unmarked "free man" (173), as Golden Gray wishes—without settling them. Since "identity for the mulatto lies between existing precepts of racial singularity," he deconstructs "essentialist notions of race" (Wilson 104) and either/or choices. Instead of fitting neatly into any identity category, the mulatto occupies an in-between position that literally renders him invisible. (Golden Gray disappears after encountering his father.)

However, disappearances leave subtle "traces" that connect the seemingly disparate pieces of the communal story. The following section that switches to Joe's remote past suggests that he may be the son of the wild woman picked up by Golden Gray and that her craziness may have "reasons" (*Jazz* 175), the savage woman lurking in the cane fields burned by the whites, confusing the fire with the golden hair of her rescuer. Joe's own quest for origins ends more ambiguously than Golden Gray's. But while he receives no "sign" of recognition from his biologic mother, he senses her unseen presence in the sunlit cave above the Treason River and in the "music the world makes" (176). We are invited to experience with Joe a liminal motherly universe, the kind that Beloved could have established were she not chased out of the previous novel. This unruly female universe that shuns the male world in an instinctive exile looks forward to the Convent women in *Paradise*.

The novel's final two chapters intersperse the "music the world makes" with the music humans make, "playing out their maple-sugar hearts, tapping it from four-hundred year old trees and letting it run down the trunk" (197). Emulating the polyphonic medium of jazz, the narrator merges her voice with those of her characters in a collective performance that helps their "greedy, reckless words, loose and infuriating" find a structure (60). These sections benefit from a "sweetheart weather" that makes trees glow with a soft but steady light (196). Even though this rejuvenation cannot last ("ash [is already] falling from the blue distance down on the streets" [198]), it allows a fresh perspective to enter the novel—that of Felice, Dorcas's confidante. When she first approaches, she appears to Violet as "another true-as-life Dorcas" (197), but Felice is a practical girl, angry with her departed friend for choosing men who treated her badly. But Felice also helps Joe and Violet understand Dorcas as a more complex figure who, far from being a mere victim, had strength and a daredevil resourcefulness.

Felice, whose name is a good harbinger for the Traces, is invited back to share their slow reawakening. The members of this supportive trio, in which Felice plays the role of the lost daughter, help each other develop a stronger sense of self that escapes traditional gender and racial biases. Felice finds "boot black" Violet pretty (205) and notes that aging Joe carries himself handsomely, like her "father when he's being a proud Pullman porter seeing the world [. . .] and not cooped up in Tuxedo Junction" (206). In Joe and Violet's company, Felice learns not to be a Dorcas, somebody's "alibi or hammer or toy" (222). Violet teaches her what she herself learned from Alice: to "make [the world] up the way you want it" so as to prevent the world from changing you (208). In turn, Felice's presence allows Violet to complete her self-clarification and achieve something few previous Morrison characters have managed: freedom from the "quiet mole" (208) of patriarchal and racial prejudice. The "me" that Violet reclaims is not arrogant or competitive: "Not like the 'me' was some tough somebody, or somebody she had put together for show. But like [. . .] somebody she favored and could count on" (210).

The novel ends with the narrator's resolve to reimagine her characters after her efforts to manipulate them toward a predictable end fail. Misreading Felice, Joe, and Violet for a mirror image of the Dorcas, Joe, and Violet triangle, the narrator expected them to succumb to new violence, "never occurring to [her] that they were thinking other thoughts, feeling other feelings, putting their lives together in ways I never dreamed of" (221). Chastened by her characters' capacity to work out problems on their own, the narrator recants her deterministic approach to historical fiction and envisions—in verbs "[c]aught midway between was and must be" (226)—Joe and Violet's comfortable tenderness in old age, freed from the cultural gaze now that "there is no stud's eye, no chippie glance to undo them" (226). Adopting the style of a "talking book" (interview with Carabi 42), the final paragraph urges readers to share in the "remaking" of history: "You are free to do it and I am free to let you because look, look. Look where your hands are. Now" (*Jazz* 229).

As Morrison has insisted, imagination is vital to good historical fiction, allowing one to move from "facts" that can exist without "human intelligence" to "truth" that cannot ("The Site of Memory" 113). By interplaying historical facts and narrative imagination, the historical novelist can present two worlds at once: "the actual and the possible" (117). The need for both perspectives is brought home dramatically in Morrison's most recent novel, *Paradise* (1998), which completes the exploration (begun in *Beloved* and *Jazz*) of gender roles in "the troubled history of black America from Reconstruction through the civil rights movement" (Allen 6). *Paradise* opens with a shocking sequence of events in 1976 that foreground the stark regime of the actual at the expense of the utopian possible. A posse of nine men from Ruby, Oklahoma—veterans of various wars armed with "rope, a palm leaf cross, handcuffs, Mace and sunglasses, along with clean, handsome guns" (3)—attack the Convent of the Sacred Cross located in the scrubland, seventeen miles away. Their target is a commune of "throwaway" women (4) who linger inside the Convent after its breakup. The men feel justified in using overwhelming force against a few defenseless women, blaming the erosion of Ruby's communal spirit on "the female malice that hides [inside the Convent]" (4). They hope this act of violent cleansing will prevent history from repeating itself, dooming their new attempt at establishing a community in Oklahoma. The first attempt at creating an all-black "dreamtown" in the wilderness was made by their grandfathers, freedmen from Louisiana and Mississippi who "stood tall in 1889 dropped to their knees in 1934 and were stomach-crawling by 1948" (5), when their Haven collapsed. The raiders are driven by a sense of mission inherited from the Old Fathers, but also by the debilitating self-doubt of second-generation settlers who witness the decline of their new settlement born in 1950 from the ashes of Haven.

From the outset, Morrison's novel throws suspicion on the raiders' motives permeated by a "nocturnal odor of righteousness" (18), contrasting their perspective with that of the rest of the town for whom the Convent women were "strange neighbors [. . .] but harmless" (11). But the revisionistic work of the novel goes deeper, exposing the ideology that motivated the assault. The erosion of the Ruby community between 1960 and 1976 is revealed to be the result of both outside pressures that reflect the political, racial, and gender divisions of the Cold War and of contradictions inherent in the patriarchal narrative of self-determination that the Ruby leaders inherited from the founders of the nineteenth-century all-back community of Haven, Oklahoma. At least indirectly, the novel also challenges "the foundational myths of the American nation. Whites here are not vigorous frontiersmen; removed to the margins of the scene, they are entirely dependent on black resources and advice" (Bold 22). The African Americans who establish the two successive communities are "founding fathers" in their own right. "Descendants of those who had been in Louisiana Territory when it was French, when it was Spanish, when it was French again, when it was sold to Jefferson and when it became a state in 1812" (193), their history *is* a history of America. As Patricia Storace has also argued, *Paradise* draws "more directly than any of Toni Morrison's other novels, [. . .] that black presence forward from the margins of imagination to the center of American literature and history. [. . .] Morrison tells a story of an African American community in the Vietnam era which is also a story about colonial America" (64–65). The story of *Paradise* is both historical, rewriting the official national myths (the narrative of Exodus, colonization, and foundation) from an African American perspective, and contemporary, re-creating the feel of Nixon's and Ford's America within the confines of a black township in rural Oklahoma.

While the "ownership of history" is one of the "things fought over in this novel that Morrison wanted to entitle 'War'" (Bold 22), the battle is more between visions of history than between white and black settlers: One vision is teleologic, patriarchal, and exclusionary, the other is more experimental, participatory, and inclusive. Rather than simply invert the race narrative, making blacks "dominant by virtue of either biology or culture" and allocating whites a subordinate role (Gilroy 191), Morrison questions purist notions of race and culture. As the author explained in an interview, in writing *Paradise* she was interested in foregrounding "the kind of violent conflict that could happen as a result of efforts to establish a Paradise. Our view of Paradise is so limited: it requires you to think of yourself as the chosen people—chosen by God that is. Which means that your job is to isolate yourself from other people" (Marcus). As a black version of the white New World myth that Morrison critiqued in *Playing in the Dark*, the Ruby community is an amalgam of utopian aspiration and prejudice, a "claim to freedom" and the "presence of the unfree within the heart of the democratic experiment" (*Playing in the Dark* 48). After the earlier settlement in Haven collapsed, the modern generation tried to repeat the experiment in Ruby but "simply couldn't sustain what the Old Fathers had created, because of the ways in which the world had changed. The Ruby elders couldn't prevent certain anxieties about drugs, about politics. And their notions [. . .] about controlling women [. . .] left them vulnerable, precisely because they had romanticized and mythologized their own history" (interview with Marcus). But while exposing the violence embedded in self-serving myths, Morrison also tries to "reimagine" Paradise in terms that are more inclusive and earthbound: "By the end of the book—the very last lines about the ships and the passengers—should suggests that an earthly Paradise is the only one we know" (interview with Marcus). *Paradise* brings to fruition Morrison's intention announced at the beginning of *Playing in the Dark* "to draw a new map, so to speak, of a critical geography and use that map to open as much space for discovery, intellectual adventure, and close exploration as did the original charting of the New World—without the mandate for conquest" (3).

The patriarchal view of history is embodied in the town's bankers, Deacon (Deek) and Steward Morgan. Descended from one of the original Oklahoma pilgrims, Zechariah, the Morgan twins control both the economic power and the power of memory: "Between them they remembered the details of everything that ever happened—things they witnessed and things they have not" (*Paradise* 13). Their sense of history is proud and partisan, commemorating the heroic struggles of their grandsires who, "[t]urned away by rich Choctaw and poor whites, chased by yard dogs, jeered at by camp prostitutes and their children" and discouraged even by the "Negro towns already being built" (13), found their destination in Indian territory guided by the inspired vision of Big Papa (Zechariah Morgan) and his son Rector. The lessons that the Morgan twins (born in 1924) learn from their predecessors are lessons of "escape and suffering, tradition, temporality, and the social organization of memory," which have animated all diasporic communities modeled after the Jewish Exodus (Gilroy 205). The Morgans' entire sense of postslave history is shaped by stories of war, "great migrations," and "tales of love deep and permanent" (14). These stories reinforce a basic division between their town and the territory "out there" that was once beckoning and free, but which now is seen as a "void where random and organized evil erupted" (16). This Manichaean vision made the Morgans and fourteen other families abandon Haven in 1948, when the town could no longer protect itself against the outside, and move deeper west to found Ruby. Thirty years later, the same polarized vision spurs the raid on an outside (female) evil deemed greater than the "Depression, the tax man and railroad" (17) that undid Haven.

Though the historical narrative preserved by the Morgans' "total memory" (107) emphasizes a combative male vision, it is the excluded female perspective that prevails in naming the new settlement after Ruby Morgan Smith, the woman who died in childbirth on the way to the new location. The men lose their bid to call the place New Haven or other names inspired by their World War II campaigns that "only the children enjoyed pronouncing" (17). The masculine bias wins in other areas, as suggested by the official version of the Convent raid that turns the assailants into heroes. Morrison's novel calls this selective and sexist memory into question by foregrounding the internal contradictions of the male perspective and by recovering the marginalized points of view. After a first chapter concerned with the men's triumphalist plot, the novel refocuses attention on the victims, recovering the stories of the refugee women in a "backward and punctured and incomplete" narration (173).

The first Convent woman to get a story is Mavis Albright, a homemaker overwhelmed by her domestic duties and terrified by her husband's moods. After she causes the accidental death of her twins by leaving them unattended in an overheated car while buying her husband's favorite dinner, Mavis steals her husband's Cadillac and drives westward to California. Robbed on the way by two other runaway girls, Mavis knocks on the first door she finds—that of the Oklahoma Convent. She is let in by Consolata, who plays the roles of gardener-caretaker-cook and surrogate mother for the refugee women. Feeling for the first time in years comfortable inside Connie's kitchen, Mavis parks away her repainted "dark as bruised blood" Cadillac and decides to stay at the Convent where she hears the laughter of her twins every night. The second Convent refugee arrives from the fabled California that Mavis was headed for. Unlike Mavis, whose car and patience simply break down in

Oklahoma, Grace (Gigi) Gibson comes to the Midwest on a quest of sorts, looking for the magical place advertised by a black man on the train, where "two trees grew in each other's arms. And if you squeezed in between them in just the right way, [...] you would feel an ecstasy no human could invent or duplicate" (66). As we learn later, Gigi had few alternatives left: Her mother is unlocatable, her father on death row, her boyfriend in prison for antisocial behavior. Gigi herself had to abandon her Berkeley activism because no one "took her seriousness seriously" (257) and "Everybody [was] dead anyway. King, another of them Kennedys, Medgar Evers, a nigger named X" (65). While the small-town mentality of Ruby repulses Gigi, she feels inspired by the "designer sky [...] bigger than everything" (75) around the Convent. Shedding her inhibitions, she fixes her eyes on K. D. (the designated Morgan heir), but when Seneca appears she "spit[s] out that K. D. person like a grape seed" (259), transferring her love to her.

If Mavis and Gigi come from the two opposite sides of the continent, Seneca hitchhikes to Ruby from a life spent in midwestern foster homes. Her sorry life is recorded in intricate maps of scars drawn on her arms and thighs. Begun accidentally while raped by her foster brother, her ritual of scarification has steadied her against predatory men and the world's misfortunes. Always the empathizer (as her tragedian name suggests), Seneca slashes new lines into her arm when King and Bobby Kennedy are killed and when Pallas returns to the Convent after a failed attempt to rejoin the world. Sixteen-year-old Pallas Truelove is the last one to join the Convent as an escapee from "the overlight of Los Angeles" (176). This suburban teenager with a life stumped by the "gnawing violence" of "rules and regulations" (254) drives off in her birthday Toyota until she is stalked and probably raped. Her "little one's story of who had hurt her" (173) ends nearby Ruby, in a hospital where she is discovered by Billie Delia and taken to the Convent. Billie tells her not to be afraid of the Convent women, "[a] little nuts, maybe, but loose, relaxed" (175–76). This advice proves redundant since what Pallas fears most is the world she left behind. After a visit with her father who fusses over her wrecked car, Pallas returns to the Convent to deliver her bastard baby in a supportive community of castaway women.

All the runaways are welcomed by Consolata Sosa, the aging Convent custodian with "smoky, sundown skin" (223). Except for the first comer Mavis, Connie can hardly tell the women apart. The "timbre of their voices told the same tale: disorder, deception and [...] drift" (221–22). She provides them with a home they never had, but on occasion she feels like snapping their necks to stop their drifting and "talk of love as if they knew anything about it" (222). Connie is the only one who has experienced deep love: first, for her adoptive mother, Mary Magna, who rescued her from poverty in Brazil, taking her along to an "asylum/boarding school for Arapaho Indian girls in some desolate part of the North American West" (223–24); then for a real man, later revealed as Deacon Morgan. Deek makes frantic love to her in the gully with the fabled interlocked fig trees. But this affair ends abruptly when Deek resumes his patriarchal role as a community builder. Returning to the solitude of the Convent, Connie undergoes a "siege of sorrow" from which she emerges with a "bat vision" that allows her "to see best in the dark" (241). Her "in-sight" (247) approaches the supernatural when she reanimates Soane Morgan's injured son, Scout, by strengthening his "pin point of light" (245) or when she intercedes with the death of her protectress. After Mary Magna passes away, Consolata becomes the new "irenic 'mother' of the Convent" (Wood 30), feeding and giving a focus to the freeloading women who drift into her plentiful kitchen and garden. The short paragraphs that remind us of Consolata's presence in the background, peeling potatoes, making hot relish, or serving rhubarb pie, bridge the self-absorbed narrative sections of the other five women. Connie encourages cooperative rather than competitive relationships that imitate the "[u]njudgemental. Tidy. Ample. Forever" (48) prairie around the Convent. Distinctions of race, class, or education become almost irrelevant as the women find a common language for their stories of pain and self-doubt. By deliberately creating confusion over questions such as the identity of the "white girl" shot in the opening sentence of the novel, Morrison averts a reading informed by conventional identity categories. She does this not by erasing difference but rather by pluralizing it to the point at which her women become figures of heterogeneity, both rebels and victims, powerful and powerless.

Working together, the women turn the Convent into a space of "blessed malelessness, like a protected domain" that allows the refugees to meet an "unbridled, authentic self [. . .] in one of the house's many rooms" (177). This is no small achievement, given the androcentric vestiges in the Convent's architecture, built by a notorious playboy-embezzler. After his arrest, the teaching Sisters take over the mansion and convert the "dining room into schoolroom, the living room into a chapel; and the game room alteration to an office." They cannot extirpate, however, the "echoes of [lewd] delight" in the building's ornamentation, which reify the female body for male pleasure. Against this burdensome patriarchal heritage, Connie teaches the women to draw their own space, tracing the contour of their naked body on the cellar floor and filling it with "half-tales and [. . .] never-dreamed" images (264). The Convent cellar functions as an enabling field for female imagination, making the Convent tenants "adult" and "unlike some people in Ruby, [. . .] no longer haunted" (266).

Lone DuPress, midwife/clairvoyant/witch, warns the Convent women of the "devilment" (269) that a posse of men is concocting back in Ruby. Though they distrust the town, the Convent squatters take no precautions. On the eve of the assault, they dance in the scented rain, "partnered, imagining each other" (179). Their imaginations converge in Connie's tale of a tropical paradise ruled over by goddesslike Piedade, "who sang but never said a word" (264). The following morning this female utopia is brutally interrupted by the arrival of the phallic raiders. "Fondling their weapons, feeling suddenly so young and good" (285), they chase and kill the dreamers of an alternative irenic world. With the Convent purged of female presence, the men observe the "lovingly drawn filth carpets on the stone floor" and are shocked by the "perversions beyond their [male] imagination" they see in them.

By the time we return to this dramatic denouement in the penultimate chapter of the novel, we know not only who the victims are but also what prodded the Ruby men to this act of unconscionable violence. The problems of the Ruby community are explored in counterpoint to the tales of the Convent women. In addition to keeping the focus on both sides of the story, this intercrossed presentation suggests the interdependence not only of the male and female narratives (several assailants have had relationships with the Convent women) but also of the two versions of utopia embodied in Ruby and the Convent. As an all-black community in the "unassigned territories" of Oklahoma, whose "edge" is also its "center" (67), Ruby functions initially as a successful model of liminality. Its self-imposed isolation seems to guard it against the "anger [that] smallpoxed other places" during the Vietnam era. By contrast to the Convent, whose residents are victims of war, civil unrest, rape, and spousal violence, Ruby appears impervious to these troubles. But its invulnerability is only an illusion. Ruby pays its dues to what Reverend Pulliam calls "Evil Times" (102): Soane and Deacon's sons perish in a war that Soane considers ironically "safer than any city in the United States" (101); Stewart's story unfolds as a sequence of losses, beginning with the forced sale of his herd in 1958 and continuing with his defeat in the 1962 elections for church secretary and discovery that he cannot have children; and Sweetie Fleetwood's baby conceived with a Vietnam veteran contaminated with Agent Orange finally dies, challenging Ruby's insolent claim to "immortality." The town is affected also by other tensions engulfing the 1960s, from civil rights militancy to generational restlessness. In spite of the older generation's effort to keep Ruby centered by building it around a communal Oven and naming its streets after the Gospel, the town falls victim to both outside intrusions and its own contradictions as suggested by the grid of streets on the two sides of Central Avenue: St. Peter, St. John, St. Luke to the east; Cross Peter, Cross John, Cross Luke, to the west.

The disagreements around such communal icons as the Oven (adorned at one point with a Black Power fist) suggest that even with "no whites (moral or malevolent) around to agitate or incense them" (102), the Ruby community experiences division. The debate between the old patriarchs headed by the Morgan twins and the Methodist Reverend Pulliam and the younger generation backed by the Baptist Reverend Misner starts from the Oven's motto but expands into an all-inclusive argument concerning the community's aspirations. The young propose to modify the inscription carved by Deacon's grandfather on the Oven ("Beware the Furrow of His Brow" [86]) to read "Be the Furrow of His Brow" (87), in order to suggest that African Americans can be agents rather than mere obeyers of God. The patriarchs cry "blasphemy," but the "wayward young" (104) meet separately to discuss renaming themselves and the Oven with proud African names. Judging from the uses that the Oven has been put to in recent years, it clearly needs renaming. By the time the final sequence of events is set in motion, this "meeting place to report on what done and what needed; on illness, births, deaths, coming and goings" (111) has lost its integrative function, witnessing the idle radio music of the young and the nocturnal plots of the disgruntled old. But the Oven is only a visible representation of cultural changes that run deeper and that include a challenge to the economic monopoly of the town's traditional bankers and to Reverend Pulliam's disciplinarian theology. Against the latter's notion of an impersonal and autocratic deity, Reverend Misner defines "God as a permanent interior engine that, once ignited, roared, purred and moved you to do your own work as well as His" (142). Interpreting the crucifixion as an emancipating experience that turned "the relationship between God and man from CEO and supplicant to one on one," promoting humans "from muttering in the wings to the principal role in the story of their lives" (146), Misner wants the Ruby community to claim a broader role in the affairs of history. History overlooks "ordinary folk" whose story is not told properly, such as the "janitor who turned the switch so the police couldn't see; the grandmother who kept all the babies so the mothers could march; the backwoods women with fresh towels in one hand and a shotgun in the other; the little children who carried batteries and food to secret meetings" (212), even though their contribution was "the spine" that made the civil rights movement possible. To prevent history from repeating this injustice, Misner urges the Ruby folk to take an active role in politics. No matter how "special they think they are," a community that turns itself into a fortress "to keep everybody in or out" (213) is doomed to failure.

Several of the Ruby women concur with this assessment. Dovey and her sister Soane watch with concern the erosion of Ruby's idealistic spirit, secretly blaming it on the arrogance of their men who turn the Oven from a "utility [into] a shrine" to their accomplishments (103). Uncomfortable with this reenshrinement of patriarchal power in the early 1970s, the Morgan women look for alternative outlets: Dovey by talking to an imaginary visitor in a foreclosed house on St. Matthew's Street; Soane by visiting Connie at the Convent and inviting the "strange feathers" Mavis, Gigi, and Seneca to a town wedding to watch in amusement the commotion they create. Other Ruby women diverge more radically from patriarchal expectations. Anna Flood violates Ruby norms not just by being an "outsider" from Detroit, with unstraightened hair and an undisguised interest in handsome Reverend Misner, but also through her impish revision of the Oven's motto: "Be the Furrow of *Her* Brow" (159). Patricia Cato-Best challenges patriarchal ideology even further, a fact acknowledged in the structure of the novel with a chapter that bears her name. Daughter of the town maverick who eventually opens a gas station and paves the dirt roads, making Ruby accessible to more than "the lost and the knowledgeable" (186), Patricia plays the role of the town's uncomfortable historian. Her secret project is a genealogical study of the Ruby community starting with the fifteen original families that made the trip from Haven to their new location. Pat's "keen imagination" opens "crevices and questions" in the "town's official history, elaborated from pulpits, in Sunday schools classes and ceremonial speeches" (188), retrieving the overlooked stories of women with "only one name [. . .] Celeste, Olive, Sorrow, Ivlin, Pansy," or with "generalized last names [. . .] Brown, Smith, Rivers, Stone, Jones [. . .] whose identity rested on the men they married" (187). She also explores the fate of the families crossed out from the history of Ruby, realizing the exclusionary power wielded by the dominant families in the name of an essentialist racial ideology. Codified by Pat as "8-R"—an "abbreviation for eight-rock, a deep deep level in the coal mines" (193)—the leading Ruby families proudly assert their racial purity and self-determination against not only the whites who forced their grandfathers out of their offices in 1875, but also their lighter-skinned brothers who "disallowed" their ancestors from settling in their communities. In trying to preserve a history of "uncorruptible worthiness" (194), the 8-R people commit their own acts of "disallowing": marginalizing Pat's father for infringing the "blood rule" (195) and marrying a wife of "racial tampering" (197); driving Menus to alcoholic insanity by forcing him to give up the sandy-haired woman he brings from Virginia; and persecuting Billie Delia for her cream-colored skin. Challenging the patriarchs' obsession with racial/cultural purity, Patricia burns her card archives and notes that tried to make sense of Ruby's genealogies, writing an atoning letter to her mother Delia who died in childbirth because the 8-R men refused to drive her to the Convent for help. Her action calls into question not only the pretensions of the 8-R men but also—to borrow Hazel V. Carby's characterization of Pauline Hopkins's fiction—the dominant culture's own "discourse of social Darwinism, undermining the tenets of 'pure blood' and 'pure race' as mythological, and implicitly exposing the absurdity of the theories of the total separation of the races" (140).

The other Ruby woman who gets a chapter named after her is Lone DuPress, the town's midwife. The 8-R men view her with suspicion because she is an outsider both in origin (as an orphan rescued by Fairy DuPress during the exodus to Ruby) and in attitude, reputed to have witchlike powers and to travel the Convent road all too often. Her clairvoyance allows her to predict Ruby's need to blame the "nasty [Convent] women" for its own decline. When Lone tries to warn the Ruby women of the "passel of menfolk planning something against the Convent" (280), she reaches only a few. With Misner and Anna out of town, most women hesitate to antagonize their husbands or fathers. Lone still manages to mobilize a few people against the impending raid: the ranchers who, living farther away, have not allowed their minds to be clouded by family relations; and one of the founding families, the DuPresses, who are descended from men who governed in the state house and of artisans who paid attention to "the ethics of the deed, the clarity of motives" (284). This fact alone warns against a hasty reading of *Paradise* along gender lines. *Paradise* maintains a tension between its conflicting perspectives (male and female, young and old, insider and outsider, racially purist and hybrid). Even those viewpoints that reflect Morrison's own humanistic-multicultural sensibilities are submitted to questioning or qualification. In response to Misner's eulogy of the cross as a symbol of unconditional love, Steward Morgan, who has seen "crosses between the titties of whores, military crosses spread for miles; crosses on fire in Negroes' yards, crosses tattooed on the forearms of dedicated killers," reflects that "A cross was no better than the bearer" (154). Upon further reflection, Misner concedes that his love-affirming position is not immune to the conflictual ethos of the Vietnam era and that Reverend Pulliam's sermon about a stern God actually "fingered a membrane enclosing a ravenous appetite for vengeance" (160) in his own soul. As Morrison explained in her interview with James Marcus, "I tend to distrust 'either/or' solutions. That's why I'm asking those kinds of questions in the book."

Either/or solutions disperse rather than stabilize a community. In the aftermath of the raid new rifts emerge between the male protagonists (while Steward is "outrageously prideful" and K. D. is "uncorrectably stupid" [298], Deacon is remorseful), between the raiders and their wives (Dovey wonders if her husband Steward thought that "just because they lived away from white law they were beyond it" [287]), or between the two Morgan sisters (while Dovey describes the Convent women as "whores" and "strange too," Soane finds that they are "Just women" [288]). People return to their houses, wondering how "could so clean and blessed a mission [like the Ruby settlement] devour itself and become the world they had escaped?" (292). The men try to contain the damage by producing two editions of the official story meant to exculpate them: "One, that nine men had gone to talk to and persuade the Convent women to leave and mend their ways; there had been a fight; the women took other shapes and disappeared into thin air. And two [. . .] that five men had gone to evict the women; that four others—the authors—had gone to restrain or stop them" (297). These versions are countered by Pat's lucid summation: "nine 8-rock murdered five harmless women (a) because the women were impure (not 8-rock); (b) because the women were unholy (fornicators at the least, abortionists at most); and (c) because they *could*" (297).

Patricia notes also a few positive transformations in the aftermath of the raid on the Convent. Deacon Morgan becomes estranged from his unrepentant brother and confesses to Misner his remorse for becoming "what the Old Fathers had cursed: the kind of man who set himself up to judge, rout and even destroy the needy, the defenseless, the different" (302). Anna and Misner visit the scene of the crime and detect an invisible "door" in the "unconquerable growth" (305) of the Convent garden, through which the Convent women may have vanished. (Their bodies were never found.) Their metaphysical insight is validated by Lone, who interprets the whole episode as a case of divine intervention: God has given Ruby a second chance by allowing its daughters to be sacrificed and recalling them to Heaven. Misner echoes this view that transposes the saving power from the crucified Son to the sacrificed Daughter, in the eulogy he gives at funeral of Sweetie's baby, Save-Marie. After rebuking the Ruby men for being "deafened by the roar of [their] own history" (306), he predicts that Save-Marie, whose name sounds like "a request (or a lament)" (295), will become Ruby's symbol of redemption.

The last chapter also imagines the possibility of the Convent women's magical return. In a coda marked off by blank space, several of the Convent women are "sighted" in places where "reprieve took years but it came" (309): Manley Gibson, who was spared the chair and works on a road crew, is visited by Gigi dressed in army cap and fatigues; Dee Dee watches her daughter Pallas retrieve a few things from her house and walk away "into a violet so ultra it broke her heart" (312), "one hand on the knapsack bottom, the other carrying a sword" (311); Sally Albright recognizes her mother Mavis and feels something "long and deep and slow and bright" developing inside her (314); Jean spots her own daughter Seneca in a parking lot, her hands bloodied by cuts, but Seneca refuses to acknowledge her. Though these sightings remain uncertain, each Convent woman dissolving out of the frame, a few "minimiracle[s]" (308) such as the reconciliation of the Poole brothers who decide to await patiently Billie Delia's choice between them, continue to occur. These minimiracles suggest that a balance between male and female interests can still be achieved.

The entire novel contributes to this readjustment of balance by disrupting the hegemonic male narratives and creating a space for women's imagination and interests. A number of chapters have a male focus initially: "Ruby" describes a community dominated by males, "Divine" begins with a religious debate between Reverend Pulliam and Reverend Misner, and the novel's title seems to commemorate the Old Fathers' establishment of a utopian all-black community. But these male references are gradually revised: The chapter entitled "Divine" is as much about the Convent women as it is about the Ruby men, and it ends with the latter applying the word "divine" to their most recent lodger, the sixteen-year-old Pallas Truelove. The title word "paradise" can be associated with the gardens that women cultivate, including the Convent's garden through which Gigi walks naked like a new Eve or the family garden where Patricia burns her Ruby genealogies; and also with Connie's vision of a tropical Eden centered around the goddess figure Piedade.

Women have functioned as "a kind of utopia" also back in Ruby, but "many of the names given girls [. . .], like the Ruby schoolgirls 'Hope, Chaste, Lovely, and Pure,'" suggest that they are cherished not for themselves but rather as "illustrations of allegorized virtues treasured and codified by men" (Storace 66). Ruby itself, whose name recalls the Proverbs 35 valuation of a good woman as "above rubies," reinforces this patriarchal view of women. At the center of the town is the womb-like Oven whose purpose is to "nourish" men's desires and "monumentalize what they had done" (*Paradise* 6–7). The Ruby men gather around the Oven to recite their history, which includes the Morgan twins' childhood recollection of an all-black town inhabited by doll-like "Negro ladies [. . .] bending their tiny waists with rippling laughter" (109). In opposition to this patriarchal utopia, the Convent women envision a place of interfemale relationships, in which, like a "feminist Pietá" (Menand 82), the replenishing Piedade supports the head of a younger woman on her shoulder, the two figures steadying each other for "the endless work they were created to do down here in Paradise" (*Paradise* 318).

Toni Morrison foregrounds the violence that emerges from the confrontation of rival utopian visions, but also resurrects the idea of a utopia that mediates otherness rather than reinforcing sameness. In this sense, paradise must remain undefinable: a place that "it's not anybody's place to define" (Morrison qtd. in Gray 68). Morrison's confessed effort in this novel is to bring paradise "off its pedestal, as a place for anyone, [opened up] for passengers and crew" (qtd. in Bold 22). Viewed from this perspective, *Paradise* crowns Morrison's career-long effort to "expand articulation, rather than to close it, to open doors, sometimes, not even closing the book—leaving the endings open for reinterpretation, revisitation, a little ambiguity" (interview with Jaffrey).

Exploring the Ignored Outskirts of the Art of Telling

INNOVATIVE FICTION AT THE TURN OF THE MILLENNIUM

> *Fiction is the art of telling. A conviction is possible, commitment to a line of reason, or morality, but the stronger that gets, the more contingencies are eliminated that are also true. As you intensify your focus on "objective truth," events on the outskirts get dimmer, events that also bear on the whole picture, and when you try to annex those suburbs the focus downtown gets dull. [. . .] You've got to move in the dark. Truth is everything included. [. . .][I]t's not constructed in language, but generated as resonances by the art of telling.*
>
> —Steve Katz, Moving Parts (73–74)

As Steve Katz suggests in *Moving Parts*, the difficult but perhaps not unattainable task of fiction is to articulate an inclusive experiential truth. The speech acts of any particular narrative are "reductive, narrow versions of the possible; only the total fiction that results from the juxtaposition of these independent articulations can suggest an adequate means of knowing the boundaries of identity" (J. Kerry Grant, "Fiction and the Facts of Life" 212). "Truth" in fiction is a matter of resonance among diverse acts of narrative and cultural articulation. Recent innovative fiction (Coover's *Pinocchio in Venice* and *John's Wife*, Pynchon's *Vineland* and *Mason & Dixon*, Federman's *To Whom It May Concern*, Sukenick's *Mosaic Man*, Morrison's *Jazz* and *Paradise*, and so on) seems more concerned with "a line of reason, or morality" (Katz, *Moving Parts* 73), more committed to a strong cultural focus. But this renewed emphasis on narrative and cultural articulation is accompanied by an awareness of the acts of exclusion and "darkening" that all narration entails. What it may have lost in experimental versatility, recent innovative fiction compensates for in critical breadth, exploring the ignored outskirts of the art of storytelling.

A literary imagination that simply reflects the power configurations of contemporary society leaves us "nerve dead" and "life dumb" (Sukenick, *Doggy Bag* 65). Therefore, innovative writers have taken upon themselves the task of revolutionizing literary imagination, returning us to the nonpolarized contingencies of life. Each of the three modes of narrative innovation discussed in this book critiques "classic representational realism characterized by a hyperrational paranoia (an attempt to prevent contingencies)," expanding fiction with forms of experimental, explorative, or "virtual realism" (Travis 13). Polysystemic fiction is interested in frame breaking and boundary crossing, seeking more inclusive sociocultural mappings that transcend polarized "world orders," old or new. Surfiction sets up a "vast coincidence" of situations and discourses through which a "sideways" narrative imagination (Sukenick, *Blown Away* 20–21) can trace new, liberating connections. Postmodern feminism rescues the "unspeakable" or repressed stories of women, radicalizing and problematizing the postmodern sense of difference. Improvising, projecting, rewriting, these narrative modes revise simultaneously the configurations of "reality" and of their own narration.

The alternative models of nonconflictive, open-ended articulation proposed by innovative fiction are just as useful in today's transitional world as they were during the Cold War era. The post–Cold War period has freed our imagination of traditional ideological divisions, but has replaced the bipolar mapping of the world with cartographies of a nationalistic and ethnocentric kind that endorse exclusionary boundaries. Much of this new ethnic fundamentalism has emerged in direct reaction to the perceived threat of globalization that erases "whole continents, histories, and experiences under the sophisticated, theoretical weight of First World" theories and practices (McLeod 82). Therefore, the input of an alternative literary imagination that emphasizes fluid, counterhegemonic mappings can play a significant corrective role in the current post–Cold War restructuring. Innovative fiction can contribute to the new "politics of location" advocated by bell hooks, "identify[ing] the space where we [can] begin the process of re-vision" (*Yearning* 145) that will replace mono-logic concepts of cultural evolution with potentially limitless mappings. It can also train our historical imagination to explore the in-between spaces, the ignored opportunities, and the potential for new conjunctures, offering us history as "reconstruction, reinvention, redemption" (Chow 53).

WORKS CITED

Abel, Elizabeth, Marianne Hirsch, and Elizabeth Langland, eds. *The Voyage In: Fictions of Female Development*. Hanover: UP of New England, 1983.

Acker, Kathy. *Blood and Guts in High School*. New York: Grove Press, 1984.

———. *Don Quixote*. New York: Grove, 1986.

———. *Empire of the Senseless*. New York: Grove Weidenfeld, 1988.

———. *My Mother: Demonology*. New York: Pantheon Books, 1993.

———. *Pussy, King of the Pirates*. New York: Grove-Atlantic, 1996.

Adams, Henry. *The Education of Henry Adams*. 1918. Boston: Houghton Mifflin, 1961.

Adorno, Theodor W. "What Does Coming to Terms with the Past Mean?" 1959. *Bitburg in Moral and Political Perspective*. Ed. Geoffrey Hartman. Bloomington: Indiana UP, 1986. 114–29.

Aldridge, John W. *The American Novel and the Way We Live Now*. New York: Oxford UP, 1983.

———. "The New American Assembly-Line Fiction: An Empty Blue Center." *The American Scholar* 1 (Winter 1990): 17–36.

Allen, Brooke. "The Promised Land." Rev. of *Paradise* by Toni Morrison. *The New York Times Book Review*, 11 January 1988: 6–7.

Alter, Robert. *Partial Magic: The Novel as a Self-Conscious Genre*. Berkeley: U of California P, 1975.

Altman, Rick. "Television/Sound." *Studies in Entertainment: Critical Approaches to Mass Culture*. Ed. Tania Modleski. Bloomington: Indiana UP, 1986. 39–54.

Amerika, Mark. "Hypertextual Consciousness, Virtual Reality, and the Avant-Pop." *The American Book Review* (Oct.-Nov. 1996): 1, 10.

Amerika, Mark, and Lance Olsen, eds. *In Memoriam to Postmodernism: Essays on the Avant-Pop*. San Diego: San Diego State UP, 1995.

Andersen, Richard. *Robert Coover*. Boston: Twayne-G. K. Hall, 1981.

Anderson, Linda, ed. *Plotting Change: Contemporary Women's Fiction*. London: Edward Arnold, 1990.

Andrews, Bruce. "Poetry as Explanation, Poetry as Praxis." Bernstein, *The Politics of Poetic Form* 23–44.

Anzaldúa, Gloria. *Borderlands/La frontera: The New Mestiza*. San Francisco: Aunt Lute, 1987.

Appadurai, Arjun. "Global Ethnoscapes: Notes and Queries for a Transnational Anthropology." *Recapturing Anthropology: Working in the Present*. Ed. Richard G. Fox. Santa Fe: School of American Research, 1991. 191–210.

Aronowitz, Stanley, and Henry Giroux. *Postmodern Education: Politics, Culture, and Social Criticism*. Minneapolis: U of Minnesota P, 1991.

Attridge, Derek. "Innovation, Literature, Ethics: Relating to the Other." *PMLA* 114.1 (Jan. 1999): 20–31.

Atwood, Margaret. *The Handmaid's Tale*. Boston: Houghton Mifflin, 1986.

———. *Lady Oracle*. New York: Avon, 1976.

———. *The Robber Bride*. New York: Nan A. Talese/Doubleday, 1993.

Bailey, Peter. "Notes on the Novel-as-Autobiography." *Novel vs. Fiction: The Contemporary Reformation*. Ed. Jackson I. Cope and Geoffrey Green. Norman, OK: Pilgrim Books, 1981. 79–93.

Bakhtin, Mikhail M. *The Dialogic Imagination.* Ed. Michael Holquist. Trans. Caryl Emerson and Michael Holquist. Austin: U of Texas P, 1981.

———. *Problems of Dostoevsky's Poetics.* Ed. and trans. Caryl Emerson. Minneapolis: U of Minnesota P, 1984.

Barnett, Pamela E. "Figurations of Rape and the Supernatural in *Beloved.*" *PMLA* 112.3 (May 1997): 418–27.

Barth, John. *Chimera.* New York: Random House, 1972.

———. *The Friday Book: Essays and Other Nonfiction.* New York: Putnam, 1984.

Barthelme, Donald. "Not-Knowing." *Voicelust: Eight Contemporary Fiction Writers on Style.* Ed. Allen Weir and Don Hendrie Jr. Lincoln: U of Nebraska P, 1985. 37–50.

Barthes, Roland. *Image-Music-Text.* Trans. Stephen Heath. New York: Hill and Wang, 1977.

———. *Sollers écrivain.* Paris: Seuil, 1979.

———. *S/Z.* Paris: Seuil, 1970. Trans. Richard Miller. New York: Hill and Wang, 1974.

Bartra, Roger. *The Imaginary Networks of Political Power.* Trans. Claire Joysmith. New Brunswick, NJ: Rutgers UP, 1992.

Bastien, Davis, and Todd Hostager. "Jazz as Social Structure, Process, and Outcome." *Jazz in Mind: Essays on the History and Meanings of Jazz.* Ed. Reginald T. Backner and Steven Weiland. Detroit: Wayne State UP, 1991. 149–65.

Baudrillard, Jean. *The Ecstasy of Communication.* Trans. Bernard and Caroline Schutze. New York: Columbia U-Sémiotext(e), 1988.

———. *The Evil Demon of Images.* Trans. Paul Patton and Paul Foss. Sydney: the Power Institute of Fine Arts, 1987.

———. *Forget Foucault.* New York: Columbia U-Sémiotext(e), 1987.

———. "Hystericizing the Millennium." *CTheory* 10 May 1994. 18 February 1998. <www.ctheory.com/ a-hystericizing_the.ht>. Trans. Charles Dudas from Baudrillard, *L'Illusion de la fin: ou La grève des événements.* Paris: Galilée, 1992.

———. *Selected Writings.* Ed. Mark Poster. Cambridge, MA: Polity, 1988.

———. *Simulations.* Trans. Paul Foss, Paul Patton, and Philip Bleitchman. New York: Columbia U-Sémiotext(e), 1983.

———. *The Transparency of Evil: Essays on Extreme Phenomena.* Trans. James Benedict. London: Verso, 1993.

Bauman, Zygmund. "The Fall of the Legislator." Docherty, *Postmodernism* 128–40.

———. *Legislators and Interpreters: On Modernity, Postmodernity and Intellectuals.* Cambridge: Polity, 1987.

Beauvoir, Simone de. *The Second Sex.* Trans. H. M. Parshley. New York: Bantam, 1961.

Beckett, Samuel. "Dante . . . Bruno, Vico . . . Joyce" *transition* 16–17 (1929): 242–53. Rpt. in *Our Egzamination Round His Factification for Incamination of "Work in Progress."* New York: New Directions, 1962. 21–22.

———. *Le dépeupleur.* Paris: Éditions de Minuit, 1970. Trans. by the author as *The Lost Ones.* New York: Grove Press, 1972.

———. *Disjecta: Miscellaneous Writings and a Dramatic Fragment.* Ed. Rudy Cohn. New York: Grove, 1984.

———. *L'innommable.* Paris: Éditions de Minuit, 1953. Trans. by the author as *The Unnamable.* New York: Grove Press, 1958.

———. *Malone meurt.* Paris: Éditions de Minuit, 1951. Trans. by the author as *Malone Dies.* New York: Grove Press, 1956.

———. *Molloy.* Paris: Éditions de Minuit, 1951.

Bellamy, Joe David. *The New Fiction: Interviews with Innovative American Writers.* Urbana: Illinois UP, 1972.

Benamou, Michel, and Charles Caramello, eds. *Performance in Postmodern Culture.* Milwaukee: Center for Twentieth Century Studies; Madison: Coda, 1977.

Benhabib, Seyla. "Epistemologies of Postmodernism: A Rejoinder to Jean-François Lyotard." Nicholson 107–31.

Benjamin, Walter. "The Storyteller: Reflections on the Works of Nikolai Leskov." *Illuminations.* Ed. with an introduction by Hannah Arendt. Trans. Harry Zohn. New York: Harcourt, Brace and World, 1968. 83–109.

Benson, Stephen. "Stories of Love and Death: Reading and Writing the Fairy Tale Romance." Sceats and Cunningham 103–13.

Benston, Kimberly. "I Yam What I Am: The Topos of (Un)naming in Afro-American Literature." Gates Jr., *Black Literature and Literary Theory* 151–72.

Bercovitch, Sacvan. "Games of Chess: A Model of Literary and Cultural Studies." Robert Newman 15–57.

Berger, James. "Ghosts of Liberalism: Morrison's *Beloved* and the Moynihan Report." *PMLA* 111.3 (May 1996): 408–20.

Bernstein, Charles. "Comedy and the Poetics of Political Form." Bernstein, *The Politics of Poetic Form* 235–44.

Bernstein, Charles, ed. *The Politics of Poetic Form: Poetry and Public Policy.* New York: Roof Books- the Segue Foundation, 1990.

Berressem, Hanjo. *Pynchon's Poetics: Interfacing Theory and Text.* Urbana: U of Illinois P, 1993.

Berry, Ellen E. "Afterword: Postmodernism East and West." *Re-Entering the Sign: Articulating New Russian Culture.* Ed. Ellen E. Berry and Anesa Miller-Pogacar. Ann Arbor: U of Michigan P, 1995. 337–58.

Berry, Ellen E., and Mikhail Epstein. *Transcultural Experiments: Russian and American Models of Creative Communication.* New York: St. Martin's Press, 1999.

Berry, Ellen E., Kent Johnson, and Anesa Miller-Pogacar. "Postcommunist Postmodernism: An Interview with Mikhail Epstein." *Common Knowledge* 2.3 (1993): 103–22.

Bertens, Hans, and Douwe Fokkema, eds. *International Postmodernism: Theory and Literary Practice.* Philadelphia: John Benjamins, 1997.

Bhabha, Homi. "DissemiNation: Time, Narrative, and the Margins of the Modern Nation." *Nation and Narration.* Ed. Homi Bhabha. New York: Routledge, 1990. 291–322.

———. *The Location of Culture.* New York: Routledge, 1994.

Blanchot, Maurice. *The Writing of the Disaster.* Trans. Ann Smock. Lincoln: U of Nebraska P, 1986.

Bleikasten, André. "Roman vrai, vrai roman ou l'indestructible récit." *Revue française d'études américaines* 31 (Jan. 1987): 7–21.

Bloom, Harold, ed. *Modern Critical Views: Toni Morrison.* New York: Chelsea House, 1990.

Bold, Christine, "An Enclave in the Wilderness." Rev. of *Paradise* by Toni Morrison. *Times Literary Supplement,* 27 March 1988: 22.

Bolling, Doug. Rev. of *The Twofold Vibration. The American Book Review* (Nov.-Dec. 1983): 7.

Boons-Grafé, M.-C. "Other/other." Trans. Margaret Whitford. *Feminism and Psychoanalysis: A Critical Dictionary.* Oxford: Blackwell, 1992. 298.

Boyers, Robert. "The Avant-Garde." Elliott, *The Columbia History of the American Novel* 726–52.

Bradbury, Malcolm. "Neorealist Fiction." Elliott, *Columbia Literary History of the United States,* 1125–41.

Brillouin, Leon. *Science and Information Theory.* New York: Academic Press, 1956.

Brooke-Rose, Christine. *Stories, Theories and Things.* Cambridge: Cambridge UP, 1991.

Brooks, Peter. *Reading for the Plot: Design and Intention in Narrative.* New York: Oxford UP, 1984.

———. "The Tale vs. the Novel." *Novel: A Forum on Fiction* 21.2-3 (Winter-Spring 1988): 285–92.

Brossard, Nicole. "Poetic Politics." Bernstein, *The Politics of Poetic Form* 73–86.

Bruner, Edward M. "Experience and Its Expressions." *The Anthropology of Experience.* Ed. Victor W. Turner and Edward M. Bruner. Chicago: U of Illinois P, 1986. 3–30.

Bruss, Paul. *Victims: Textual Strategies in Recent American Fiction.* Lewisburg: Bucknell UP, 1981.

Buell, Frederick. *National Culture and the Global System.* Baltimore: Johns Hopkins UP, 1994.

Burroughs, William S. *Electronic Revolution, 1970–1971.* Cambridge: Blackmoor Head Press, 1971.

Butler, Judith. *Bodies that Matter: On the Discursive Limits of "Sex."* New York: Routledge, 1993.

———. "Poststructuralism and Postmarxism." *Diacritics* 23.4 (Winter 1993): 3–11.

———. "Subversive Bodily Acts." *Gender Trouble: Feminism and the Subversion of Identity.* London: Routledge, 1990. 127–43.

Byerman, Keith. *Fingering the Jagged Grain: Tradition and Form in Recent Black Fiction.* Athens: U of Georgia P, 1985.

Calinescu, Matei. *Five Faces of Modernity: Modernism, Avant-Garde, Decadence, Kitsch, Postmodernism.* Durham: Duke UP, 1987.

Callinicos, Alex. *Against Postmodernism: A Marxist Critique.* New York: St. Martin's, 1989.

———. *Theories and Narratives: Reflections on the Philosophy of History.* Durham: Duke UP, 1995.

Campbell, Joseph. *The Hero with a Thousand Faces.* 1948. Princeton: Princeton UP, 1968.

Cannon, Elizabeth M. "Following the Traces of Female Desire in Toni Morrison's *Jazz.*" *African American Review* 31.2 (1997): 235–47.

Carabi, Angels. "Toni Morrison." An Interview. *Belle Lettres* 10.2 (1995): 40–43.

Caramello, Charles. "On Styles of Postmodern Writing." Benamou and Caramello 221–34.

———. *Silverless Mirrors: Book, Self, and Postmodern American Fiction.* Tallahassee: UP of Florida, 1983.

Carby, Hazel V. *Reconstructing Womanhood: The Emergence of the Afro-American Woman Novelist.* New York: Oxford UP, 1987.

Cârneci, Magda. *Arta anilor '80: Texte despre postmodernism.* Bucharest: Litera, 1996.

Caruth, Cathy. Introduction to *Psychoanalysis, Culture and Trauma.* Spec. issue of *American Imago* 48.1 (Spring 1991): 1–12.

———. *Unclaimed Experience: Trauma, Narrative, and History.* Baltimore: Johns Hopkins UP, 1996.

Chénetier, Mark. "Écriture engagée: pléonasme ou oxymore." *Révue française d'études américaines* 9.29 (May 1986): 215–38.

Chow, Rey. *Ethics after Idealism: Theory—Culture—Ethnicity—Reading.* Bloomington: Indiana UP, 1998.

Christgau, Robert. "What Pretentious White Men Are Good For." *Village Voice Literary Supplement* 44 (April 1986): 7–8.

Christian, Barbara. "Fixing Methodologies: *Beloved.*" *Cultural Critique* 24 (1993): 5–15.

———. "Layered Rhythms: Woolf and Morrison." Peterson, *Toni Morrison* 19–36.

———. "The Race for Theory." *Cultural Critique* 6 (1987): 51–63.

Cirlot, Juan Eduardo. *A Dictionary of Symbols.* New York: Philosophic Library, 1974.

Cixous, Hélène. "The Laugh of the Medusa." Trans. Keith and Paula Cohen. *Signs* 1 (Summer 1976): 875–99. Rpt. in *New French Feminisms: An Anthology.* Ed. Elaine Marks and Isabelle de Courtivron. Amherst: U of Massachusetts P, 1980. 245–64.

Clerc, Charles, ed. *Approaches to Gravity's Rainbow.* Columbus: Ohio UP, 1983.

Connor, Steven. *The English Novel in History: 1950–1995.* New York: Routledge, 1996.

———. *Postmodernist Culture: An Introduction to Theories of the Contemporary.* Oxford: Basil Blackwell, 1989.

Coover, Robert. *Briar Rose.* New York: Grove, 1997.

———. "The End of Books." *New York Times Book Review,* 21 June 1992: 1, 25.

———. *Gerald's Party.* New York: Simon & Schuster, 1986.

———. *Ghost Town.* New York: Henry Holt, 1998.

———. *John's Wife: A Novel.* New York: Simon & Schuster, 1996.

———. *A Night at the Movies: Or, You Must Remember This.* New York: Linden Press-Simon & Schuster, 1987.

———. "On Reading 300 American Novels." *New York Times Book Review*, 18 March 1984: 37–38.

———. *The Origin of the Brunists*. New York: G. P. Putnam, 1966.

———. *Pinocchio in Venice*. 1991. New York: Grove, 1997.

———. *Pricksongs and Descants: Fictions*. New York: E. P. Dutton, 1969.

———. *The Public Burning*. New York: Viking, 1977.

———. *Spanking the Maid*. New York: Grove, 1982.

———. *The Universal Baseball Association, Inc., J. Henry Waugh, Prop*. New York: Random House, 1968.

Cope, Jackson I. *Robert Coover's Fictions*. Baltimore: Johns Hopkins UP, 1986.

Cornell, Drucilla. *Beyond Accommodation: Ethical Feminism, Deconstruction, and the Law*. New York: Routledge, 1991.

Cornis-Pope, Marcel. "Critical Theory and the *Glasnost* Phenomenon: Ideological Reconstruction in Romanian Literary and Political Culture." *College Literature* 21.1 (1994): 131–56.

———. "'Going to BEthiCKETT on the Way to Heaven': The Politics of Self-Reflection in Postmodern Fiction." *Beckett and the Political*. Ed. Henry Sussman. Albany: State U of New York P, 2000. 83–111.

———. "Narrative Innovation and Cultural Rewriting: The Pynchon-Morrison-Sukenick Connection." *Narrative and Culture*. Ed. Janice Carlisle and Daniel R. Schwarz. Athens: U of Georgia P, 1994. 216–37.

———. "Postmodernism Beyond Self-Reflection: Radical Mimesis in Recent Fiction." *Mimesis, Semiosis and Power*. Ed. Ronald Bogue. Philadelphia: John Benjamins, 1991. 127–55.

———. "Rethinking Postmodern Liminality: Marginocentric Characters and Projects in Thomas Pynchon's Polysystemic Fiction." *Symploke* 5.1–2 (1997): 27–47.

———. "Rewriting the Encounter with the Other: Narrative and Cultural Transgression in *The Public Burning*." *Critique: Studies in Contemporary Fiction* 42.1 (Fall 2000): 40–50.

———. "Self-Referentiality." Bertens and Fokkema 257–64.

———. "Systemic Transgression and Cultural Rewriting in Pynchon's Fiction." *Pynchon Notes* 28–29 (Spring-Fall 1991): 77–90.

———. *The Unfinished Battles: Romanian Postmodernism Before and After 1989*. Iaşi: Polirom Press, 1996.

Coste, Didier. *Narrative as Communication*. Minneapolis: U of Minnesota P, 1989.

Couturier, Maurice. *Textual Communication: A Print-Based Theory of the Novel*. New York: Routledge, 1991.

Cowart, David. "Attenuated Postmodernism: Pynchon's *Vineland*." Green, Greiner, and McCaffery 3–13.

———. *Thomas Pynchon: the Art of Allusion*. Carbondale: Southern Illinois UP, 1980.

Creamer, Mark. "A Cognitive Processing Formulation of Posttrauma Reactions." *Beyond Trauma: Cultural and Societal Dynamics*. Ed. Rolf Kleber, Charles R. Figley, and Berthold P. R. Gersons. New York: Plenum, 1995. 55–74.

Crouch, Stanley. *Notes of a Hanging Judge: Essays and Reviews 1979–1989*. New York: Oxford UP, 1990.

Dainotto, Roberto Maria. "The Excremental Sublime: The Postmodern Literature of Blockage and Release." *Postmodern Culture* 3.3 (May 1993): 28 pp. 28 October 1996. <http://jefferson.village.virginia.edu/pmc/issue.593/ dainotto.593.html>.

D'Amico, Robert. "The Meaning of Pornography." *Humanities in Society* 7.1–2 (Winter-Spring 1984): 87–101.

Darling, Marsha. "In the Realm of Responsibility: A Conversation with Toni Morrison." *Women's Review of Books* 5 (March 1988): 4–5.

Davis, Cynthia A. "Self, Society, and Myth in Toni Morrison's Fiction." *Contemporary Literature* 23.3 (Summer 1982): 323–42. Rpt. in Bloom 7–25.

Davis, Johanna. *Life Signs*. New York, Atheneum, 1973.

Debord, Guy. *La société du spectacle.* 1967. Paris: Gallimard, 1992.

De Certeau, Michel. *Heterologies: Discourse on the Other.* Minneapolis: U of Minnesota P, 1986.

Dekoven, Marianne. "Postmodernism and Post-Utopian Desire in Toni Morrison and E. L. Doctorow." Peterson, *Toni Morrison* 111–30.

De Lauretis, Teresa. "Strategies of Coherence: the Poetics of Film Narrative." Phelan, *Reading Narrative* 186–206.

———. *Technologies of Gender.* Bloomington: Indiana UP, 1987.

Deleuze, Gilles, and Felix Guattari. *Anti-Oedipus: Capitalism and Schizophrenia.* Trans. Robert Hurley, Mark Seem, and Helen R. Lane. Minneapolis: U of Minnesota P, 1983.

DeLillo, Don. *White Noise.* New York: Viking, 1985.

De Man, Paul. "Conclusions: Walter Benjamin's 'The Task of the Translator.'" *The Resistance to Theory.* Minneapolis: U of Minnesota P, 1986. 73–105.

———. "Literary History and Literary Modernity." *Blindness and Insight: Essays in the Rhetoric of Contemporary Criticism.* Minneapolis: U of Minnesota P, 1983. 142–65.

Derrida, Jacques. "Des Tours de Babel." *Difference in Translation.* Ed. Joseph F. Graham. Ithaca: Cornell UP, 1985. 209–48; English trans. 165–207.

———. *Of Grammatology.* Trans. Gayatri Chakravorty Spivak. Baltimore: Johns Hopkins UP, 1976.

———. *Specters of Marx: The State of the Debt, the Work of Mourning, and the New International.* Trans. Peggy Kamuf. New York: Routledge, 1994.

———. *Spurs: Nietzsche's Styles.* Trans. Barbara Harlow. Chicago: U of Chicago P, 1979.

———. *Writing and Difference.* Trans. Alan Bass. Chicago: U of Chicago P, 1978.

Desai, Gaurav. "The Invention of Invention." *Cultural Critique* 24 (1993): 119–42.

D'haen, Theo, and Hans Bertens, eds. *"Closing the Gap": American Postmodern Fiction in Germany, Italy, Spain, and the Netherlands.* Amsterdam: Rodopi, 1997.

Didion, Joan. *Democracy.* New York: Simon & Schuster, 1984.

Dienstfrey, Harris. "The Choice of Inventions." *fiction international* 2–3 (1974): 147–50.

Dimić, Milan V. "Friedrich Schlegel's and Goethe's Suggested Models of Universal Poetry and World Literature and Their Relevance for Present Debates about Literature as System." *Proceedings of the XIth Congress of the International Comparative Literature Association, Paris 1985.* Ed. G. Gillespie. New York: Peter Lang, 1991. 39–50.

Dixon, Melvin. *Ride Out the Wilderness: Geography and Identity in Afro-American Literature.* Urbana: U of Illinois P, 1987.

Doane, Mary Ann, Patricia Mellencamp, and Linda Williams, eds. *Re-vision: Essays in Feminist Film Criticism.* Frederick, Maryland: U Publications of America and the American Film Institute, 1984.

Docherty, Thomas. *After Theory: Postmodernism/Postmarxism.* New York: Routledge, 1990.

———. *Reading (Absent) Character: Towards a Theory of Characterization in Fiction.* New York: Oxford UP, 1983.

Docherty, Thomas, ed. *Postmodernism: A Reader.* New York: Columbia UP, 1993.

Doctorow, E. L. "Ragtime Revisited." *Neiman Reports* (Summer-Autumn 1977): 3–8.

———. *The Waterworks.* New York: Random House, 1994.

Dowling, David. "Raymond Federman's America: Take It or Leave It." *Contemporary Literature* 30.3 (Fall 1989): 348–69.

Drabble, Margaret. *The Waterfall.* 1969. New York: Fawcett, 1977.

Du Bois, W. E. B. *The Souls of Black Folk.* 1903. New York: Bantam, 1989.

Dugdale, John. *Thomas Pynchon: Allusive Parables of Power.* New York: St. Martin's, 1990.

DuPlessis, Rachel Blau. *Writing beyond the Ending: Narrative Strategies of Twentieth-Century Women Writers.* Bloomington: Indiana UP, 1985.

DuPlessis, Rachel Blau, and Members of Workshop 9. "For the Etruscans: Sexual Difference and Artistic Production—The Debate Over a Female Aesthetic." Rpt. in *The New Feminist Criticism: Essays on Women, Literature, and Theory.* Ed. Elaine Showalter. New York: Pantheon, 1985. 271–91.

Durand, Régis. "Image, récit: notes sur la rupture." *Fabula* 1 (March 1983): 93–107.

———. "Le continu de la fiction." *Fabula* 1 (March 1983): 125–28.

Eagleton, Terry. "Capitalism, Modernism and Postmodernism." *New Left Review* 52 (July-Aug. 1985): 60–72. Rpt. in *Against the Grain*. London: Verso, 1986. 131–48.

———. *Saints and Scholars*. London and New York: Verso, 1987.

Ebert, Teresa L. *Ludic Feminism and After: Postmodernism, Desire, and Labor in Late Capitalism*. Ann Arbor: U of Michigan P, 1996.

Eco, Umberto. *Foucault's Pendulum*. Trans. William Weaver. San Diego: Harcourt Brace Jovanovich, 1989.

———. "Reflections on *The Name of the Rose*." *Encounter* 64.4 (April 1985): 7–19.

———. *The Role of the Reader: Explorations in the Semiotics of Texts*. Bloomington: U of Indiana P, 1979.

Eisler, Riane. "Violence and Male Dominance: The Ticking Time Bomb." *Humanities in Society* 7.1–2 (Winter-Spring 1984): 3–18.

Elliott, Emory, ed. *The Columbia History of the American Novel*. New York: Columbia UP, 1991.

———. *Columbia Literary History of the United States*. New York: Columbia UP, 1988.

Ellison, Ralph. *Invisible Man*. New York: Modern Library, 1992.

———. *Shadow and Act*. 1953. New York: Random, 1964.

Epstein, Mikhail. *After the Future: The Paradoxes of Postmodernism and Contemporary Russian Culture*. Amherst: U of Massachusetts P, 1995.

Erdrich, Louise. *Tracks*. New York: Henry Holt, 1988.

Ermarth, Elizabeth Deeds. *Sequel to History: Postmodernism and the Crisis of Representational Time*. Princeton: Princeton UP, 1991.

Even-Zohar, Itamar. "Interference in Dependent Literary Polysystems" (1981). Rpt. in rev. form. *Poetics Today* 11.1 (Spring 1990): 79–83.

———. "Introduction" to *Polysystem Studies*. Spec. issue of *Poetics Today* 11.1 (Spring 1990): 1–8.

———. "Polysystem Theory." *Poetics Today* 1.1–2 (1979): 287–310. Rpt. in rev. form in *Poetics Today* 11.1 (Spring 1990): 9–26.

Fairbanks, Lauren. *Sister Carrie*. Normal, IL: Dalkey Archive Press, 1993.

Fanon, Franz. *Toward the African Revolution*. Trans. H. Chevalier. London: Pelican, 1970.

Federman, Raymond. *Amer Eldorado*. Paris: Éditions Stock, 1974.

———. "And I Followed My Shadow." Ms. New York, 1956–57.

———. "The Art of Genocide." *The American Book Review* (March-April 1986): 1.

———. "Avant-Pop: You're Kidding! or the Real Begins where the Spectacle Ends." Amerika and Olsen 175–78.

———. *Critifiction: Postmodern Essays*. Albany: State U of New York P, 1993.

———. *Double or Nothing: A Real Fictitious Discourse*. Chicago: Swallow, 1971. Redesigned 3rd ed. Boulder: Fiction Collective Two, 1992.

———. "Federman: Voices within Voices." Benamou and Caramello 159–98.

———. *La fourrure de ma tante Rachel: Roman improvisé en fourire*. Strasbourg: Circé, 1996.

———. "From the Book of Sarah." *Formations* 3.1 (Spring 1986): 4–14.

———. "Imagination as Plagiarism." *New Literary History* 7 (Spring 1976): 563–78.

———. "The Impossibility of Saying the Same Old Thing the Same Old Way—Samuel Beckett's Fiction since *Comment c'est*." *L'Esprit créateur* 11 (Fall 1971): 21–43.

———. "In." Rev. of *Out* by Ronald Sukenick. *Partisan Review* 41.1 (1974): 137–42.

———. "Inside the Thing." Interview. *Cream City Review* 5 (1979): 90–94.

———. *Journey to Chaos: Samuel Beckett's Early Fiction*. Berkeley: U of California P, 1965.

———. "The Last Stand of Literature." *ANQ: A Quarterly Journal of Short Articles, Notes, and Reviews* 5.4 (Oct. 1992): 190–92.

———. *Loose Shoes: A Life Story of Sorts*. 4 May 2000. <http://epc.buffalo.edu/authors/federman/shoes/>

———. "Playgiarism: A Spatial Displacement of Words." *SubStance* 16 (1977): 107–12.

————. "Samuel Beckett, the Gift of Words." *Fiction International* 19.1 (Fall 1990): 180–83.

————. "Self-Reflexive Fiction." Elliott, *Columbia Literary History of the United States* 1142–57. Rpt. in rev. form in *Critifiction* 17–34.

————. *Smiles on Washington Square (A Love Story of Sorts).* New York: Thunder's Mouth, 1985.

————. "Surfiction—Four Propositions in Form of an Introduction." Federman, *Surfiction* 5–15.

————. *Take It or Leave It: An Exaggerated Second-Hand Tale to Be Read Aloud Either Standing or Sitting.* New York: Fiction Collective, 1976.

————. *To Whom It May Concern: A Novel.* Boulder: Fiction Collective Two, 1990.

————. *The Twofold Vibration.* Bloomington: Indiana UP, 1982.

————. "A Version of My Life—The Early Years." *Contemporary Authors Autobiography Series.* Ed. Mark Zadrezny. Vol. 8. Detroit: Gale Research, 1989. 63–81.

————. *The Voice in the Closet/La voix dans le cabinet de débarras.* Madison, WI: Coda, 1979.

————. "Why Maurice Roche?" *Visual Literature Criticism: A New Collection.* Ed. Richard Kostelanetz. Carbondale: Southern Illinois UP, 1979. 129–33.

————. "The Writer as Self-Translator." Friedman, Rossman, and Sherzer 7–16.

Federman, Raymond, ed. *Surfiction: Fiction Now . . . and Tomorrow.* 1975. 2nd ed., enlarged. Chicago: Swallow Press, 1981.

Federman, Raymond, and Ronald Sukenick. "The New Innovative Fiction." *Antaeus* 20 (1976): 138–49.

Felman, Shoshana, and Dori Laub. *Testimony: Crises of Witnessing in Literature, Psychoanalysis, and History.* New York: Routledge, 1992.

Felski, Rita. "Fin du Siècle, Fin du Sexe: Transexuality, Postmodernism, and the Death of History." Robert Newman 225–37.

Ferguson, Mary Anne. "The Female Novel of Development and the Myth of Psyche." Abel, Hirsch, and Langland 228–43.

Fetterley, Judith. *The Resisting Reader: A Feminist Approach to American Fiction.* Bloomington: Indiana UP, 1978.

Fielding, Helen. *Bridget Jones's Diary.* London: Picador, 1996.

Fish, Stanley. "Consequences." *Against Theory: Literary Studies and the New Pragmatism.* Ed. W. J. T. Mitchell. Chicago: U of Chicago P, 1983. 106–31.

Flaubert, Gustave. *Correspondances. Supplément.* Ed. R. Dumesnil, J. Pommier, and C. Digeon. Vol. 2. Paris: Conard, 1954.

Flax, Jane. *Thinking Fragments: Psychoanalysis, Feminism, and Postmodernism in the Contemporary West.* Berkeley: U of California P, 1990.

Fleischman, Suzanne. *Tense and Narrativity: From Medieval Performance to Modern Fiction.* Austin: U of Texas P, 1990.

Fludernik, Monika. "Hänsel und Gretel, and Dante: The Coordinates of Hope in Pynchon's *Gravity's Rainbow. Arbeiten aus Anglistik und Amerikanistik* 14.1 (1989): 39–55.

————. "Second-Person Fiction: Narrative *You* as Addressee and/or Protagonist." *Arbeiten aus Anglistik und Amerikanistik* 18.2 (1993): 217–47.

Fokkema, Dowe W. *Literary History, Modernism and Postmodernism.* Philadelphia: John Benjamins, 1984.

Foster, Hal. *Recodings: Art, Spectacle, Cultural Politics.* Port Townsend, WA: Bay Press, 1985.

Foucault, Michel. *Language, Counter-Memory, Practice: Selected Essays and Interviews.* Trans. Donald F. Bouchard and Sherry Simon. Ithaca: Cornell UP, 1977.

————. *The Order of Things: Archaeology of the Human Sciences.* New York: Pantheon Books, 1970.

————. *Power, Truth, Strategy.* Ed. Meaghan Morris and Paul Patton, trans. Paul Foss and Meaghan Morris. Sydney: Feral, 1979.

————. "The Subject and Power." *Critical Inquiry* 8.4 (1982): 777–95.

Fowler, Douglas. *A Reader's Guide to "Gravity's Rainbow."* Ann Arbor: Ardis, 1980.

Francese, Joseph. *Narrating Postmodern Time and Space.* Albany: State U of New York P, 1997.

Fraser, Nancy, and Nicholson, Linda J. "Social Criticism without Philosophy: An Encounter between Feminism and Postmodernism." Nicholson 19–38.

Freud, Sigmund. *The Standard Edition of the Complete Psychological Works.* Trans. and ed. James Strachey. 24 vols. London: Hogarth, 1953. Rpt. with corrections 1958, 1978.

Frick, Daniel E. "Coover's Secret Sharer? Richard Nixon in *The Public Burning.*" *Critique: Studies in Contemporary Fiction* 37.2 (Winter 1996): 82–91.

Friedan, Betty. *The Feminine Mystique.* New York: Norton, 1963.

Friedman, Alan Warren, Charles Rossman, and Dina Sherzer, eds. *Beckett Translating/Translating Beckett.* University Park: Pennsylvania State UP, 1987.

Friedman, Ellen G. "Where Are the Missing Contents: (Post)Modernism, Gender, and the Canon." *PMLA* 108.2 (March 1993): 240–52.

Friedman, Ellen G., and Miriam Fuchs, eds. *Breaking the Sequence: Women's Experimental Fiction.* Princeton: Princeton UP, 1989.

Friedman, Melvin J. "Making the Best of Two Worlds: Raymond Federman, Beckett, and the University." *The American Writer and the University.* Ed. Ben Siegel. Newark: U of Delaware P, 1989. 136–45.

Friedman, Susan Stanford. "Lyric Subversion of Narrative in Women's Writing: Virginia Woolf and the Tyranny of Plot." Phelan, *Reading Narrative* 162–85.

———. *Mappings: Feminism and the Cultural Geographies of Encounter.* Princeton: Princeton UP, 1998.

Frye, Joanne. *Living Stories, Telling Lives: Women and the Novel in Contemporary Experience.* Ann Arbor: U of Michigan P, 1986.

Fukuyama, Francis. *The End of History and the Last Man.* New York: Free Press, 1992.

Gado, Frank. *First Person: Conversations on Writers and Writing.* Schenectady, NY: Union College P, 1973.

Gass, William H. *Fiction and the Figures of Life.* New York: Knopf, 1970.

———. *Habitations of the Word: Essays.* New York: Simon and Schuster, 1985.

Gates, Henry Louis, Jr. "African American Criticism." *Redrawing the Boundaries: The Transformation of English and American Literary Studies.* Ed. Steven Greenblatt and Giles Gunn. New York: MLA, 1992. 303–19.

———. "Criticism in the Jungle." Gates, *Black Literature and Literary Theory* 1–24.

———. *Figures in Black: Words, Signs, and the Racial Self.* New York: Oxford UP, 1987.

———. *The Signifying Monkey: A Theory of African-American Literary Criticism.* New York: Oxford UP, 1988.

Gates, Henry Louis, Jr., ed. *Black Literature and Literary Theory.* New York: Methuen, 1984.

Genette, Gérard. *Narrative Discourse: An Essay in Method.* Trans. Jane E. Lewin. Ithaca: Cornell UP, 1980.

———. *Narrative Discourse Revisited.* Trans. Jane E. Lewin. Ithaca: Cornell UP, 1988.

Gibson, Donald B. "Text and Countertext in Toni Morrison's *The Bluest Eye.*" *Literature, Interpretation, Theory* 1.1–2 (1989): 19–32.

Gibson, William. *Neuromancer.* New York: Ace Books, 1984.

Gibson, William, and Bruce Sterling. *The Difference Engine.* New York: Bantam Books, 1991.

Gilroy, Paul. *The Black Atlantic: Modernity and Double Consciousness.* Cambridge: Harvard UP, 1993.

Girard, René. "Innovation and Repetition." *SubStance* 19.2–3 (1990): 7–20.

Giroux, Henry A. "Post-Colonial Ruptures and Democratic Possibilities: Multiculturalism as Anti-Racist Pedagogy." *Cultural Critique* 21 (Spring 1992): 5–39.

Godwin, Gail. *The Odd Woman.* New York: Warner, 1974.

———. *Violet Clay.* New York: Warner, 1978.

Grant, J. Kerry. *A Companion to "The Crying of Lot 49."* Athens: U of Georgia P, 1994.

———. "Fiction and the Facts of Life: Steve Katz's *Moving Parts.*" *Critique: Studies in Contemporary Fiction* 34.4 (Summer 1993): 206–14.

Grant, Robert. "Absence into Presence: The Thematics of Memory and 'Missing' Subjects in Toni Morrison's *Sula.*" McKay 90–102.

Graves, Robert. *The White Goddess: A Historical Grammar of Poetic Myth.* New York: Farrar, Strauss & Giroux, 1966.

Gray, Paul. "Paradise Found." Rev. of *Paradise* by Toni Morrison. *Time* 19 January 1988: 63–68.

Green, Geoffrey, Donald J. Greiner, and Larry McCaffery, eds. *The Vineland Papers: Critical Takes on Pynchon's Novel.* Normal, IL: Dalkey Archive, 1994.

Greenblatt, Stephen J. *Renaissance Self-Fashioning: From More to Shakespeare.* Chicago: U of Chicago P, 1980.

Greene, Gayle. *Changing the Story: Feminist Fiction and the Tradition.* Bloomington: Indiana UP, 1991.

Gregory, Philippa. "Love Hurts." Sceats and Cunningham 139–48.

Griffin, Farrah J. *The African-American Migration Narrative.* New York: Oxford UP, 1995.

Grosz, Elizabeth. *Volatile Bodies: Toward a Corporeal Feminism.* Bloomington: Indiana UP, 1994.

Hagen, W. M. Rev. of Thomas Pynchon, *Mason & Dixon. World Literature Today* 71.4 (Autumn 1997): 788–89.

Handlin, Oscar. *Truth in History.* Cambridge: Harvard UP, 1979.

Haraway, Donna. "A Cyborg Manifesto: Science, Technology, and Socialist-Feminism in the Late Twentieth Century." *Simians, Cyborgs, and Women: The Reinvention of Nature.* London: Free Association Books, 1991. 149–81.

Harding, Sandra. "Feminism, Science, and the Anti-Enlightenment Critiques." Nicholson 83–106.

Harlow, Barbara. "Drawing the Line: Cultural Politics and the Legacy of Partition." *Polygraph* 5 (1992): 84–111.

Harris, Joel Chandler. *Uncle Remus: His Songs and Sayings.* New York: D. Appleton & Co., 1880. Rev. ed., New York: Grosset and Dunlap, 1908.

Hart, Roderick P. *Seducing America: How Television Charms the Modern Voter.* New York: Oxford UP, 1994.

Hassan, Ihab. "The Aura of a New Man." *Salmagundi* 67 (Summer 1985): 163–70.

———. "Joyce, Beckett, and the Postmodern Imagination." *TriQuarterly* 34 (1975): 179–200.

———. *The Postmodern Turn: Essays in Postmodern Theory and Culture.* Columbus: State U of Ohio P, 1987.

Hauser, Marianne. "About My Life So Far." *Contemporary Authors Autobiography Series.* Ed. Mark Zadrezny. Vol. 11. Detroit: Gale Research Inc., 1990. 123–38.

———. *Dark Dominion.* New York: Random House, 1947.

———. "Literary Cross-Dressing." *AWP Chronicle* March-April 1990: 3.

———. *Me & My Mom.* Los Angeles: Sun & Moon Press, 1993.

———. *The Memoirs of the Late Mr. Ashley: An American Comedy.* Los Angeles: Sun & Moon Press, 1986.

———. *Prince Ishmael.* New York: Stein and Day, 1963.

———. *The Talking Room: A Novel.* New York: Fiction Collective, 1976.

Hawthorne, Nathaniel. *Twice-Told Tales. The Centenary Edition of the Works of Nathaniel Hawthorne.* Vol. 9. Ed. William Charvat, Roy Harvey Pearce, and Claude M. Simpson. Columbus: Ohio State UP, 1974.

Hayles, Katherine N. "'Who Was Saved?' Families, Snitches, and Recuperation in Pynchon's *Vineland.*" Green, Greiner, and McCaffery 14–30.

Hayman, David. "Beckett: Impoverishing the Means—Empowering the Matter." Friedman, Rossman, and Sherzer 109–19.

Heinze, Denise. *The Dilemma of "Double-Consciousness": Toni Morrison's Novels.* Athens: U of Georgia P, 1993.

Henderson, Katherine Usher. "Joan Didion: The Bond between Narrator and Heroine in *Democracy.*" *American Women Writing Fiction: Memory, Identity, Family, Space.* Ed Mickey Pearlman. Lexington: U of Kentucky P, 1989. 69–93.

Henderson, Mae G. "Toni Morrison's *Beloved:* Re-Membering the Body as Historical Text." *Contemporary American Identities: Race, Sex, and Nationality in the Modern Text.* Ed. Hortense J. Spillers. New York: Routledge, 1991. 62–86.

Hertzel, Leo J. "An Interview with Robert Coover." *Critique* 11.3 (1969): 25–29.

Higgins, Lynn A., and Brenda R. Silver. "Introduction: Rereading Rape." *Rape and Representation.* Ed. Higgins and Silver. New York: Columbia UP, 1991. 1–11.

Hite, Molly. *Ideas of Order in the Novels of Thomas Pynchon.* Columbia: Ohio State UP, 1983.

———. *The Other Side of the Story: Structures and Strategies of Contemporary Feminist Narrative.* Ithaca: Cornell UP, 1989.

———. "Postmodern Fiction." Elliott, *The Columbia History of the American Novel* 697–725.

Hogue, W. Lawrence. *Race, Modernity, Postmodernity: A Look at the History and Literatures of People of Color Since the 1960s.* Albany: State U of New York P, 1996.

Holston, James. "Spaces of Insurgent Citizenship." *Polygraph* 8 (1996): 65–78.

Homans, Margaret. "Feminist Fictions and Feminist Theories of Narrative." *Narrative* 2.1 (Jan. 1994): 3–16.

hooks, bell. "Representing Whiteness in the Black Imagination." *Cultural Studies.* Ed. Lawrence Grossberg and Paula Treichler. New York: Routledge, 1992. 338–46.

———. *Talking Back: Thinking Feminist, Thinking Black.* Boston: South End Press, 1989.

———. *Yearning: Race, Gender, and Cultural Politics.* Boston: South End Press, 1990.

Hovet, Grace Ann, and Barbara Lounsberry. "Flying as Symbol and Legend in Toni Morrison's *The Bluest Eye, Sula,* and *Song of Solomon.*" *The Journal of Children's Literature* 27 (Dec. 1983): 119–40.

Howe, Irving. "The Idea of the Modern." *Literary Modernism.* Ed. Irving Howe. New York: Fawcett Publications, 1967. 11–40.

Hunt, Erica. "Notes for an Oppositional Poetics." Bernstein, *The Politics of Poetic Form* 197–212.

Hutcheon, Linda. *Narcissistic Narrative: The Metafictional Paradox.* Waterloo, Ontario: Wilfred Laurier UP, 1980.

———. *A Poetics of Postmodernism: History, Theory, Fiction.* New York: Routledge, 1988.

———. *The Politics of Postmodernism.* New York: Routledge, 1989.

———. *A Theory of Parody: The Teachings of Twentieth-Century Art Forms.* New York: Methuen, 1985.

Huyssen, Andreas. *After the Great Divide: Modernism, Mass Culture, Postmodernism.* Bloomington: Indiana UP, 1986.

———. "Mapping the Postmodern." *New German Critique* 33 (1984): 5–52.

Inman, P. "One to One." Bernstein, *The Politics of Poetic Form* 221–25.

Iorgulescu, Mircea. "The Resilience of Poetry." *The Times Literary Supplement* 19–24 January 1990: 61.

Irigaray, Luce. *Speculum of the Other Woman.* Trans. Gillian C. Gill. Ithaca: Cornell UP, 1985.

———. *This Sex Which Is Not One.* Trans. Catherine Porter with Carolyn Burke. Ithaca: Cornell UP, 1985.

Iser, Wolfgang. "Representation: A Performative Act." Krieger 217–32.

Jacobus, Mary. *Women Writing and Writing about Women.* New York: Barnes and Noble, 1979.

Jaffrey, Zia. "The Salon Interview: Toni Morrison." 2 February 1998. 2 June 1998. <http//www.salonmagazine.com/books/int/1998/02/cov_si_02int.html>

Jameson, Fredric. *Postmodernism, or, the Cultural Logic of Late Capitalism.* Durham: Duke UP, 1991.

———. *The Prison-House of Language: A Critical Account of Structuralism and Russian Formalism.* Princeton: Princeton UP, 1974.

Jardine, Alice. *Gynesis: Configurations of Women and Modernity.* Ithaca: Cornell UP, 1985.

Jeffords, Susan. *The Remasculinization of America: Gender and the Vietnam War.* Bloomington: Indiana UP, 1989.

Jencks, Charles. *Postmodernism.* New York: Rizzoli International, 1987.

Jong, Erica. "Blood and Guts: The Tricky Problem of Being a Woman Writer in the Late Twentieth Century." *The Writer on Her Work.* Ed. Janet Sternburg. New York: Norton, 1980. 169–79.

———. *Fear of Flying.* New York: Signet, 1973.

———. *How to Save Your Own Life.* New York: Holt, Rinehart, 1977.

Joplin, Patricia Klindienst. "The Voice of the Shuttle Is Ours." *Stanford Literary Review* 1 (1984): 25–53.

Joyce, James. *Ulysses. The Corrected Text.* Ed. Hans Walter Gabler et al. New York: Random House, 1986.

Katz, Steve. *The Exagggerations of Peter Prince.* New York: Holt, Rinehart &Winston, 1968.

———. *Florry of Washington Heights.* Los Angeles: Sun & Moon Press, 1987.

———. *Moving Parts.* New York: Fiction Collective, 1977.

———. *Swanny's Way.* Los Angeles: Sun & Moon Press, 1995.

———. *Wier & Pouce.* Los Angeles: Sun & Moon Press, 1984.

Kaufman, Sue. *Diary of a Mad Housewife.* New York, Random House, 1967.

Kenner, Hugh. *Samuel Beckett.* Berkeley: U of California P, 1968.

Kiernan, Robert F. "American Literature." *The Encyclopedia of World Literature in the 20th Century.* Vol. 5. Ed. Steven R. Serafin. New York: Continuum, 1993. 23–30.

Kingston, Maxine Hong. *The Woman Warrior: Memoirs of Girlhood among Ghosts.* 1976. New York: Vintage, 1977.

Klinkowitz, Jerome. "Experimental Realism in Recent American Painting and Fiction." Mc-Caffery, *Postmodern Fiction* 63–77.

———. "The Extra-Literary in Contemporary American Fiction." *Contemporary American Fiction.* Ed. Malcolm Bradbury and Sigmund Ro. London: Edward Arnold, 1987. 19–38.

———. *The Life of Fiction.* Urbana: U of Illinois P, 1977.

———. *Literary Disruptions: The Making of Post-Contemporary American Fiction.* 2nd ed. Urbana: U of Illinois P, 1980.

———. *The Self-Apparent Word: Fiction as Language/Language as Fiction.* Carbondale: Southern Illinois UP, 1984.

Kosinski, Jerzy. *Being There.* 1971. New York: Bantham, 1972.

———. *The Devil Tree.* New York: Harcourt Brace Jovanovich, 1973.

Krieger, Murray, ed. *The Aims of Representation: Subject/Text/History.* New York: Columbia UP, 1987.

Kristeva, Julia. *Revolution in Poetic Language.* Trans. Margaret Waller. New York: Columbia UP, 1984.

———. "The Subject in Signifying Practice." *Semiotext(e)* 1.3 (1975): 22.

———. "The System and the Speaking Subject." *The Kristeva Reader.* Ed. Toril Moi. Oxford: Basil Blackwell, 1986. 24–33.

———. *Tales of Love.* Trans. Leon S. Roudiez. New York: Columbia UP, 1998.

———. "Women's Time." *Signs* 7.1 (1981): 13–35. Rpt. in *The Kristeva Reader* 187–213.

Kroker, Arthur, Marilouise Kroker, and David Cook. *Panic Encyclopedia: The Definitive Guide to the Postmodern Scene.* New York: St. Martin's, 1989.

Kuehl, John. *Alternate Worlds: A Study of Postmodern Antirealistic American Fiction.* New York: New York UP, 1989.

Kutnik, Jerzy. *The Novel as Performance: The Fiction of Ronald Sukenick and Raymond Federman.* Carbondale: Southern Illinois UP, 1986.

Lacan, Jacques. *Le Séminaire. Livre III. Les psychoses 1955–1956.* Ed. Jacques-Alain Miller. Paris: Seuil, 1973.

Landon, Brooks. "No Slipping Out." Rev. of *In Memoriam to Identity* by Kathy Acker. *The American Book Review* (Oct.-Nov. 1991): 7.

Lanzmann, Claude. "The Obscenity of Understanding: An Evening with Claude Lanzmann." *American Imago* 48.4 (Winter 1991): 473–96.

Laplanche, Jean, and Serge Leclaire. "The Unconscious: A Psychoanalytic Study." *Yale French Studies* 48 (1972): 118–75.

Lasch, Christopher. *The Culture of Narcissism: American Life in the Age of Diminishing Expectations.* New York: W. W. Norton, 1978.

Laub, Dori, and Nanette C. Auerhahn. "Knowing and Not Knowing Massive Psychic Trauma: Forms of Traumatic Memory." *The International Journal of Psycho-Analysis* 74.2 (1993): 287–302.

Lauzen, Sarah E. "Men Wearing Macintoshes, the Macguffin in the Carpet (Aunt Martha—Still?—on the Stairs)." *Chicago Review* 33.3 (1982): 57–79.

———. "This New Text." Introduction to "A Special Section on {In/Re} novative Fiction in Two Parts." *Chicago Review* 33.2 (1982): 4–8.

Lazarus, Neil. "Doubting the New World Order: Marxism, Realism, and the Claims of Post-modern Social Theory." *Differences* 3.3 (1991): 94–138.

LeClair, Tom. *The Art of Excess: Mastery in Contemporary American Fiction.* Urbana: U of Illinois P, 1989.

———. *In the Loop: Don DeLillo and the Systems Novel.* Urbana: U of Illinois P, 1987.

———. "Postmodern Mastery." McCaffery, *Postmodern Fiction* 117–28.

LeClair, Tom, and Larry McCaffery, eds. *Anything Can Happen: Interviews with Contemporary American Novelists.* Urbana: U of Illinois P, 1983.

Le Clézio, J. M. G. *The Book of Flights.* Trans. Simon Watson Taylor. London: Cape, 1971.

Lefebvre, Henri. *The Production of Space.* Oxford: Basil Blackwell, 1991.

Le Guin, Ursula K. *The Left Hand of Darkness.* New York: Walker, 1969.

———. "Writing without Conflict." *Harper's Magazine* (March 1989): 35–39.

Leitch, Vincent B. *Postmodernism: Local Effects, Global Flows.* Albany: State U of New York P, 1996.

Lejeune, Philippe. *Je est un autre.* Paris: Seuil, 1980.

Leland, John. "MTV: A Salute to America's Most Wasted." *Newsweek,* 19 July 1993: 61.

Lessing, Doris. *The Golden Notebook.* 1962. New York: Bantam, 1981.

Lester, Rosemarie K. "An Interview with Toni Morrison." Hessian Radio Network, Frankfurt, Germany. Rpt. in McKay 47–56.

Levi, Neil. "The Subject of History: Gadamer, Lacoue-Labarthe, and Lyotard." *Textual Practice* 5 (1991): 40–54.

Lippard, Lucy. *Get the Message? A Decade of Art for Social Change.* New York: E. P. Dutton, 1984.

Locke, Alain, ed. *The New Negro: An Interpretation.* New York: Albert and Charles Boni, 1925. Rpt. New York: Atheneum, 1968.

Lorde, Audre. *Zami: A New Spelling of My Name.* Watertown, MA: Persephone Press, 1982.

Lotman, Iurij. *Analysis of the Poetic Text.* Trans. D. Barton Johnson. Ann Arbor: Ardis, 1976.

———. "On the Semiotic Mechanism of Culture." *New Literary History* 9.2 (1978): 211–32.

Luhmann, Niklas. *Soziale systeme.* 2nd ed. Frankfurt/M.: Suhrkamp, 1985.

Lukács, Georg. *The Historical Novel.* Trans. Hannah and Stanley Mitchell. London: Merlin Press, 1962.

Lurie, Alison. *The War between the Tates.* New York, Random House, 1974.

Lyotard, Jean-François. "Answering the Question: What Is Postmodernism?" Trans. Régis Durand. Lyotard, *The Postmodern Condition* 71–82.

———. *The Differend: Phrases in Dispute.* Trans. George Van Den Abbeele. Minneapolis: U of Minnesota P, 1988.

———. "Discussion avec Richard Rorty." *Critique* 456 (1985): 581–84.

———. *Heidegger and "the Jews."* Trans. Andreas Michel and Mark S. Roberts, foreword David Carroll. Minneapolis: U of Minnesota P, 1990.

———. *The Postmodern Condition: A Report on Knowledge.* Trans. Geoff Bennington and Brian Massumi. Minneapolis: U of Minnesota P, 1984.

Macherey, Pierre. *A Theory of Literary Production.* 1966. Trans. Geoffrey Wall. London: Routledge & Kegan Paul, 1978.

Mackey, Louis. "Paranoia, Pynchon, and Preterition." *SubStance* 30 (1981): 16–30.

———. "Representation and Reflection: Philosophy and Literature in *Crystal Vision* by Gilbert Sorrentino." *Contemporary Literature* 28.2 (Summer 1987): 206–22.

Madsen, Deborah L. *The Postmodern Allegories of Thomas Pynchon.* New York: St. Martin's, 1991.

Magnusson, Magnus, and Herman Pálsson, trans. *The Vineland Sagas: The Norse Discovery of America.* Middlesex: Penguin, 1987.

Major, Clarence. *All-Night Visitors.* New York: Olympia Press, 1969. Rpt. Evanston: Northeastern UP, 1998.

————. *The Dark and Feeling: Black American Writers and Their Work.* New York: The Third Press, 1974.

————. *Dirty Bird Blues.* New York: Mercury House, 1996.

————. "Making Up Reality." *fiction international* 2.3 (1974): 151–54.

————. "A Meditation: Space and Time in Bamism." McCaffery, *Postmodern Fiction* 163–77.

————. *No.* New York: Emerson Hall, 1973.

————. *Painted Turtle: Woman with Guitar.* Los Angeles: Sun & Moon Press, 1988.

————. *Reflex and Bone Structure.* New York: Fiction Collective, 1975.

————. "Self Interview: On Craft." Major, *The Dark and Feeling* 125–32.

————. *Such Was the Season.* San Francisco: Mercury House, 1987.

Malmgren, Carl Darryl. *Fictional Space in the Modernist and Postmodernist American Novel.* Lewisburg: Bucknell UP, 1985.

Maltby, Paul. *Dissident Postmodernists: Barthelme, Coover, Pynchon.* Philadelphia: U of Pennsylvania P, 1991.

————. "Postmodern Thoughts on the Visionary Moment." *Centennial Review* 41.1 (Winter 1997): 119–41.

Marcus, James. "This Side of Paradise." The *Amazon.Com* Interview with Toni Morrison. 1997. 3 July 1998. <http// www.amazon.com/obidos/subst/c . . . re/morrison.interview/ 002-2468042-1767452M>

Marshall, Brenda K. *Teaching the Postmodern: Fiction and Theory.* New York: Routledge, 1992.

Martin, Valerie. *Mary Reilly.* New York: Doubleday, 1990.

Mazurek, Raymond A. "Metafiction, the Historical Novel, and Coover's *The Public Burning.*" *Critique* 23.3 (1982): 29–42.

McCaffery, Larry. "Avant-Pop: Still Life after Yesterday's Crash." Editor's Introduction. *After Yesterday's Crash: The Avant-Pop Anthology.* Ed. Larry McCaffery. Harmondsworth: Penguin, 1995. xi-xxxi.

————. "The Avant-Pop Phenomenon." *ANQ* 5.4 (Oct. 1992): 215–19.

————. "The Fictions of the Present." Elliott, *The Columbia Literary History of the United States* 1161–77.

————. "New Rules of the Game: *Take It or Leave It.*" *Chicago Review* 29.1 (Summer 1977): 144–49.

————. "Reconfiguring the Logic of Hyperconsumer Capitalism." *The American Book Review* (Aug.-Sept. 1996): 1, 12.

————. "Robert Coover." *Dictionary of Literary Biography.* Vol. 2. Ed. Jeffrey Helterman and Richard Layman. Detroit: Gale Research, 1978. 106–21.

McCaffery, Larry, ed. *Postmodern Fiction: A Bio-Bibliographical Guide.* Westport: Greenwood P, 1986.

————, ed. *Some Other Frequency: Interviews with Innovative American Authors.* Philadelphia: U of Pennsylvania P, 1996.

McCaffery, Larry, Thomas Hartl, and Doug Rice, eds. *Federman A to X-X-X-X: A Recyclopedic Narrative.* San Diego: San Diego State UP; 1998.

McEwan, Neil. *Perspective in British Historical Fiction Today.* Wolfboro, NH: Longman Academic, 1987.

McHale, Brian. *Constructing Postmodernism.* New York: Routledge, 1992.

————. *Postmodern Fiction.* New York: Methuen, 1987.

————. "Postmodernism or the Anxiety of Master Narratives." *Diacritics* 22.1 (Spring 1992): 17–33.

————. "You Used to Know What These Words Mean: Misreading *Gravity's Rainbow.*" *Language and Arts* 18.1 (Winter 1985): 93–118. Rpt. in *Constructing Postmodernism* 87–114.

McHoul, Alec, and David Wills. *Writing Pynchon.* Urbana: U of Illinois P, 1990.

McKay, Nellie Y., ed. *Critical Essays on Toni Morrison.* Westport, CT: Greenwood Press, 1988.

McLaughlin, Robert L. Rev. of *Mason & Dixon* by Thomas Pynchon. *Review of Contemporary Fiction* 17.3 (Fall 1997): 216–17.

McLeod, Bruce. "Collegiate Maneuvers: The University of Chicago, Postmodern Geographies and Dislocating the University." *Polygraph* 8 (1996): 79–94.

McLuhan, Marshall. *Understanding Media: The Extensions of Man.* New York: McGraw-Hill, 1964.

Meese, Elizabeth A. *(Ex)tensions: Re-Figuring Feminist Criticism.* Chicago: U of Illinois P, 1990.

Menand, Louis. "The War between Men and Women." Rev. of *Paradise* by Toni Morrison. *The New Yorker* 12 January 1998: 78–82.

Mendelson, Edward. "Gravity's Encyclopedia." *Mindful Pleasures: Essays on Thomas Pynchon.* Ed. G. Levine and D. Leverenz. Boston: Little, Brown, 1971. 161–95.

———. "The Sacred, the Profane, and *The Crying of Lot 49.*" Mendelson, *Pynchon* 112–46.

Mendelson, Edward, ed. *Pynchon: A Collection of Critical Essays.* Englewood Cliffs, NJ: Prentice-Hall, 1978.

Mepham, John. "The Intellectual as Heroine: Reading and Gender." Sceats and Cunningham 17–28.

Meyer, Charlotte M. "An Interview with Ronald Sukenick." *Contemporary Literature* 23.2 (Spring 1982): 129–44.

Michaels, Walter Benn. "'You who never was there': Slavery and the New Historicism, Deconstruction and the Holocaust." *Narrative* 4.1 (1996): 1–16.

Miller, J. Hillis. *Ariadne's Thread: Story Lines.* New Haven: Yale UP, 1992.

———. *Topographies.* Stanford: Stanford UP, 1995.

Miyoshi, Masao, and H. D. Harootunian. Introduction to *Postmodernism and Japan.* Spec. issue of *The South Atlantic Quarterly* 87.3 (Summer 1988): 387–99.

Mobley, Marilyn Sanders. *Folk Roots and Mythic Wings in Sarah Orne Jewett and Toni Morrison: The Cultural Function of Narrative.* Baton Rouge: Louisiana State UP, 1991.

"Modern Library/100 Best Novels." 31 May 1999. <www.randomhouse.com/modernlibrary/100best/novels.html>

Moers, Ellen. *Literary Women.* New York: Oxford UP, 1985.

Moglen, Helene. "Redeeming History: Toni Morrison's *Beloved.*" *Cultural Critique* 24 (Spring 1993): 17–40.

Mohanty, Satya P. "The Epistemic Status of Cultural Identity: On *Beloved* and the Postcolonial Condition." *Cultural Critique* 24 (Spring 1993): 41–80.

Moi, Toril. *Feminist Literary Theory.* London: Methuen, 1985.

Molesworth, Charles. "Culture, Power, and Society." Elliott, *Columbia Literary History of the United States* 1023–45.

Moore, Thomas. *The Style of Connectedness: Gravity's Rainbow and Thomas Pynchon.* Columbia: U of Missouri P, 1987.

Morace, Robert A. "Robert Coover, the Imaginative Self, and the 'Tyrant Other.'" *Papers on Language and Literature* 21.2 (Spring 1985): 192–209.

Moraru, Christian. "Reading the Onomastic Text: 'The Politics of the Proper Name' in Toni Morrison's *Song of Solomon.*" *Names* 44.3 (Sept. 1996): 189–204.

Moreland, Richard C. "'He Wants to Put His Story Next to Hers': Putting Twain's Story Next to Hers in Morrison's *Beloved.*" Peterson, *Toni Morrison* 154–79.

Morrison, Toni. "The Art of Fiction." Interview with Elissa Schappel and Claudia Brodsky Lacour. *Paris Review* 128 (1993): 83–125.

———. *Beloved.* New York: Knopf, 1987.

———. *The Bluest Eye.* 1970. New York: Pocket Books, 1972.

———. "City Limits, Village Values: Concepts of the Neighborhood in Black Fiction." *Literature and the Urban Experience.* Ed. Michael C. Jaye and Ann Chalmers Watts. New Brunswick: Rutgers UP, 1981. 35–43.

———. "Friday on the Potomac." *Race-ing Justice and En-gendering Power: Essays on Anita Hill, Clarence Thomas, and the Construction of Social Reality.* Ed. Toni Morrison. New York: Pantheon, 1992. vii-xxx.

———. *Jazz.* 1992. New York: Penguin-Plume, 1993.

———. "Nobel Lecture 1993." Peterson, *Toni Morrison.* 267–73.

———. *Paradise.* New York: Knopf, 1998.

———. *Playing in the Dark: Whiteness and the Literary Imagination.* Cambridge: Harvard UP, 1992.

———. "Rootedness: The Ancestor as Foundation." *Black Women Writers (1950–1980): A Critical Evaluation.* Ed. Marie Evans. Garden City, NY: Anchor Press-Doubleday, 1984. 339–45.

———. "The Site of Memory." *Inventing the Truth: The Art and Craft of Memoir.* Ed. William Zinsser. New York: Houghton-Mifflin, 1987. 101–24.

———. *Song of Solomon.* 1977. New York: Penguin-Plume, 1987.

———. *Sula.* 1973. New York: New American Library-Plume, 1982.

———. *Tar Baby.* Boston: G. K. Hall, 1981.

———. "Unspeakable Things Unspoken: The Afro-American Presence in American Literature." *Michigan Quarterly Review* 28 (Winter 1989): 1–34. Rpt. in Bloom, *Modern Critical Views* 201–30.

Morrissette, Bruce. *Novel and Film: Essays in Two Genres.* Chicago: U of Chicago P, 1985.

Morson, Gary Saul. *Narrative and Freedom: The Shadow of Time.* New Haven: Yale UP, 1995.

Motte, Warren F., Jr. *The Poetics of Experiment: A Study of the Work of Georges Perec.* Lexington, KY: French Forum Publishers, 1984.

Mouffe, Chantal. "The Sex/Gender System and the Discursive Construction of Women's Subordination." *Rethinking Ideology: A Marxist Debate.* Ed. Sakari Hanninen and Leena Paldan. Berlin: Argument Verlag; New York: International General-IMMRC, 1993. 139–43.

Mukherjee, Bharati. *The Holder of the World.* New York: Knopf, 1993.

———. *Jasmine.* New York: Grove Weidenfeld, 1989.

Mulvay, Laura. "Visual Pleasure and Narrative Cinema." *Screen* 16 (1975): 6–18.

Nadel, Alan. *Containment Culture: American Narratives, Postmodernism, and the Atomic Age.* Durham: Duke UP, 1995.

Nash, Christopher. *World-Games: The Tradition of Anti-Realist Revolt.* New York: Methuen,1987.

Natoli, Joseph. *Mots d'Ordre.* Albany: State U of New York P, 1992.

Naylor, Gloria, and Toni Morrison. "A Conversation." *Southern Review* 21 (1985): 567–93.

Nelson-Born, Katherine A. "Trace of a Woman: Narrative Voice and Decentered Power in the Fiction of Toni Morrison, Margaret Atwood, and Louise Erdrich." *Literature, Interpretation, Theory* 7.1 (1996): 1–12.

Newman, Charles. *The Post-Modern Aura: The Act of Fiction in an Age of Inflation.* Evanston: Northwestern UP, 1985.

Newman, Robert, ed. *Centuries' Ends, Narrative Means.* Stanford: Stanford UP, 1996.

Nicholson, Linda J., ed. *Feminism/Postmodernism.* New York: Routledge, 1990.

Nietzsche, Friedrich. "Vom Nutzen und Nachteil der Historie für das Leben" [Of the Use and Misuse of History for Life]. *Werke in drei Bänden.* Vol. 1. Ed. Karl Schlecta. München: Hanser, 1956. 212–45.

———. *The Will to Power.* Trans. Walter Kaufmann and R. J. Hollingdale. New York: Vintage, 1968.

Nohrnberg, James. "Pynchon's Paraclete." Mendelson, *Pynchon* 147–61.

O'Kane, John. "Cultural Pessimism and the Anti-Intellectual American Tradition." *Salmagundi* 67 (Summer 1985): 171–82.

Olsen, Lance, and Mark Amerika. "Smells Like Avant-Pop: An Introduction of Sorts." Amerika and Olsen 1–31.

Olson, Gary A., and Elizabeth Hirsh, eds. *Women Writing Culture.* Albany: State U of New York P, 1995.

Olster, Stacey. "When You're a (Nin)jette, You're a (Nin)jette All the Way—or Are You?: Female Filmmaking in *Vineland*." Green, Greiner, and McCaffery 119–34.

Ong, Walter J. *Orality and Literacy: The Technologizing of the Word*. 1982. New York: Methuen, 1986.

Ordóñez, Elizabeth J. "Narrative Texts by Ethnic Women: Rereading the Past, Reshaping the Future." *MELUS* 9 (1982): 19–28.

Ormiston, Gayle L., and Raphael Sassower. *Narrative Experiments: The Discursive Authority of Science and Technology*. Minneapolis: U of Minnesota, 1989.

Ornatowski, Cezar M., and Jerzy Durczak. "Kosinski: The Final Chapter." *The American Book Review* (Oct.-Nov. 1996): 12–13.

Orr, Leonard. *Problems and Poetics of the Nonaristotelian Novel*. Lewisburg, PA: Bucknell UP, 1991.

Orwell, George. "Inside the Whale" (1940). Rpt. in *The Collected Essays, Journalism and Letters of George Orwell*. Vol. 1. Ed. Sonia Orwell and Ian Angus. New York: Harcourt, Brace and World, 1968. 493–527.

Paulson, William R. *The Noise of Culture: Literary Texts in a World of Information*. Ithaca: Cornell UP, 1988.

Pearce, Richard. *The Novel in Motion: An Approach to Modern Fiction*. Columbus: Ohio State UP, 1983.

Pearce, Richard, ed. *Essays on Thomas Pynchon*. Boston: G. K. Hall, 1981.

Perec, Georges. *La Disparition*. Paris: Les Lettres nouvelles, 1969.

———. *La Vie, mode d'emploi*. Paris: Hachette littérature, 1978. Trans. by David Bellos as *Life, a User's Manual: Fictions*. London: Collins Harvill, 1987.

Pérez-Torres, Rafael. "Knitting and Knotting the Narrative Thread—*Beloved* as Postmodern Novel." Peterson, *Toni Morrison* 91–109.

Peterson, Nancy J. "'Say Make Me, Remake Me': Toni Morrison and the Reconstruction of African-American History." Peterson, *Toni Morrison* 201–21.

Peterson, Nancy J., ed. *Toni Morrison: Critical and Theoretical Approaches*. Baltimore: Johns Hopkins UP, 1997.

Phelan, James. "Toward a Rhetorical Reader-Response Criticism: The Difficult, the Stubborn, and the Ending of *Beloved*. Peterson, *Toni Morrison* 225–44.

Phelan, James, ed. *Reading Narrative: Form, Ethics, Ideology*. Columbus: Ohio State UP, 1989.

Piercy, Marge. *Braided Lives*. New York: Fawcett Crest, 1982.

———. *Small Changes*. New York: Fawcett Crest, 1973.

Piombino, Nick. "Cultivate Your Own Wildness." Bernstein, *The Politics of Poetic Form* 232–34.

Plath, Sylvia. *The Bell Jar*. London: Faber and Faber, 1966.

Popper, Karl R. *The Poverty of Historicism*. Boston: Beacon Press, 1957.

Pratt, Mary Louise. "Comparative Literature and Global Citizenship." *Comparative Literature in the Age of Multiculturalism*. Ed. Charles Bernheimer. Baltimore: Johns Hopkins UP, 1995. 58–65.

Probyn, Elspeth. *Outside Belongings*. New York: Routledge, 1996.

Przybylowicz, Donna. "Toward a Feminist Cultural Criticism." *The Construction of Gender and Modes of Social Division II*. Spec. issue of *Cultural Critique* 14 (Winter 1989–90): 259–301.

Przybylowicz, Donna, Nancy Hartsock, and Pamela McCallum. Introduction to *The Construction of Gender and Modes of Social Division*. Spec. issue of *Cultural Critique* 13 (Fall 1989): 5–14.

Pynchon, Thomas. *The Crying of Lot 49*. 1966. New York: Harper & Row-Perennial Library, 1986.

———. *Gravity's Rainbow*. New York: Viking, 1973.

———. "Is It O.K. to Be a Luddite?" *The New York Times Book Review* 24 October 1984: 40–41.

———. *Mason & Dixon*. New York: Henry Holt, 1997.

———. *Slow Learner*. Boston: Little, Brown, 1984.

———. *V.* Philadelphia: J. B. Lippincott, 1963.

———. *Vineland*. Boston: Little, Brown, 1990.

Quartermain, Peter. "Trusting the Reader." *Chicago Review* 32.2 (Autumn 1980): 65–74.

Rabinow, Paul. "Representations Are Social Facts: Modernity and Post-Modernity in Anthropology." *Writing Culture: The Poetics and Politics of Ethnography.* Ed. James Clifford and George E. Marcus. Berkeley: U of California P, 1986. 224–61.

Radhakrishnan, R. "Feminist Historiography and Post-Structuralist Thought: Intersections and Departures." *The Difference Within: Feminism and Critical Theory.* Ed. Elizabeth Meese and Alice Parker. Philadelphia: John Benjamins, 1988. 189–205.

Radway, Janice A. *Reading the Romance: Woman, Patriarchy and Popular Literature.* London: Verso, 1987.

Reed, Ishmael. *Mumbo Jumbo.* 1972. New York: Bard, 1978.

Rhys, Jean. *Wide Sargasso Sea.* London: Deutsch, 1966.

Ricardou, Jean. "Nouveau roman, Tel Quel." Federman, *Surfiction: Fiction Now . . . and Tomorrow* 104–33.

Rich, Adrienne. *Of Woman Born: Motherhood as Experience and Institution.* New York: Bantam, 1977.

———. *On Lies, Secrets, and Silence: Selected Prose, 1966–1978.* New York: W. W. Norton, 1979.

Richardson, Brian. "The Poetics and Politics of Second-Person Narrative." *Genre* 24 (Fall 1991): 309–30.

———. "Remapping the Present: The Master Narrative of Modern Literary History and the Lost Forms of Twentieth-Century Fiction." *Twentieth Century Literature* 43.3 (Fall 1977): 291–309.

Robins, Kevin. "Prisoners of the City: Whatever Could a Postmodern City Be." *Space and Place: Theories of Identity and Location.* Ed. Erica Carter, James Donald, and Judith Squires. London: Lawrence and Wishart, 1993. 303–30.

Robinson, Lou. "Reverse Everything." *The American Book Review* (Oct.-Nov. 1991): I, II, 12.

Roche, Maurice. *Codex; roman.* Paris: Seuil, 1974.

Rodrigues, Eusebio L. "Experiencing *Jazz.*" Peterson, *Toni Morrison* 245–66.

Rooney, Ted. "The Fiction of Fiction: An Interview with Ronald Sukenick." *Poetry Flash* May-June 1995: 1–5, 6, 8–11.

Rorty, Richard. *Contingency, Irony, and Solidarity.* New York: Cambridge UP, 1989.

Rosenbaum, Jonathan. "Pynchon's Prayer." Rev. of *Vineland* by Thomas Pynchon. *The Reader: Chicago's Free Weekly* 9 March 1990: 8, 29–31.

Rosenfelt, Deborah, and Judith Stacey. "Second Thoughts on the Second Wave." *Feminist Studies* 13.2 (Summer 1987): 341–61.

Rothenberg, Jerome. "Ethnopoetics and Politics/The Politics of Ethnopoetics." Bernstein, *The Politics of Poetic Form* 1,21.

Ruas, Charles. *Conversations with American Writers.* New York: Knopf, 1985.

Russ, Joanna. *The Female Man.* Boston: Gregg Press, 1977.

———. *To Write Like a Woman: Essays in Feminism and Science Fiction.* Bloomington: Indiana UP, 1995.

Russell, Charles. "Individual Voice in the Collective Discourse: Literary Innovations in Postmodern American Fiction." *SubStance* 27 (1980): 29–39.

———. *Poets, Prophets, and Revolutionaries: The Literary Avant-Garde from Rimbaud through Postmodernism.* New York: Oxford UP, 1985.

———. "Subversion and Legitimation: The Avant-Garde in Postmodern Culture." *Chicago Review* 33.2 (1982): 54–59.

Russell, Sandi. "It's OK to Say OK." Interview with Toni Morrison. *Women's Review,* 5 March 1986: 22–24. Rpt. in McKay 43–47.

Ruthrof, Horst. "Narrative and the Digital: On the Syntax of the Postmodern." *AUMLA: Journal of the Australian Universities Language and Literature Association* 74 (Nov. 1990): 185–200.

Ryan, Judylyn S. "Contested Visions/Double-Vision in *Tar Baby.*" Peterson, *Toni Morrison* 63–87.

Saintsbury, George. *A Short History of English Literature.* London: Macmillan, 1924.

Samuels, Wilfred D., and Clenora Hudson-Weems. *Toni Morrison.* Boston: Twayne, 1990.

Sayre Henry M. "The Avant-Garde and Experimental Writing." Elliott, *Columbia Literary History of the United States* 1178–99.

Sceats, Sarah, and Gail Cunningham, eds. *Image and Power: Women in Fiction in the Twentieth Century.* New York: Longman, 1996.

Schaub, Thomas Hill. *American Fiction in the Cold War.* Madison: U of Wisconsin P, 1991.

Schmidt, Siegfried J. "Literary Systems as Self-Organizing Systems." *Empirical Studies of Literature: Proceedings of the Second IGEL Conference, Amsterdam 1989.* Ed. E. Ibsch, D. Schram, and G. Steen. Amsterdam: Rodopi, 1991. 413–24.

Scholes, Robert. *Structuralism in Literature.* New Haven: Yale UP, 1974.

Schwarzbach, F. S. "A Matter of Gravity." Mendelson, *Pynchon* 56–67.

Shange, Ntozake. "Interview with Toni Morrison." *American Rag* (Nov. 1978): 48–52.

Shohat, Ella, and Robert Stam. *Unthinking Eurocentrism: Multiculturalism and the Media.* New York: Routledge, 1994.

Showalter, Elaine. "Piecing and Writing." *The Poetics of Gender.* Ed. Nancy K. Miller. New York: Columbia UP, 1986. 222–47.

Silko, Leslie Marmon. *Almanac of the Dead.* New York: Simon & Schuster, 1991.

Sklar, Robert. "An Anarchist Miracle: The Novels of Thomas Pynchon." Mendelson, *Pynchon* 87–96.

Sloan, De Villo. "The Decline of American Postmodernism." *SubStance* 16.3 (1987): 29–43.

Soja, Edward. "Inside Exopolis." *Variations of a Theme Park: The New American City and the End of Public Space.* Ed. Michael Sorkin. New York: Hill and Wang, 1992. 94–122.

———. *Postmodern Geographies: The Reassertion of Space in Critical Social Theory.* London: Verso, 1989.

———. *Thirdspace: Journeys to Los Angeles and Other Real-and-Imagined Spaces.* London: Basil Blackwell, 1996.

Sokolov, Raymond A. "Ga-Ga-Ga-Gug." Rev. of *Up* by Ronald Sukenick. *The New York Times Book Review* 14 July 1968: 34.

Sollers, Philippe. *Writing and the Experience of Limits.* Trans. Philip Barnard with David Hayman. New York: Columbia UP, 1983.

Somerson, Wendy. "Becoming Rasta: Recentering White Masculinity in the Era of Transnationalism." *The Comparatist* 23 (1999): 128–40.

Sorrentino, Gilbert. *Crystal Vision.* San Francisco: North Point Press, 1981.

———. *Something Said. Essays by Gilbert Sorrentino.* San Francisco: North Point Press, 1984.

Southgate, Beverley. *History: What and Why? Ancient, Modern, and Postmodern Perspectives.* New York: Routledge, 1996.

Spanos, William. "Rethinking the Postmodernity of the Discourse of Postmodernism." Bertens and Fokkema 65–74.

Spariosu, Mihai. *Dionysus Reborn: Play and the Aesthetic Dimension in Modern Philosophical and Scientific Discourse.* Ithaca: Cornell UP, 1989.

———. *God of Many Names: Play, Poetry and Power in Hellenic Thought from Homer to Aristotle.* Durham: Duke UP, 1991.

———. *Literature, Mimesis and Play: Essays in Literary Theory.* Tübingen: Gunter Narr Verlag, 1982.

———. *The Wreath of Wild Vine: Play, Liminality, and the Study of Literature.* Albany: State U of New York P, 1997.

Spiegel, Gabrielle. "History and Post-Modernism." *Past and Present* 135 (1992): 194–208.

Spiller, Robert E., et al. *Literary History of the United States.* 1948. 3rd ed., rev. New York: Macmillan Company, 1963.

Spillers, Hortense J. "A Hateful Passion, a Lost Love." *Feminist Issues in Literary Scholarship.* Ed. Shari Benstock. Bloomington: Indiana UP, 1987. 181–207.

Spiro, Rand J., et al. "Knowledge Acquisition for Application: Cognitive Flexibility and Transfer in Complex Content Domains." *Executive Control Processes in Reading.* Ed. Bruce K. Britton and Shawn M. Glynn. Hillsdale, NJ: Lawrence Erlbaum Associates, 1987. 181–92.

Spivak, Gayatri Chakravorty. "Can the Subaltern Speak?" *Marxism and the Interpretation of Culture.* Ed. Cary Nelson and Lawrence Grossberg. Urbana: U of Illinois P, 1988. 271–313.

Stanzel, Franz K. *A Theory of Narrative.* Trans. Charlotte Goedsche. Cambridge: Cambridge UP, 1984.

Stepto, Robert B. *From Behind the Veil: A Study of Afro-American Narrative.* Urbana: U of Illinois P, 1979.

———. "'Intimate Things in Place': A Conversation with Toni Morrison." *Chant of Saints.* Ed. Robert B. Stepto and Michael Harper. Urbana: Illinois UP, 1979. 213–29.

Stern, J. P. *On Realism.* London: Routledge & Kegan Paul, 1973.

Stevick, Philip. *Alternative Pleasures: Postrealist Fiction and the Tradition.* Urbana: U of Illinois P, 1981.

Stimpson, Catharine R. "Literature as Radical Statement." Elliott, *Columbia Literary History of the United States* 1160–76.

Stonehill, Brian. *The Self-Conscious Novel: Artifice in Fiction from Joyce to Pynchon.* Philadelphia: U of Pennsylvania P, 1988.

Storace, Patricia. "The Scripture of Utopia." *The New York Review of Books,* 11 June 1988: 64–69.

Strauss, Walter A. Rev. of *Journey to Chaos: Samuel Beckett's Early Fiction* by Raymond Federman. *Modern Language Journal* 50.7 (Nov. 1966): 505–6.

Strehle, Susan. "Pynchon's 'Elaborate Game of Doubles' in *Vineland.*" Greene, Greiner, and McCaffery 113–16.

Sukenick, Ronald. "Autogyro: My Life in Fiction." *Contemporary Authors Autobiography Series.* Vol. 8. Ed. Mark Zadrezny. Detroit: Gale Research. 283–95.

———. "aVANT-pOP, sUR-fICTION, hYPER-fICTION." Amerika and Olsen 48–50.

———. *Blown Away.* Los Angeles: Sun & Moon Press, 1986.

———. *The Death of the Novel and Other Stories.* New York: Dial Press, 1969.

———. *Doggy Bag: Hyperfictions.* Boulder: Black Ice Books, 1994.

———. *Down and In: Life in the Underground.* New York: Beech Tree Books-William Morrow, 1987.

———. *The Endless Short Story.* New York: Fiction Collective, 1986.

———. *In Form: Digressions on the Act of Fiction.* Carbondale: Southern Illinois UP, 1985.

———. "Introduction: The Dirty Secret." *Experimental Fiction.* Spec. issue of *Witness* 3.2–3 (Summer-Fall 1989): 7–9.

———. Introductory Note to section 8 of *Out. Massachusetts Review* 14 (Spring 1973): 352–53.

———. *Long Talking Bad Conditions Blues.* New York: Fiction Collective, 1979.

———. "Love Conks Us All." Introduction to the sixth edition of *98.6.* Boulder, CO: Fiction Collective Two, 1994. i–vii.

———. *Mosaic Man.* Normal: Illinois State U-Fiction Collective Two, 1999.

———. *Narralogues: Truth in Fiction.* Albany: State U of New York P, 2000.

———. "The New Tradition." *Partisan Review* 39 (1972): 580–88. Rpt. in *In Form* 202–13.

———. *98.6.* New York: Fiction Collective, 1975.

———. *Out.* Chicago, Swallow Press, 1973.

———. "Post Modern Fiction and Oppositional Art." *Descrizioni e iscrizioni: politiche del discorso.* Ed. Carla Locatelli and Giovanna Covi. Trento: Editrice Universitá degli Studi di Trento, 1998. 173–81.

———. "Refugee from the Holocaust." *New York Times Book Review* 1 October 1972: 40–41.

———. "The Rival Tradition." Interview with Jack Foley. *FlashPoint* (Summer 1996): 1–12.

———. "Unwriting: Talmudic Fiction." *The American Book Review* (Dec. 1991-Jan. 1992): 4, 26–27.

———. *Up.* New York: Dial Press, 1968.

———. "Up from the Garret: Success Then and Now." *The New York Times Book Review* 27 January 1985, late ed., sec. 7:1.

———. *Wallace Stevens: Musing the Obscure.* New York: New York UP, 1967.

Suleiman, Susan Robin. *The Female Body in Western Culture: Contemporary Perspectives.* Cambridge: Harvard UP, 1986.

————. "The Politics of Postmodernism after the Wall." Bertens and Fokkema 51–64.

————. *Subversive Intent: Gender, Politics, and the Avant-Garde.* Cambridge: Harvard UP, 1990.

Szilard, Leo. "On the Decrease of Entropy in a Thermodynamic System by the Intervention of Intelligent Beings." 1929. *The Collected Works of Leo Szilard.* Vol. 1, *Scientific Papers.* Ed. Bernard T. Feld and Gertrude Weiss Szilard. Cambridge: MIT P, 1972. 120–29.

Tabbi, Joseph. "Pynchon's Groundward Art." Green, Greiner, and McCaffery 89–100.

Tanner, Tony. "V. and V-2." Mendelson, *Pynchon* 16–55.

Tate, Claudia. "Toni Morrison." Interview. *Black Women Writers at Work.* Ed. Claudia Tate. New York: Continuum, 1983. 117–31.

Tatham, Campbell. E-mail rev. of *Critifiction* by Raymond Federman. Rpt. in McCaffery, Hartl, and Rice 86–88.

————. "Mythotherapy and Postmodern Fictions: Magic Is Afoot." Benamou and Caramello 137–57.

Taylor, Charles. "The Politics of Recognition." *Multiculturalism: Examining the Politics of Recognition.* Ed. Amy Gutmann. Princeton: Princeton UP, 1994. 25–73.

Thomas, Brook. "Turner's 'Frontier Thesis' as a Narrative of Reconstruction." Robert Newman 117–37.

Toolan, Michael J. *Narrative: A Critical Linguistic Introduction.* 1988. New York: Routledge, 1995.

Tötösy de Zepetnek, Steven. "Systemic Approaches to Literature—An Introduction with Selected Bibliographies." *Canadian Review of Comparative Literature* 19.1–2 (March-June 1992): 21–93.

Trachtenberg, Stanley. "The Way That Girl Pressed Against You in the Subway: Ronald Sukenick's Real Act of Imagination." *The Journal of Narrative Technique* 12.1 (Winter 1982): 57–71.

Travis, Molly Abel. *Reading Cultures: The Construction of Readers in the Twentieth Century.* Carbondale: Southern Illinois UP, 1998.

Turner, Victor. "Liminal to Liminoid in Play Flow and Ritual." *Rice University Studies* 60.3 (1974): 53–92.

————. *On the Edge of the Bush: Anthropology as Experience.* Tucson: U of Arizona P, 1985.

————. *The Ritual Process.* Chicago: Aldine, 1969.

Unspeakable Practices: A Celebration of Iconoclastic American Fiction. Symposium held at Brown University, April 1988. *Critique: Studies in Contemporary Fiction.* 21.4 (Summer 1990): 233–75.

Van Der Zee, James, Owen Dodson, and Camille Billops. *The Harlem Book of the Dead.* Foreword by Toni Morrison. Dobbs Ferry, NY: Morgan and Morgan, 1978.

Varsava, Jerry A. *Contingent Meanings: Postmodern Fiction, Mimesis, and the Reader.* Tallahassee: Florida State UP, 1990.

Vattimo, Gianni. "The End of (Hi)story." *Zeitgeist in Babel: The Post-modernist Controversy.* Ed. Ingeborg Hoesterey. Bloomington: Indiana UP, 1991. 132–41.

Verdery, Katherine. *National Ideology under Socialism: Identity and Cultural Politics in Ceauşescu's Romania.* Berkeley: U of California P, 1991.

Vickroy, Laurie. "*Beloved* and *Shoah*: Witnessing the Unspeakable." *The Comparatist* 22 (1998): 123–44.

Virilio, Paul. *The Art of the Motor.* Minneapolis: U of Minnesota P, 1995.

Walker, Alice. *The Color Purple.* New York: Harcourt Brace Jovanovich, 1982.

————. *Meridian.* New York: Washington Square Press, 1976.

————. *In Search of Our Mothers' Gardens: Womanist Prose.* New York: Harvest-HBJ, 1983.

Walsh, W. H. *An Introduction to the Philosophy of History.* 3rd ed. London: Hutchinson, 1967.

Washington, Mary Helen. *Invented Lives: Narratives of Black Women, 1860–1960.* Garden City, NY: Doubleday, 1987.

Weber, Max. *The Protestant Ethic and the Spirit of Capitalism.* 1904. Trans. Talcott Parsons, introduction Randall Collins. Los Angeles: Roxbury, 1996.

Weiman, Robert. "History, Appropriation, and the Uses of Representation in Modern Narrative." Krieger 175–215.

White, Curtis. "An Essay-Simulacrum on Avant-Pop." Amerika and Olsen 70–80.

———. "Jameson Out of Touch?" *The American Book Review* (Dec.-Jan. 1993): 21, 30.

White, Hayden. *Metahistory: The Historical Imagination in Nineteenth-Century Europe.* Baltimore: Johns Hopkins UP, 1973.

———. "Storytelling: Historical and Ideological." Robert Newman 58–78.

———. *Tropics of Discourse: Essays in Cultural Criticism.* Baltimore: Johns Hopkins UP, 1978.

Wiegman, Robyn. "Negotiating AMERICA: Gender, Race, and the Ideology of the Interracial Male Bond." *Cultural Critique* 13 (Fall 1989): 89–117.

Wielgosz, Anne-Katrin. "Displacement in Raymond Federman's *Double or Nothing* or: 'noodles and paper coincide.'" *Journal of Narrative Technique* 25.1 (1995): 91–107.

Wilde, Alan. *Middle Grounds: Studies in Contemporary American Fiction.* Philadelphia: U of Pennsylvania P, 1987.

———. "'Strange Displacements of the Ordinary': Apple, Elkin, Barthelme, and the Problem of the Excluded Middle." *boundary 2* 10.2 (Winter 1982): 177–99.

Wilden, Anthony. *System and Structure: Essays in Communication and Exchange,* 2nd ed. New York: Tavistock, 1980.

Willis, Susan. "Eruptions of Funk: Historicizing Toni Morrison." Gates, *Black Literature and Literary Theory* 263–83.

Wills, Gary. *The Second Civil War: Arming for Armageddon.* New York: NAL, 1968.

Wilson, Kimberly A. C. "The Function of the 'Fair' Mulatto: Complexion, Audience, and Mediation in Frances Harper's *Iola Leroy*." *Cimarron Review* 106 (1994): 104–13.

Wood, James. "The Color Purple." Rev. of *Paradise* by Toni Morrison. *The New Republic* 2 March 1988: 29–32.

Woolf, Virginia. *A Room of One's Own.* 1929. New York: HBJ, 1957.

Wyatt, Jean. "Giving Body to the Word: The Maternal Symbolic in Toni Morrison's *Beloved*." *PMLA* 108.3 (May 1993): 474–88.

Zabel, M. D. Introduction to *Tales of Heroes and History.* By Joseph Conrad. Garden City, NY: Doubleday, 1960.

Zavarzadeh, Mas'ud. *The Mythopoetic Reality—the Postwar American Nonfiction Novel.* Urbana: Illinois UP, 1976.

Zimmerman, Bonnie. "Exiting from Patriarchy: The Lesbian Novel of Development." Abel, Hirsch, and Langland 244–67.

———. "Feminist Fiction and the Postmodern Challenge." McCaffery, *Postmodern Fiction.* 175–88.

Žižek, Slavoj. *The Sublime Object of Ideology.* London: Verso, 1989.

INDEX

Printed in the United States
By Bookmasters